SCHAUM'S
outlines
™

Statistics for Engineers

Larry J. Stephens, Ph.D.

Professor of Mathematics
University of Nebraska at Omaha

WITHDRAWN

Schaum's Outline Series

Mc
Graw
Hill

New York   Chicago   San Francisco   Lisbon   London
Madrid   Mexico City   Milan   New Delhi   San Juan
Seoul   Singapore   Sydney   Toronto

LARRY J. STEPHENS is professor of mathematics at the University of Nebraska at Omaha, where he has taught since 1974. He has also taught at the University of Arizona, Gonzaga University, and Oklahoma State University. He has worked for NASA, Lawrence Berkeley National Laboratory, and Los Alamos National Laboratory. He has consulted widely and spent 10 years as a consultant for the engineering group at 3M in Valley, Nebraska. Starting in 2008, he now conducts a monthly statistical seminar for Omaha Public Power District (OPPD) employees in the pricing and forecasting services department at OPPD. He has over 35 years of experience in teaching statistical methodology, engineering statistics, and mathematical statistics. He has over 50 publications in professional journals and has published books in the Schaum's outline series as well as books in the Utterly Confused and Demystified series published by McGraw-Hill.

Schaum's Outline of STATISTICS FOR ENGINEERS

Copyright © 2011 by The McGraw-Hill Companies, Inc. All rights reserved. Printed in the United States of America. Except as permitted under the Copyright Act of 1976, no part of this publication may be reproduced or distributed in any form or by any means, or stored in a database or retrieval system, without the prior written permission of the publisher.

1 2 3 4 5 6 7 8 9 10 11 12 13 14 15   ROV/ROV   1 9 8 7 6 5 4 3 2 1

ISBN 978-0-07-173646-6 (print book)
MHID 0-07-173646-8

ISBN 978-0-07-173647-3 (e-book)
MHID 0-07-173647-6

This publication is designed to provide accurate and authoritative information in regard to the subject matter covered. It is sold with the understanding that neither the author nor the publisher is engaged in rendering legal, accounting, securities trading, or other professional services. If legal advice or other expert assistance is required, the services of a competent professional person should be sought.
        —*From a Declaration of Principles Jointly Adopted by a Committee of the American Bar Association and a Committee of Publishers and Associations*

Library of Congress Cataloging-in-Publication Data is on file with the Library of Congress.

Trademarks: McGraw-Hill, the McGraw-Hill Publishing logo, Schaum's, and related trade dress are trademarks or registered trademarks of The McGraw-Hill Companies, Inc. and/or its affiliates in the United States and other countries, and may not be used without written permission. All other trademarks are the property of their respective owners. The McGraw-Hill Companies, Inc. is not associated with any product or vendor mentioned in this book.

MINITAB® is a registered trademark of Minitab, Inc.

McGraw-Hill books are available at special quantity discounts to use as premiums and sales promotions, or for use in corporate training programs. To contact a representative, please e-mail us at bulksales@ mcgraw-hill.com.

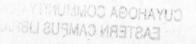

Preface

To create *Schaum's Outline of Statistics for Engineers*, I consulted many different engineering statistics texts as well as the Internet in creating the solved problems and the supplementary problems for this text. I hope that these problems will give the student a broader picture of how statistics is used in solving different engineering problems.

When "engineering statistics" is entered as a search topic on the Internet, the following is given from Wikipedia, the free encyclopedia.

Engineering statistics is a branch of statistics that has several subtopics which are particular to engineering:

1. Design of Experiments (DOE) uses statistical techniques to test and construct models of engineering components and systems.

2. Quality control and process control use statistics as a tool to manage conformance to specifications of manufacturing processes and their products.

3. Time and methods engineering use statistics to study repetitive operations in manufacturing in order to set standards and find optimum (in some sense) manufacturing procedures.

4. Reliability engineering which measures the ability of a system to perform for its intended function (and time) and has tools for improving performance.

5. Probabilistic design involving the use of probability in product and system design.

Also, statistics.com is a source for online courses in statistics. If you follow the path **Home > Our courses > Engineering**, you are referred to a course by Dr. Thomas P. Ryan. This is but one example of online courses available in engineering statistics.

Statistical software plays a large role in this *Schaum's Outline of Statistics for Engineers*. I would like to thank the following people and the companies they are associated with for the use of their software in the book:

MINITAB – Linda Holderman, Coordinator of the author assistance program that Minitab sponsors. Portions of the input and output contained in this publication/book are printed with the permission of Minitab, Inc. All material remains the exclusive property and copyright of Minitab, Inc. The web address for MINITAB is *www.minitab.com.*

SPSS: An IBM Company – Jill Crist, Account Manager, Publications, IBM. The following quote is from Wikipedia: SPSS is a computer program used for statistical analysis. Between 2009 and 2010 the premier software for SPSS was called PASW (Predictive Analytics Soft Ware) Statistics. The company announced on July 28, 2009, that it was being acquired by IBM for US$1.2 billion. As of January 2010, it became "SPSS: An IBM Company." The web address for SPSS is *www.spss.com.*

STATISTIX – Dr. Gerard Nimis, President, Analytical Software. I quote from the website: "If you have data to analyze, but you're a researcher not a statistician, Statistix is designed for you. You'll be up and running in minutes without programming or using a manual. This easy-to-learn and simple-to-use software saves you valuable time and money. Statistix combines all the basic and advanced statistics and powerful data manipulation tools you need in a single, inexpensive package." The web address for STATISTIX is *www.statistix.com.*

SAS – Sandy Varner, Marketing Operations Manager, SAS Publishing, Cary, NC, USA. Created with SAS software. Copyright 2005. SAS Institute Inc. I quote from the website: "SAS is a leader in business intelligence software and services. Over 30 years, SAS has grown from seven employees to nearly 10,000 worldwide, from a few customer sites to more than 40,000, and has been profitable every year." The web address for SAS is *www.sas.com.*

MAPLE – Dr. Tom Lee, Ph.D., Vice President Market Development and Executive Product Director, Maplesoft. I quote from the website: "Technical professionals, like engineers, research scientists and financial analysts, use Maplesoft products to save time and costs through increased productivity, rapid solution development and deployment, and easy technical knowledge capture." The web address for MAPLE is *www.maplesoft.com.*

EXCEL – Microsoft Excel has been around since 1985. It is available to almost all college students and professional engineers. It is widely used in this Schaum's Outline.

I would like to thank Shruti Vasishta at Glyph International for her help as the project manager of this book.

I wish to thank my wife Lana, for her understanding when I was working on the book in my mind's eye instead of giving her my undivided attention.

LARRY J. STEPHENS

References for the Solved Problems and Supplementary Problems

1. Howard Gitlow, Shelly Gitlow, Alan Oppenheim, and Rosa Oppenheim, *Tools and Methods for the Improvement of Quality,* Irwin, Boston, MA 02116, 1989.
2. William J. Kolarik, *Creating Quality, Concepts, Systems, Strategies, and Tools,* McGraw-Hill, New York, NY, 1995.
3. Irwin Miller and Marylees Miller, *Statistical Methods for Quality with Applications to Engineering and Management,* Prentice Hall, Englewood Cliffs, NJ 07632, 1995.
4. William Navidi, *Statistics for Engineers and Scientists,* McGraw-Hill, New York, NY, 2006.
5. Thomas P. Ryan, *Statistical Methods for Quality Improvement,* John Wiley and Sons, New York, NY, 1989.
6. Thomas Pyzdek, *What Every Engineer Should Know About Quality Control,* Marcel Dekker, Inc., New York, NY, 1989.
7. Richard L. Scheaffer, Madhuri S. Mulekar, and James T. McClave, *Probability and Statistics for Engineers,* Brooks/Cole, New York, NY, Fifth edition, 2011.
8. Geoffrey Vining and Scott Kowalski, *Statistical Methods for Engineers,* Brooks/Cole, Boston, MA 02210, Third edition, 2011.

Contents

CHAPTER 1

Treatment of Data Using Statistical Software

Graphical Displays of Data

This first chapter will deal with the treatment of data and introduce you to five of the six software packages that will be used in the book. EXCEL, MINITAB, SAS, SPSS, and STATISTIX are the primary statistical packages that will be referenced. MAPLE is a package that is more mathematically oriented. It will be used to assist with the calculus aspects of the course. The official Internet website for MAPLE is www.maplesoft.com.

Output from different software will be contrasted to allow you to compare the output and see the similarities and differences in the packages. The comparisons of the software will help you learn the concepts. Downloads of MINITAB, SPSS, and STATISTIX for approximately 30 days are available on the Internet at www.minitab.com, www.spss.com, and www.statistix.com respectively. Another website where much of the software is available at reasonable rents is www.e-academy.com. EXCEL is readily available on university computers and on most home computers. SAS is available on most university computers. These six sources of statistical software will be used in the text to solve statistical problems. I am indebted to the software manufacturers for granting me permission to use output in the book. When an engineer or a scientist goes to work, he or she will use the software to help solve problems.

Figure 1.1 is the MINITAB worksheet with the data from Table 1.1 in column C1. This data can either be entered directly from the keyboard or read from a file called mpg.mtw. A **histogram** is a graphical device

Figure 1.1   MINITAB worksheet containing the data from Table 1.1.

Table 1.1   mpg Values for 100 Pickup Trucks in an Engineering Study

19.7	17.9	15.7	16.2	15.9	15.7	17.9	20.2	15.2	18.4
20.8	18.8	17.4	17.1	16.3	16.9	12.6	16.1	17.5	16.6
22.9	20.7	20.1	18.8	18.0	13.5	15.3	17.9	16.3	12.6
17.8	16.3	16.2	16.0	15.3	16.4	16.2	18.9	19.8	14.7
14.4	18.1	19.1	14.2	21.1	18.8	11.4	18.0	17.5	19.0
14.0	14.4	18.2	15.5	15.4	19.5	15.2	17.6	16.6	16.3
16.1	16.3	19.9	15.6	15.3	18.6	18.9	14.6	19.7	20.1
19.6	20.8	15.5	19.3	17.1	14.8	17.0	21.2	14.1	15.3
17.8	15.5	17.7	10.4	18.8	18.1	20.5	16.0	18.3	17.0
19.2	12.9	19.5	12.5	14.7	14.9	13.1	19.0	16.8	21.5

that uses rectangles to show the distribution of data. The heights of the rectangles give the frequency of the classes and the bases represent the classes. It is one of the oldest constructs in use which shows distributions. The pulldown **Graph ⇒ Histogram** gives a dialog box in which a simple histogram is chosen. In the simple histogram dialog box, it is requested that a histogram for the variable mgp be prepared. The result is shown in Figure 1.2.

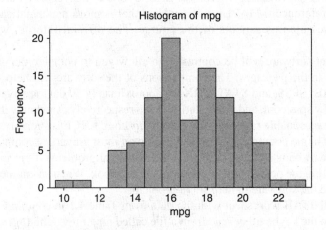

Figure 1.2   MINITAB histogram for the data in Table 1.1.

The histogram shows the engineer or scientist how the data is distributed. In this case, the miles per gallon data has a distribution we call the **normal distribution**. Most of the data is near the middle. There are a few values that are down in the 10 and 11 range and a few that are up near the 22 and 23 range. Most of the data is concentrated in the middle, between 13 and 21 miles per gallon.

     Another graphical picture called a dot plot is constructed using the software package MINITAB. **Dot plots** place dots atop numerical values that are placed along a horizontal axis. They are particularly useful with small data sets. The pulldown **Graph ⇒ Dot plot** gives the graph shown in Figure 1.3. The mpg values are plotted along the horizontal axis and dots are plotted above the numbers on the horizontal axis. The dot plot shows that most of the mpg values are between 12 and 20 with the minimum value being 10.4 and the maximum value being 22.9.

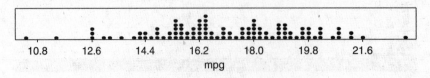

Figure 1.3   MINITAB dot plot for the data in Table 1.1.

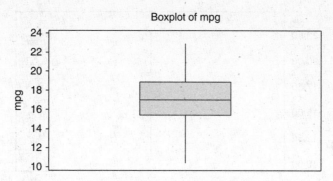

Figure 1.4   MINITAB box plot for the data in Table 1.1.

Figure 1.4 gives a box plot for the data in Table 1.1. The pulldown **Graph ⇒ Box plot** gives the graph shown in Figure 1.4.

In its simplest form a **box plot** shows a **box** which contains the middle 50% of the data values. It also shows two **whiskers** that extend from the box to the maximum value and from the box to the minimum value. In Figure 1.4, the box extends from a number between 15 and 16 to a number between 18 and 19. The whiskers extend from a number a little larger than 10 to the box and from the box to a number a little less than 23. We shall have more to say about box plots as the text develops.

Figure 1.5 shows the worksheet for the software package SPSS containing the mpg data.

mpg.sav [DataSet1] - SPSS Data Editor

File Edit View Data Transform Analyze Graphs Utilities Add-ons Window Help

1 : mpg   19.7

	mpg	var	var	var	var	var	var	var	var	var
1	19.70									
2	20.80									
3	22.90									
4	17.80									
5	14.40									
6	14.00									
7	16.10									
8	19.60									
9	17.80									
10	19.20									
11	17.90									
12	18.80									
13	20.70									
14	16.30									
15	18.10									
16	14.40									
17	16.30									
18	20.80									
19	15.50									
20	12.90									
21	15.70									
22	17.40									

Data View / Variable View /

SPSS Processor is ready

start   Eng-Stat-Demystified   chp1.doc - Microsoft ...   mpg.sav [DataSet1]-...   2:56 PM

Figure 1.5   SPSS worksheet containing the data from Table 1.1.

Figure 1.6 gives an SPSS created box plot. The pulldown **Graph ⇒ Box plot** is used to create the box plot in SPSS. The box plot gives a box that covers the middle 50% of the data. The box in Figure 1.6 extends from 15.5 to 18.8. The whiskers extend to 10.4 on one end of the box and to 22.9 on the other end. If the data has a bell-shaped histogram, it will have a box plot like the one shown in Figure 1.6. The middle of the data is shown by the heavy line in the middle of the box and divide the box in half. Half of the data will be below the middle line and half above the middle line. The two whiskers will be roughly equal in length for a normal distribution.

The SAS worksheet for the data in Table 1.1 is given in Figure 1.7. The SAS pulldown **Graph ⇒ Histogram** gives the histogram of the mpg values and is shown in Figure 1.8.

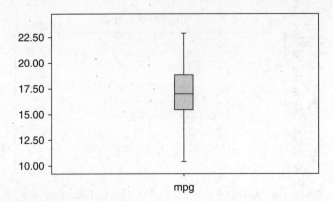

Figure 1.6   SPSS generated box plot for the data in Table 1.1.

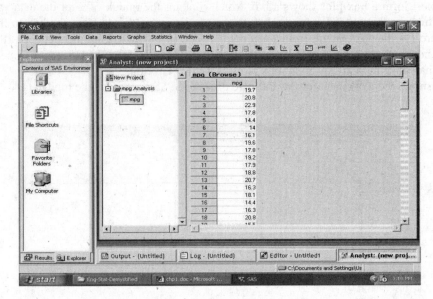

Figure 1.7   SAS worksheet with the data in Table 1.1.

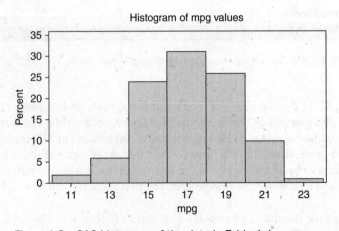

Figure 1.8   SAS histogram of the data in Table 1.1.

```
Statistix 8.0
Stem and Leaf Plot of mpg
       Leaf Digit Unit = 0.1              Minimum   10.400
       10   4   represents  10.4          Median    17.000
                                          Maximum   22.900

              Stem  Leaves
        1   10   4
        1   10
        2   11   4
        2   11
        2   12
        6   12   5669
        7   13   1
        8   13   5
       13   14   01244
       18   14   67789
       25   15   2233334
       32   15   5556779
       45   16   0011222333334
       49   16   6689
       (5)  17   00114
       46   17   556788999
       37   18   0011234
       30   18   6888899
       23   19   00123
       18   19   5567789
       11   20   112
        8   20   5788
        4   21   12
        2   21   5
        1   22
        1   22   9

      100 cases included    0 missing cases
```

Figure 1.9   STATISTIX stem-and-leaf plot of the data in
             Table 1.1.

A **stem-and-leaf plot** breaks the data down into stems and leaves. For example, 45 consists of stem 4 and leaf 5 or 4 tens and 5 ones. Figure 1.9 gives a STATISTIX created stem-and-leaf plot of the data in Table 1.1. The smallest number is 104 and when multiplied by 0.1 becomes 10.4 mpg. Note that if the stem-and-leaf plot in Figure 1.9 is rotated 90 degrees, the shape of the histogram is shown. In addition to the shape being shown, the actual numbers that comprise the stem-and-leaf plot are also discernable. The pulldown **Statistics ⇒ Summary Statistics ⇒ Stem-and-leaf plot** will give a stem-and-leaf plot. For this stem-and-leaf plot, the first row represents 10.4, the third row represents 11.4, the sixth row represents 12.5, 12.6, 12.6, and 12.9, the seventh row represents 13.1, and the eighth row represents 13.5. The first column represents the cumulative count. The leaves (in the third column) for any row are 0 through 4 or 5 through 9. (See the following from the upper portion of the stem-and-leaf plot in Figure 1.9.)

```
          1      10      4
          1      10
          2      11      4
          2      11
          2      12
          6      12      5669
          7      13      1
          8      13      5
```

The largest number is 22.9. The number in the third row from the bottom of the stem-and-leaf plot is 21.5, the next two numbers in the fourth row from the bottom are 21.1 and 21.2. The next four numbers are 20.5, 20.7, 20.8, and 20.8. The first column is the cumulative counts from the bottom. (See the following from the lower portion of the stem-and-leaf plot in Figure 1.9.)

8	20	5788
4	21	12
2	21	5
1	22	
1	22	9

The class that contains the middle number contains the frequency for that class in parenthesis. In this case the middle number is 17.0. Note that the stem is the number of ones and is in the second column and the leaf is the number of tens and is in the third column.

Figure 1.10 shows the STATISTIX worksheet with the mpg data. The pulldown **Statistics ⇒ Summary Statistics ⇒ Histogram** is used to generate the histogram for the data in Table 1.1. This histogram is shown in Figure 1.11. Note that the classes used to form the histograms by the various packages differ. MINITAB uses 10, 11, . . . , 23 as the center of the classes in Figure 1.2. SAS uses 11, 13, 15, 17, 19, 21, and 23 as the center of the classes in Figure 1.8. STATISTIX uses 10.3, 11.2, 12.1, . . . , 22.9 as the class limits in Figure 1.11. However, the distribution shape is the same with all the packages.

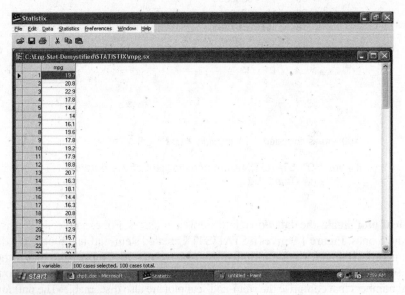

Figure 1.10   STATISTIX worksheet with data from Table 1.1.

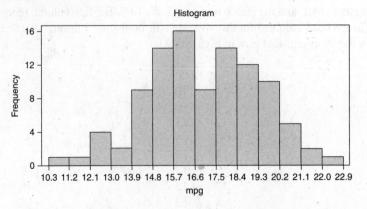

Figure 1.11   STATISTIX histogram of the data for Table 1.1.

All of the statistical software given in this section reveals a set of data that is distributed symmetrically. A few values are low and a few are high, but most are in the center of the distribution. Such a distribution is said to have a **normal** or **bell-shaped** distribution. Statistical graphics helps the engineer form some idea of the characteristic with which he or she is dealing. How variable is it? What type of distribution does it have? Are there some extreme values that occur? In all the spreadsheets shown, the data may be entered directly by the engineer or it may be read from a file.

The histogram, dot plot, box plot, and stem-and-leaf plot are used with quantitative data. **Quantitative engineering data** is obtained from measuring. **Qualitative engineering data** gives classes or categories for your data values. If your data is qualitative, other techniques are used to graphically display your data. Suppose daily inspection of some product results in rejection due to scratches, dents, smudges, failure to meet specifications, and other causes. Suppose the data shown in Table 1.2 was gathered during a given shift at a large company.

Table 1.2   Fifty Causes for Rejection of Items

Scratch	Spec	Dent	Spec	Dent
Dent	Scratch	Scratch	Spec	Dent
Smudge	Other	Scratch	Other	Scratch
Spec	Dent	Dent	Scratch	Scratch
Dent	Spec	Other	Spec	Dent
Scratch	Other	Spec	Dent	Scratch
Dent	Spec	Other	Spec	Other
Smudge	Dent	Dent	Spec	Spec
Spec	Smudge	Dent	Scratch	Dent
Other	Spec	Spec	Smudge	Dent

A **pie chart** is a plot of qualitative data in which the size of the piece of pie is proportional to the frequency of a category. The angles of the pieces of pie that correspond to the categories add up to 360 degrees. For example, the piece of pie that represents dents has an angle equal to 0.30(360 degrees) = 108 degrees. The piece of pie that represents smudges has an angle equal to 0.08(360 degrees) = 28.8 degrees. The data in Table 1.2 may be summarized as follows:

Category	Frequency	Percent
Dent	15	30
Other	7	14
Scratch	10	20
Smudge	4	8
Spec	14	28

EXCEL may be used to form a pie chart of the data. The data, in summarized form, is entered into the worksheet. The **chart wizard** is used to form the following pie chart. Figure 1.12 gives the EXCEL pie chart for the data in Table 1.2.

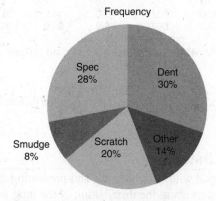

Figure 1.12   EXCEL pie chart of data
in Table 1.2.

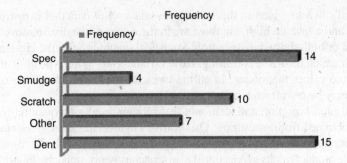

Figure 1.13   EXCEL horizontal bar chart of data in Table 1.2.

A **bar chart** is a plot of qualitative data in which the frequency or percent of the data is the height of a rectangle and each category is the base of the rectangle. Figure 1.13 is an EXCEL horizontal bar chart illustrating the data in Table 1.2.

A **Pareto chart** is a special form of a bar chart. It is a tool used by the engineers to prioritize problems for solution. It was named after the Italian economist Vilfredo Pareto (1848–1923). The bars of the Pareto chart are arranged in descending order. A Pareto chart for the data in Table 1.2 is shown in Figure 1.14. The curve above the histogram is the cumulative curve of frequencies. The curve starts at a count of 15 or 30% (15 out of 50 is 30%) above "Dent." The curve continues and above the tick mark for "Spec" it passes through 29 or 58%. This means that 29 or 58% of the defects are either dents or specs. The curve continues in this manner. The pulldown **Stat ⇒ Quality Tools ⇒ Pareto Chart** is used to produce the Pareto chart shown in Figure 1.14.

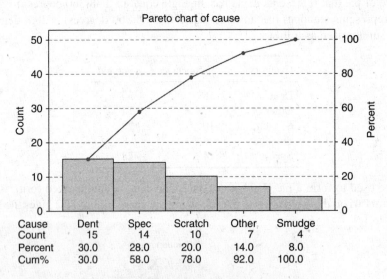

Figure 1.14   MINITAB Pareto chart of the data in Table 1.2.

In this first section, we have considered various graphs from the different software packages for illustrating the distribution of the data.

Descriptive Measures of Data

Descriptive statistics is concerned with summarizing and presenting data concisely. The graphical techniques in the previous section tell us about the distribution of the mpg values in Table 1.1. But suppose we wanted one number that would, in a sense, represent all the data in the table. There are three descriptive statistical measures that we will discuss. They are mean, median, and mode. The **mean** is the sum of all the data divided by the number of observations. The **median** is the middle of the sorted data. The **mode** is the most frequently occurring value in the data set. The sum of the mpg values is 1704.9 and when this sum is divided by 100, the result is 17.049. If the data is sorted, the two values closest to the middle are 17 and 17. They are the fiftieth and fifty-first values in the sorted array. They are averaged to obtain the median, 17. The most frequently occurring value is 16.3 mpg. It has a frequency of 5. We find that the mean, median, and mode are 17.049, 17, and 16.3 respectively. These values are found using EXCEL by entering the numbers in A2:A101. The label mpg is entered into A1. The pulldown **Tools ⇒ Data Analysis** is used to obtain the dialog box. **Descriptive Statistics** is selected from the box. The results are as follows:

EXCEL Output for Descriptive Statistics

mpg

Mean	**17.049**
Standard Error	0.239699
Median	**17**
Mode	**16.3**
Standard Deviation	2.396989
Sample Variance	5.745555
Kurtosis	–0.13851
Skewness	–0.17413
Range	12.5
Minimum	10.4
Maximum	22.9
Sum	1704.9
Count	100

The values of the mean, median, and mode are shown in bold.

When MINITAB is used, the pulldown is **Stat ⇒ Basic Statistics ⇒ Display Descriptive Statistics**. The results are as follows:

MINITAB Output for Descriptive Statistics

Descriptive Statistics: mpg

Variable	Mean	StDev	Minimum	Q1	Median	Q3	Maximum
mpg	17.049	2.397	10.400	15.425	17.000	18.875	22.900

A measure of variation or dispersion that is simple to calculate and easy to understand is the **range**. The range is the largest value minus the smallest value in the data set. By consulting the EXCEL or MINITAB output above, we see that the maximum value is 22.9 and the minimum value is 10.4. The range is therefore 22.9–10.4 or 12.5. If a data set is high in variability, the range is large. If a data set is low in variability, the range is small. The range therefore is representative of the variability of a set of data.

The SAS, SPSS, and STATISTIX output for the descriptive statistics are as follows. The SAS pulldown is **Statistics ⇒ Descriptive ⇒ Summary Statistics**.

SAS Output for Descriptive Statistics

The MEANS Procedure
Analysis Variable : mpg

Mean	Std Dev	N	Minimum	Maximum
17.0490000	2.3969886	100	10.4000000	22.9000000

The STATISTIX pulldown is **Statistics** $\Rightarrow$ **Summary Statistics** $\Rightarrow$ **Descriptive Statistics.**

STATISTIX Output for Descriptive Statistics

Descriptive Statistics	
	mpg
N	100
Mean	17.049
SD	2.3970
Variance	5.7456
Minimum	10.400
Median	17.000
Maximum	22.900

The SPSS pulldown is **Analyze** $\Rightarrow$ **Descriptive Statistics** $\Rightarrow$ **Descriptive**.

SPSS Output for Descriptive Statistics

Descriptive Statistics

	N	Range	Minimum	Maximum	Sum	Mean	Std. Deviation	Variance
mpg	100	12.50	10.40	22.90	1704.90	17.0490	2.39699	5.746
Valid N (listwise)	100							

The variance and standard deviation is easiest to understand if we select a smaller data set before finding these measures on a larger set, such as the mpg values. Suppose we wished to find the variance/standard deviation of a set of golf scores for engineer Jane Doe. The **variance** is defined to be $S^2 = \dfrac{\sum(X - \bar{X})^2}{n-1}$ or $\dfrac{\sum X^2 - \dfrac{(\sum X)^2}{n}}{n-1}$. The two forms can be shown to be algebraically equivalent. ($\bar{X}$ is the symbol that is used to represent the mean of a sample.)

Suppose Jane shot the following on 5 different occasions: 85, 90, 95, 88, and 92. First note that the mean is 90. Using the first form of the equation for variance, we have the following:

X, score	$X - \bar{X}$, deviation	$[X - \bar{X}]^2$, square
85	−5	25
90	0	0
95	5	25
88	−2	4
92	2	4
	$\sum[X - \bar{X}] = 0$	$\sum[X - \bar{X}]^2 = 58$

The middle column in the table above illustrates that the sum of the deviations about the mean is always 0. The third column gives the numerator for the variance. When it is divided by $(n-1)$, the variance is obtained. The variance is $S^2 = 58/4 = 14.5$. The standard deviation is the square root of 14.5 or 3.81.

The variance and standard deviation may also be found using the equivalent form of the formula, $S^2 = \dfrac{\sum X^2 - \dfrac{(\sum X)^2}{n}}{n-1}$. When using this formula, the sum and the sum of squares are found first. $\sum X = 85 + 90 + 95 + 88 + 92 = 450$ and $\sum X^2 = 7225 + 8100 + 9025 + 7744 + 8464 = 40{,}558$. Then, $S^2 = \dfrac{40558 - 40500}{4} = \dfrac{58}{4} = 14.5$. Today, no one finds the variance or standard deviation by hand. A software package is used to find them.

The variance and standard deviation are given for the 100 mpg values by the five descriptive statistics routines above. The variance is seen to be 5.75 in all of the five outputs and the standard deviation is 2.40. When looking over the descriptive statistics above, it is seen that the mean is 17.049, the median is 17.000, the minimum is 10.40, the maximum is 22.90, the range is 12.50, the variance is 5.75, and the standard deviation is 2.40. There are also other quantities in the descriptive statistics that we shall encounter later.

For data that is normally distributed, as is the data in Table 1.1, there is a special interpretation for mean and standard deviation. The **empirical rule** states that for any normal distribution the following relationships approximately hold true: (1) Approximately 68% of the distribution is between the (mean − standard deviation) and the (mean + standard deviation). (2) Approximately 95% of the distribution is between the (mean − 2 standard deviations) and the (mean + 2 standard deviations). (3) Approximately 99.7% of the distribution is between the (mean − 3 standard deviations) and the (mean + 3 standard deviations). The data in Table 1.1 is given in sorted form in Table 1.3.

Table 1.3   Sorted mpg Values for 100 Pickup Trucks

10.4	14.2	15.3	15.7	16.3	17.0	17.9	18.6	19.2	20.1
11.4	14.4	15.3	15.9	16.3	17.1	17.9	18.8	19.3	20.2
12.5	14.4	15.3	16.0	16.3	17.1	17.9	18.8	19.5	20.5
12.6	14.6	15.3	16.0	16.3	17.4	18.0	18.8	19.5	20.7
12.6	14.7	15.4	16.1	16.4	17.5	18.0	18.8	19.6	20.8
12.9	14.7	15.5	16.1	16.6	17.5	18.1	18.9	19.7	20.8
13.1	14.8	15.5	16.2	16.6	17.6	18.1	18.9	19.7	21.1
13.5	14.9	15.5	16.2	16.8	17.7	18.2	19.0	19.8	21.2
14.0	15.2	15.6	16.2	16.9	17.8	18.3	19.0	19.9	21.5
14.1	15.2	15.7	16.3	17.0	17.8	18.4	19.1	20.1	22.9

One standard deviation below the mean is $17.049 - 2.397 = 14.652$ and one standard deviation above the mean is $17.049 + 2.397 = 19.446$. There are 68% of the values in Table 1.3 between 14.652 and 19.446. Two standard deviations below the mean is $17.049 - 4.794 = 12.255$ and two standard deviations above the mean is $17.049 + 4.794 = 21.843$. Of the data in Table 1.3, 97% is between these two values. All the data is within 3 standard deviations of the mean. The percents 68, 97, and 100 are approximately equal to 68, 95, and 99.7.

Measures of Relative Standing

A large group of engineering students took a standardized test in a course entitled Calculus for Engineers. The scores attained by 225 students are shown in Table 1.4.

Joe was told that his raw score was 650. This meant very little to Joe. What Joe really wanted to know was how did he "stack up" with the rest of the group. What percent of the scores was lower than his? Percentiles are the answer to Joe's question. There are 99 percentiles, 9 deciles, and 3 quartiles. For example, the 90th percentile is a score such that 90% of the scores are less than that score and only 10% exceed that score. The *p*th **percentile** is the number that divides the lower $p\%$ of a set of data from the upper $(100 - p)\%$ of the data.

The 1st decile is the same as the 10th percentile, the 2nd decile is the same as the 20th percentile, etc. The 1st quartile is the same as the 25th percentile, the 2nd quartile is the same as the 50th percentile, and the 3rd quartile is the same as the 75th percentile. The notation $P_1, \ldots, P_{99}$ represents percentiles, the notation $D_1, D_2, \ldots, D_9$ represents deciles, and $Q_1, Q_2,$ and Q_3 represents quartiles. We have the following equalities: $Q_1 = P_{25}, Q_2 = P_{50} = \text{median}$, and $Q_3 = P_{75}$.

If the data in Table 1.4 is entered into cells A1:O15 of the EXCEL worksheet and the command = PERCENTILE(A1:O15,0.25) is entered into any empty cell, the number 441 is returned. Similarly, the command =PERCENTILE(A1:O15,0.50) gives 517 and =PERCENTILE(A1:O15,0.75) gives 580. We know

Table 1.4   Scores on a Standardized Test in Calculus for Engineers

526	600	492	577	480	528	304	525	522	549	386	358	565	565	574
488	397	327	448	420	624	606	417	487	539	417	467	444	620	498
608	546	502	573	449	438	688	468	613	506	517	518	532	427	491
493	502	488	453	546	517	721	775	310	532	437	340	446	413	502
535	656	426	319	527	397	457	311	298	719	594	709	521	441	487
553	468	435	666	305	365	543	486	472	525	522	378	483	379	539
370	506	536	496	606	539	661	444	520	676	662	404	449	413	417
716	546	447	612	479	359	584	533	460	503	478	498	581	542	537
724	570	438	587	342	407	446	622	605	553	529	637	384	495	504
619	500	327	460	586	314	569	562	588	375	645	381	455	580	434
431	525	642	537	615	712	519	529	630	536	409	629	566	526	576
593	404	471	482	364	582	416	452	480	619	495	267	401	792	388
622	798	488	425	512	702	406	507	460	559	426	658	565	274	664
679	534	645	579	530	381	667	477	433	472	632	552	641	440	574
477	590	489	402	326	651	619	355	445	577	743	521	607	574	599

that Joe, who scored 650 on the test, scored higher than 75% of those taking the test since the third quartile is 580. The ninth decile, given by the command =PERCENTILE(A1:O15,0.9), gives 645. This means that Joe's score is greater than Q_3 as well as D_9. The command =PERCENTILE(A1:O15,0.91) gives 655.2. Joe scored between D_9 and P_{91}.

If MINITAB is used to find some of these same measures, the data is put into column C1 and the descriptive statistics command gives basically the same values for the first, second, and third quartiles as shown below.

Descriptive Statistics: Score

Variable	Mean	StDev	Minimum	Q1	Median	Q3	Maximum
score	512.42	104.46	267.00	440.50	517.00	580.50	798.00

If Proc Univariate of SAS is used, the following percentiles are given: 100% Max is the largest value in the data set, $P_{99} = 775$, $P_{95} = 688$, $P_{90} = 645$, $Q_3 = 580$, $Q_2 = 517$, $Q_1 = 441$, $P_{10} = 379$, $P_5 = 327$, $P_1 = 298$, and 0% Min is the smallest number in the data set.

The UNIVARIATE Procedure

Variable: Quantile	Score Estimate
100% Max	798
99%	775
95%	688
90%	645
75% Q3	580
50% Median	517
25% Q1	441
10%	379
5%	327
1%	298
0% Min	267

When SPSS is used and the pulldown **Analyze ⇒ Descriptive Statistics ⇒ Frequencies** is given, the following output is obtained. $P(i)$ represents the ith percentile.

Statistics

Score		
N	Valid	225
	missing	0
		P(i)
	1	280.2
	5	327.0
	10	378.6
	25	440.5
	50	517.0
	75	580.5
	90	647.4
	95	697.8
	99	787.5

When STATISTIX is used, the pulldown **Statistics ⇒ Summary Statistics ⇒ Percentiles** is used.

Percentiles

Variable	Cases	1.0	5.0	25.0	50.0	75.0
score	225	280.24	327.00	440.50	517.00	580.50

It is seen that software packages give about the same values for the percentiles. However, differences do exist because the software programs do use slightly different techniques to find the percentiles.

One technique for finding percentiles "by hand" is the following: **(1) Sort or order the n observations from smallest to largest. Suppose we are calculating the pth percentile. (2) Calculate $np/100$. If $np/100$ is not an integer, round it up to the next integer and find the corresponding ordered value. If $np/100$ is an integer, say k, calculate the mean of the kth and $(k+1)$th ordered observations.**

For example, suppose we are looking for the 25th percentile of the data in Table 1.4. Then $np/100$ is $225(25)/100 = 56.25$. Round this up to 57. The 57th observation when the 225 scores are sorted is 441. This is basically the same as given by all the software packages.

Another measure of dispersion is defined that depends on understanding percentiles. That measure is the **inter-quartile range**. The inter-quartile range (IQR) is defined to be

$$IQR = Q_3 - Q_1.$$

The IQR measures the spread of the middle 50% of the data. The IQR for the spread in the calculus scores is $580.5 - 440.5 = 140$ points. The IQR defines the width of the box in a box plot.

Another measure of relative standing for a measurement is the Z-score for a measurement. The **Z-score** for a measurement X is defined as

$$\text{Z-score} = \frac{X - \bar{X}}{S}.$$

For example, using the data in Table 1.4, the mean is 512.4 and the standard deviation is 104.5. Joe's score was 650. The Z-score is $\frac{650 - 512.4}{104.5} = 1.32$. Joe's score is 1.32 standard deviations above the mean.

Using the rather simple techniques given in this first chapter, engineers may analyze a set of data. They may find the distribution of the data by using software packages to construct a histogram, a box plot, a dot plot, and a stem-and-leaf plot. They may also find a representative measure such as mean, median, or mode. They may also find a measure that is reflective of the variability of the data. In addition, measures of location may be found.

SOLVED PROBLEMS

Graphical displays of data

1.1. The data in Table 1.5 is the inside diameters of piston rings sampled from a forge.

Table 1.5   Inside Diameters of Fifty Piston Rings

14.619	14.829	14.985	14.788	14.966
14.867	14.939	15.023	14.922	14.879
15.187	14.854	15.056	15.089	14.995
14.838	15.093	15.187	14.652	15.017
15.113	14.950	14.929	15.001	14.850
14.941	14.829	14.818	14.795	14.774
14.771	15.058	14.898	14.951	14.909
14.967	15.096	14.841	14.815	14.627
14.809	14.978	14.809	15.140	15.042
14.971	14.764	14.732	15.089	14.810

A STATISTIX box plot is given. By considering the graphic, give the lower end of the lower whisker, Q_1, Q_2, Q_3, and the upper end of the upper whisker.

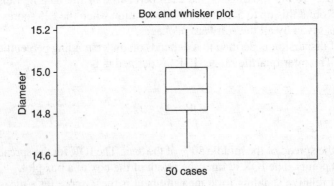

SOLUTION

Lower end of the lower whisker = minimum = 14.619

$$Q_1 = 14.814 \qquad Q_2 = 14.926 \qquad Q_3 = 15.019$$

Upper end of the upper whisker = 15.187

1.2. Use MINITAB to construct a dot plot for the inside diameters of the piston rings. By considering the graphic, answer the following: How many of the diameters are less than 14.72? How many of the diameters are greater than 14.96?

SOLUTION

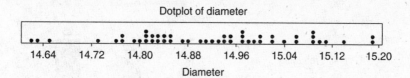

Dotplot of diameter

Three of the 100 diameters are less than 14.72. Twenty of the 100 diameters are greater than 14.96.

1.3. Construct a SPSS histogram for the inside diameters given in Problem 1.1 and identify the shape that is assigned to this histogram.

SOLUTION

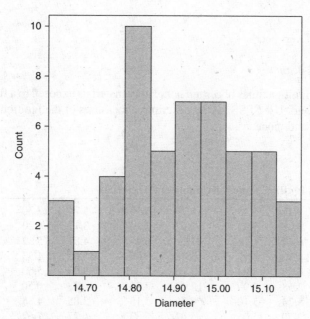

The shape is normal or bell-shaped.

1.4. Seventy-five drivers were asked to judge the maneuverability of a car that new engineering aspects were built into. The following EXCEL pie chart resulted:

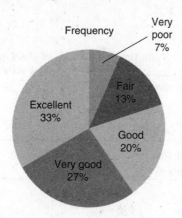

(a) How many of the drivers gave a judgment of excellent?
(b) How many gave a judgment of very poor or fair?

SOLUTION

(a) $0.33(75) = 25$ (b) $(0.07 + 0.13)(75) = 15$

1.5. Consider the following bar chart in regards to Problem 1.3.

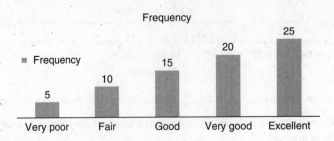

(a) Which judgment was given the most often?
(b) Which judgment was given the least often?

SOLUTION

(a) excellent   (b) very poor

Descriptive measures of data

1.6. Table 1.6 gives the ignition times of certain upholstery materials exposed to a flame, given to the nearest hundredth of a second. Use SPSS to find descriptive measures of the ignition times. In particular find the median, mean, and mode.

Table 1.6   Ignition Times of Upholstery Materials

7.25	7.73	5.38	6.79	5.10	5.70	2.30	4.61
5.26	6.08	5.92	2.92	4.20	3.17	4.20	6.75
3.70	6.47	4.14	7.11	6.45	4.87	5.32	5.56
4.64	14.89	3.52	3.48	5.56	5.08	4.33	6.81
5.40	3.09	3.04	3.16	7.28	5.03	5.60	2.85
14.18	3.58	3.99	12.08	7.71	6.38	6.50	6.48
4.45	5.25	5.10	7.08	5.15	3.85	4.85	2.69
4.24	2.79	5.05	5.02	5.33	6.27	7.67	7.23
5.22	6.78	4.36	5.65	4.36	1.99	7.57	7.62
5.94	3.25	5.49	6.03	4.10	11.78	1.04	15.75

SOLUTION

The data from Table 1.6 is entered into the first column of the data editor. The first column is given the name times. The pulldown **Analyze $\Rightarrow$ Descriptive Statistics $\Rightarrow$ Frequencies** is given. The SPSS output consists of N, mean, median, and mode.

SPSS Output

Statistics
Times

N	Valid	80
	Missing	1
Mean		5.6324
Median		5.2550
Mode		4.20[a]

[a]Multiple modes exist. The smallest value is shown.

The mean and the median are unique. However the mode is only one of four. Such a distribution is said to be multimodal. The modes are 4.20, 4.36, 5.10, and 5.56. Each occurs twice.

1.7. Find the variance and standard deviation of the data in Table 1.6 using EXCEL.

SOLUTION

A	B	C	D	E	F	G	H
7.25	7.73	5.38	6.79	5.1	5.7	2.3	4.61
5.26	6.08	5.92	2.92	4.2	3.17	4.2	6.75
3.7	6.47	4.14	7.11	6.45	4.87	5.32	5.56
4.64	14.89	3.52	3.48	5.56	5.08	4.33	6.81
5.4	3.09	3.04	3.16	7.28	5.03	5.6	2.85
14.18	3.58	3.99	12.08	7.71	6.38	6.5	6.48
4.45	5.25	5.1	7.08	5.15	3.85	4.85	2.69
4.24	2.79	5.05	5.02	5.33	6.27	7.67	7.23
5.22	6.78	4.36	5.65	4.36	1.99	7.57	7.62
5.94	3.25	5.49	6.03	4.1	11.78	1.04	15.75

Go to any open cell and enter the command =VAR(A1:H10). You will get 6.80088 for the variance of the data.

Also, go to any open cell and enter the command =STDEV(A1:H10). You will get 2.6079 for the standard deviation of the 80 ignition values.

1.8. Use MINITAB to draw a histogram of the data in Table 1.6. Describe the shape of the distribution.

SOLUTION

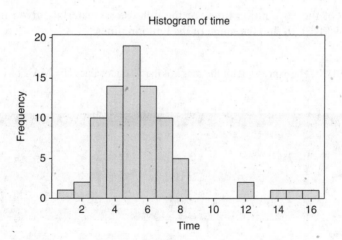

The histogram shows that the distribution is **skewed to the right**.

1.9. $S^2 = \dfrac{\sum (X - \bar{X})^2}{n-1}$ or $\dfrac{\sum X^2 - \dfrac{(\sum X)^2}{n}}{n-1}$. The built-in function =VAR(A1:H10) in EXCEL gives 6.80088 for the variance. The sample mean is $\bar{X} = 5.6324$. (See solved Problem 1.6.) Show that you get the variance if you perform the equation

$$S^2 = \frac{\sum (X - \bar{X})^2}{n-1}.$$

SOLUTION

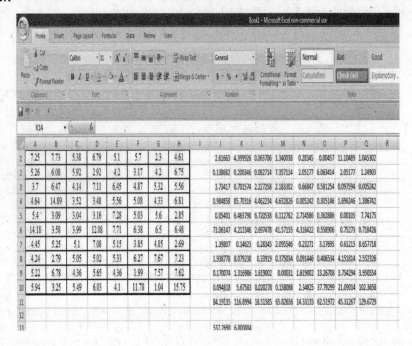

In the screen capture above, the 80 data values are shown in A1:H10. The terms of $\Sigma(X - \bar{X})^2$ are shown in J1:Q10. The terms are added and given in J13 as 537.2698.

The division by $n - 1 = 79$ is shown in K13 as 6.80088. This is the variance. This example shows that the built-in function =VAR(A1:H10) does give the variance. The standard deviation is obtained if we take the square root of the variance.

1.10. Find the range of the 80 ignition times of the upholstery materials given in Table 1.6 of solved problems. Use EXCEL to find the range of the ignition times.

SOLUTION

The data is in A1:H10. The expression for the range of the 80 data values is = MAX (A1:H10) – MIN(A1:H10) = 14.71 seconds.

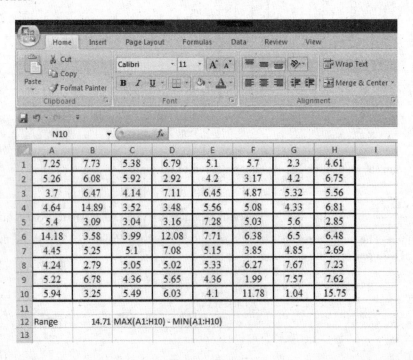

Measures of relative standing

1.11. For the 80 ignition times, find P_{10}, Q_1, median, Q_3, and P_{95} using STATISTIX.

SOLUTION

Enter the data from Table 1.6 into the worksheet of STATISTIX. Give the pulldown **Statistics ⇒ Summary Statistics ⇒ Percentiles**. The STATISTIX output is as follows:

Statistix 8.1
7/8/2009, 2:33:42 PM
Percentiles

Variable	Cases	10.0	25.0	50.0	75.0	95.0
times	80	3.0450	4.1550	5.2550	6.4950	12.065

$$P_{10} = 3.045 \qquad Q_1 = 4.155 \qquad \text{median} = 5.255 \qquad Q_3 = 6.495 \qquad P_{95} = 12.065$$

1.12. Different packages have different algorithms for computing statistical measures such as percentiles. It is not easy to see the algorithm that is used by the various statistical software to compute the percentiles. In Problem 1.11, the statistical software STATISTIX gave: $P_{10} = 3.045$, $Q_1 = 4.155$, median = 5.255, $Q_3 = 6.495$, and $P_{95} = 12.065$.

Suppose we use EXCEL to find these same percentiles using the algorithm that we discuss, that is, the technique for finding percentiles "by hand" is the following: (1) Sort or order the n observations from smallest to largest. Suppose we are calculating the pth percentile. (2) Calculate $np/100$. If $np/100$ is not an integer, round it up to the next integer and find the corresponding ordered value. If $np/100$ is an integer, say k, calculate the mean of the kth and $(k+1)$th ordered observations. Compute these percentiles by hand and compare them with what was obtained by STATISTIX.

SOLUTION

First order or sort the 80 ignition times and get the following using EXCEL.

1.04	1.99	2.30	2.69	2.79	2.85	2.92	3.04	3.09
3.16	3.17	3.25	3.48	3.52	3.58	3.70	3.85	3.99
4.10	4.14	4.20	4.20	4.24	4.33	4.36	4.36	4.45
4.61	4.64	4.85	4.87	5.02	5.03	5.05	5.08	5.10
5.10	5.15	5.22	5.25	5.26	5.32	5.33	5.38	5.40
5.49	5.56	5.56	5.60	5.65	5.70	5.92	5.94	6.03
6.08	6.27	6.38	6.45	6.47	6.48	6.50	6.75	6.78
6.79	6.81	7.08	7.11	7.23	7.25	7.28	7.57	7.62
7.67	7.71	7.73	11.78	12.08	14.18	14.89	15.75	

When finding P_{10} calculate $np/100 = 80(10)/100 = 8$. Find the 8th and 9th observations in the sorted array and average the two values. $(3.04 + 3.09)/2 = 3.065$. This compares with 3.045 using STATISTIX in Problem 1.11.

When finding Q_1 or P_{25}, calculate $np/100 = (80)(25)/100 = 20$. Find the 20th and 21st observations in the sorted array and average the two values. $(4.14 + 4.20)/2 = 4.17$. This compares with 4.155 using STATISTIX in Problem 1.11.

When finding Q_2 = median = P_{50}, calculate $np/100 = (80)(50)/100 = 40$. Find the 40th and the 41st observations in the sorted array and average the two values. $(5.25 + 5.26)/2 = 5.255$. This compares with 5.255 using STATISTIX in Problem 1.11.

When finding Q_3 or P_{75}, calculate $np/100 = (80)(75)/100 = 60$. Find the 60th and 61st values in the sorted array and average the two values. $(6.48 + 6.50) = 6.49$. This compares with 6.495 in Problem 1.11.

When finding P_{95}, calculate $np/100 = (80)(95)/100 = 76$. Find the 76th and the 77th observations in the sorted array and average the two values. $(11.78 + 12.08)/2 = 11.93$. This compares with 12.065 in Problem 1.11.

1.13. Use the built-in functions of EXCEL to find P_{10}, Q_1, median, Q_3, and P_{95} of the ignition times. Compare these values with those found in Problems 1.11 and 1.12.

SOLUTION

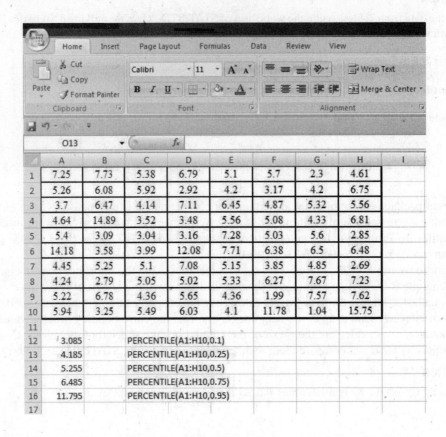

	A	B	C	D	E	F	G	H	I
1	7.25	7.73	5.38	6.79	5.1	5.7	2.3	4.61	
2	5.26	6.08	5.92	2.92	4.2	3.17	4.2	6.75	
3	3.7	6.47	4.14	7.11	6.45	4.87	5.32	5.56	
4	4.64	14.89	3.52	3.48	5.56	5.08	4.33	6.81	
5	5.4	3.09	3.04	3.16	7.28	5.03	5.6	2.85	
6	14.18	3.58	3.99	12.08	7.71	6.38	6.5	6.48	
7	4.45	5.25	5.1	7.08	5.15	3.85	4.85	2.69	
8	4.24	2.79	5.05	5.02	5.33	6.27	7.67	7.23	
9	5.22	6.78	4.36	5.65	4.36	1.99	7.57	7.62	
10	5.94	3.25	5.49	6.03	4.1	11.78	1.04	15.75	
11									
12	3.085		PERCENTILE(A1:H10,0.1)						
13	4.185		PERCENTILE(A1:H10,0.25)						
14	5.255		PERCENTILE(A1:H10,0.5)						
15	6.485		PERCENTILE(A1:H10,0.75)						
16	11.795		PERCENTILE(A1:H10,0.95)						
17									

The built-in functions are found by entering your data into an array and using the function =PERCENTILE(data array, The percentile you want to find entered as a decimal). This is shown in the copy of the EXCEL sheet above.

1.14. Compare the results for P_{10}, Q_1, median, Q_3, and P_{95} of the ignition times using STATISTIX, EXCEL using the By Hand algorithm, and the built-in functions of EXCEL.

SOLUTION

Percentile	STATISTIX	EXCEL By Hand	EXCEL Built-in function
P_{10}	3.045	3.065	3.085
Q_1	4.155	4.170	4.185
Median	5.255	5.255	5.255
Q_3	6.495	6.490	6.485
P_{95}	12.065	11.93	11.795

Note that slightly different values are obtained for the percentiles depending on which statistical software is used. However, they are all close.

1.15. Dr. Stephens' students obtained the following scores in their final exam of engineering statistics.

Student	score	Z-score
Alvin	60	−0.07936
Joe	85	1.464151
Sam	56	−0.34244
Shirley	64	0.146776
Lana	53	−0.52581
Tina	64	0.159442
Sally	55	−0.43351
Scott	35	−1.6805
Stan	90	1.778616
Mike	54	−0.48889
average	61.72	
	15.9	

(a) Explain what the Z-score for Stan means.
(b) What is the average for the Z-scores?
(c) What is the standard deviation for the Z-scores?

SOLUTION

(a) Stan's Z-score, 1.78, means he scored 1.78 standard deviations above the mean of the class.
(b) The average of the Z-scores is 0.
(c) The standard deviation of the Z-scores is 1.

SUPPLEMENTARY PROBLEMS

1.16. A computer engineer, trying to optimize system performance, collected data on the time between requests for a particular process service. Table 1.7 contains 50 measurements collected in the study.

Table 1.7   Time Between Requests in Microseconds

0.390	2.793	0.535	0.129	0.574
1.922	0.935	0.274	0.956	0.184
0.157	5.417	0.378	0.070	0.975
1.614	2.152	1.868	2.103	0.006
0.035	0.000	0.293	5.448	0.115
0.272	5.296	2.250	2.525	0.391
0.347	0.902	0.771	0.619	1.530
3.192	1.114	1.305	1.376	0.208
0.222	0.568	0.266	1.588	1.240
1.109	0.592	0.486	0.013	0.026

Construct a dot plot for the data in Table 1.7. What shape does the distribution have? How many times exceed 4.8 microseconds? How many of the times are less than 0.8 microseconds?

1.17. Construct a stem-and-leaf plot for the data in Table 1.7. Looking at the stem-and-leaf plot, what are the three largest times?

1.18. Sort the times in Table 1.7. Find the mean and standard deviation. What percent of the data is within 3 standard deviations of the mean? That is, what percent of the data is between (mean − 3 standard deviations) and (mean + 3 standard deviations)?

1.19. Construct a box plot for the data in Table 1.7. Explain the structure of the box plot.

1.20. Construct a histogram of the data in Table 1.7.

1.21. Give the SPSS descriptive statistics for the data in Table 1.7.

1.22. Use EXCEL to find the range of the data in Table 1.7.

1.23. Find the percent of the data in Table 1.7 that lies within 1 standard deviation of the mean. That is, what percent of the data is between (mean − 1 standard deviation) and (mean + 1 standard deviation)?

1.24. Give the descriptive statistics provided by EXCEL.

1.25. Contrast finding the median using the built-in function =**Median(array)** and the technique of sorting an array and averaging the two values nearest to the middle of the array using EXCEL. Use the data in Table 1.7.

1.26. Compressive strength was measured on 30 specimens of a new aluminum alloy that was undergoing development as a material for airplanes. The measurements are shown in Table 1.8.

Table 1.8   Compressive Strength Measurements on an Aluminum Alloy

55.5	61.4	63.8
53.1	49.8	60.8
56.4	61.4	55.7
63.1	60.2	60.3
61.2	53.2	56.7
70.9	52.7	56.0
56.7	58.0	66.1
59.3	64.6	53.0
63.5	60.5	67.9
59.5	54.8	59.2

Use the software STATISTIX to find P_{10}, P_{20}, P_{30}, P_{40}, and P_{50}.

1.27. Use SPSS on the data in Table 1.8 to find P_{10}, P_{20}, P_{30}, P_{40}, and P_{50}.

1.28. Use EXCEL on the data in Table 1.8 to find P_{10}, P_{20}, P_{30}, P_{40}, and P_{50}.

1.29. Use MINITAB to find the mean and standard deviation of the strengths in Table 1.8.

1.30. Use MINITAB to find the median and the mean of the strengths in Table 1.8.

CHAPTER 2

Probability

Sample Space, Events, and Operations on Events

Before discussing probability, some definitions are needed. An **experiment** is any operation or procedure whose outcome cannot be predicted with certainty. The **sample space** associated with an experiment consists of all possible outcomes associated with the experiment. An **event** is a subset of the sample space.

EXAMPLE 2.1 The simplest example of an experiment is flipping a coin. The sample space is head or tail or in the short hand of mathematics $S = \{H, T\}$. Another example is the toss of a die. The sample space for this experiment may be represented as $S = \{1, 2, 3, 4, 5, 6\}$.

EXAMPLE 2.2 Consider the following experiment. A student is to be selected from a course entitled Differential Equations for Engineers to be interviewed concerning the engineering program at Midwestern University. The students in the course may be classified according to their sex and their area of engineering. The following table gives the makeup of the class according to sex and major.

	Chemical engineering	Electrical engineering	Civil engineering	Industrial engineering
Male	Scott Hassan	Todd Jim Sam	Tom Joe	Ted
Female	Eve Heather Sue	Lana Beverly	Jody Brittany Sandy Lisa	Kristi Courtney Marie

Define the following events on the experiment of selecting one student from this class. Event A is that a female is selected and event B is that the student is majoring in chemical or electrical engineering. These events may be represented in the form of a **Venn diagram**. In a Venn diagram, the sample space is represented as a rectangle and events are rectangles inside the sample space. Figure 2.1 shows the sample space S and events A and B. The events A and B are composed of the following outcomes:

$A = \{$Eve, Heather, Sue, Lana, Beverly, Jody, Brittany, Sandy, Lisa, Kristi, Courtney, Marie$\}$

$B = \{$Scott, Hassan, Todd, Jim, Sam, Eve, Heather, Sue, Lana, Beverly$\}$

Operations on the sets may be then defined. The main operations are intersection, union, and complement.

The **intersection** of A and B are the outcomes they have in common. The intersection is written as $A \cap B$. The intersection is $A \cap B = \{$Eve, Heather, Sue, Lana, Beverly$\}$. It is shown in Figure 2.1 in gray.

The **union** of two events, A and B, is the set of outcomes in A or B or both. It is represented as $A \cup B$. In Figure 2.1, $A \cup B = \{$Jody, Brittany, Sandy, Lisa, Kristi, Courtney, Marie, Eve, Heather, Sue, Lana, Beverly, Scott, Hassan, Todd, Jim, Sam$\}$

The **complement** of A is all the outcomes in S that are not in A. It is represented as A^c. In Figure 2.1, $A^c = \{$Scott, Todd, Sam, Hassan, Jim, Tom, Joe, Ted$\}$. The complement of B is $B^c = \{$Jody, Sandy, Kristi, Marie, Brittany, Lisa, Courtney, Tom, Joe, Ted$\}$.

Two events, A and B, are **mutually exclusive** if $A \cap B$ is empty.

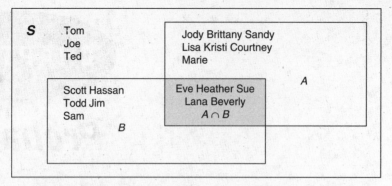

Figure 2.1   The figure shows events *A*, *B* and the intersection of events *A* and *B* in gray.

EXAMPLE 2.3   The following may be seen by considering Figure 2.1. (The reader should verify these.)

$$A \cap B^c = \{\text{Jody, Sandy, Kristi, Marie, Brittany, Lisa, Courtney}\}$$

$$B \cap A^c = \{\text{Scott, Todd, Sam, Hassan, Jim}\}$$

$$(A \cup B)^c = \{\text{Tom, Joe, Ted}\}$$

The general Venn diagram representation of the intersection of events *A* and *B* is shown in Figure 2.2.

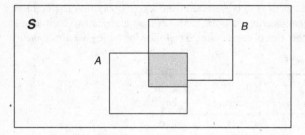 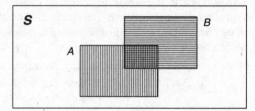

Figure 2.2   Intersection of events *A* and *B* is shown in gray.

Figure 2.3   Union of events *A* and *B* are lined horizontal (*B*), vertical (*A*), and cross-hatched (*A* and *B*).

The union is shown in Figure 2.3. Event *A* is shaded vertically, event *B* is shaded horizontally, and $A \cap B$ is crosshatched.

The complement of *A*, represented as A^c, is the gray portion of Figure 2.4.

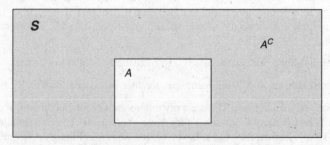

Figure 2.4   Complement of event *A* is shown in gray.

Union means the same as the word *or*. Intersection means the same as the word *and*, and the complement is the same as the word *not*. That is, the intersection of two events is the outcomes that are in *A* and *B*, the union is the outcomes that are in *A* or *B*. The complement of *A* is the outcomes that are in *S* that are not in *A*.

Venn diagrams are often used to verify relationships among sets.

EXAMPLE 2.4 For example, suppose we wished to show that $(A \cap B)^c = A^c \cup B^c$. First, draw a Venn diagram representing $(A \cap B)^c$. This is everything outside of the intersection of A and B as shown in Figure 2.5 in gray.

In Figure 2.6, B^c is shaded horizontally.

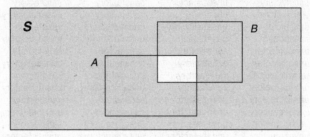

Figure 2.5 $(A \cap B)^c$ is shaded gray.

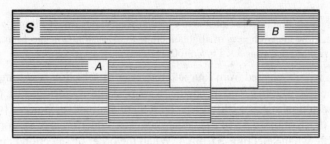

Figure 2.6 B^c is shaded horizontally.

In Figure 2.7, A^c is shaded vertically.

When the union of the set in Figure 2.6 and the set in Figure 2.7 is created, the result is Figure 2.5. Hence, it is seen that $(A \cap B)^c = A^c \cup B^c$.

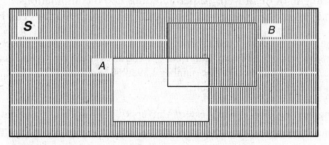

Figure 2.7 A^c is shaded vertically.

The Multiplicative Rule, Permutations, and Combinations

Suppose an engineer is asked to rate the new Toyota Yaris according to steering, shifting, handling, and comfort. Each characteristic may be rated as satisfactory or unsatisfactory. A tree diagram may be used to list the number of different ratings that the engineer can give the Yaris. This tree diagram can be built using an EXCEL worksheet. The results are shown in Figure 2.8.

There are two choices for each category. Each set of four is called a **branch** of the tree. This tree diagram has $2 \times 2 \times 2 \times 2 = 16$ branches.

Suppose there are k sets of elements, n_1 in the first set, n_2 in the second set, . . . , and n_k in the kth set. A sample of k elements is formed by taking one element from each of the k sets. The number of different samples that can be formed is the product $n_1 n_2 \cdots n_k$. This is called the **multiplicative rule**.

steering	shifting	handling	comfort
satisfactory	satisfactory	satisfactory	satisfactory
satisfactory	satisfactory	satisfactory	unsatisfactory
satisfactory	satisfactory	unsatisfactory	satisfactory
satisfactory	satisfactory	unsatisfactory	unsatisfactory
satisfactory	unsatisfactory	satisfactory	satisfactory
satisfactory	unsatisfactory	satisfactory	unsatisfactory
satisfactory	unsatisfactory	unsatisfactory	satisfactory
satisfactory	unsatisfactory	unsatisfactory	unsatisfactory
unsatisfactory	satisfactory	satisfactory	satisfactory
unsatisfactory	satisfactory	satisfactory	unsatisfactory
unsatisfactory	satisfactory	unsatisfactory	satisfactory
unsatisfactory	satisfactory	unsatisfactory	unsatisfactory
unsatisfactory	unsatisfactory	satisfactory	satisfactory
unsatisfactory	unsatisfactory	satisfactory	unsatisfactory
unsatisfactory	unsatisfactory	unsatisfactory	satisfactory
unsatisfactory	unsatisfactory	unsatisfactory	unsatisfactory

Figure 2.8   Possible satisfactory/unsatisfactory ratings for the Yaris.

EXAMPLE 2.5   There are three flights from Houston to Chicago, four flights from Chicago to Memphis and five flights from Memphis to Atlanta. How many choices of flights include the Houston-Chicago-Memphis-Atlanta connection? The multiplicative rule gives the answer as $3 \times 4 \times 5$ or 60.

Suppose a sample of two is selected from four and the order of selection is not important. For simplicity, assume the items are the letters a, b, c, and d. There are six possible choices. {(a,b), (a,c), (a,d), (b,c), (b,d), (c,d)}. If the order of selection makes a difference, then there are twelve choices: {(a,b), (b,a), (a,c), (c,a), (a,d), (d,a), (b,c), (c,b), (b,d), (d,b), (c,d), (d,c)}. When order of selection is of no concern, we say there are 6 **combinations** possible when selecting 2 from 4. When the order of selection is important, we say there are 12 **permutations** possible when selecting 2 from 4.

Generally, when r objects are selected from a set of n distinct objects, there are $P_r^n = \dfrac{n!}{(n-r)!}$ permutations possible. There are $\binom{n}{r} = \dfrac{n!}{r!(n-r)!}$ combinations possible. Applying this formula to the above example, $n = 4$ and $r = 2$. The number of permutations possible is $P_2^4 = \dfrac{4!}{(4-2)!} = \dfrac{24}{2} = 12$ and the number of combinations possible is $\binom{n}{r} = \dfrac{n!}{r!(n-r)!} = \dfrac{4!}{2!(4-2)!} = \dfrac{24}{2 \times 2} = 6$. The answers obtained above when all the combinations and permutations were worked out by listing all of them were 12 permutations and 6 combinations.

When n and r become larger, the number of combinations and permutations can be quite large.

EXAMPLE 2.6   Suppose a poker hand is dealt to an engineer at a casino and the engineer would like to know how many poker hands are possible when 5 cards are dealt from 52. The answer can be worked out using EXCEL. Enter the expression =COMBIN(52,5) in any empty cell. The number 2,598,960 is returned. This means that there are 2,598,960 different poker hands that the engineer could be dealt.

EXAMPLE 2.7   The Engineering club at Midwestern University is to select a president, a vice president, and a secretary/treasurer from the 38 members of the club. Note that order is important in this case. The answer is $P_3^{38} = \dfrac{38!}{35!} = 38(37)(36) = 50,616$. The EXCEL solution is obtained by entering the expression =PERMUT(38,3) in any cell. The number 50,616 is returned.

EXAMPLE 2.8   A calibration study needs to be conducted to see if 20 scales are giving the same weights. How many ways may be selected to perform a preliminary study? The answer is =COMBIN(20,5). The value 15,504 is returned.

EXAMPLE 2.9   Industrial engineers are timing workers to rank their speed at assembling computers. There are 10 workers. How many different permutations are possible? When the rankings are finished, there will be a number 1 (fastest), a number 2, . . . , a number 10 (slowest). The answer is $P_{10}^{10} = \dfrac{10!}{0!} = 10! = 3,628,800$. This is given in EXCEL by =PERMUT(10,10).

Probability

The probability that an event will occur is the likelihood of the occurrence of the event. When walking on campus, the likelihood of spotting a student holding a cell phone is very high. The likelihood that a student has been on the Internet within the last 24 hours is high. The likelihood that a student has been to a BMW dealership within the last 24 hours is low. The likelihood that the student is majoring in computer engineering is low. Probability is a way of assigning a numerical value to these events. There are many different ways of defining probability. We shall discuss three of these. These three definitions are the **classical** definition, the **relative frequency** definition, and the **personal or subjective** definition. There are some definitions such as the measure theoretic approach that are very mathematical and require more mathematics background than most of the readers of this book might have.

The **classical definition** might also be called the equally likely outcomes definition. Suppose an experiment has n equally likely outcomes and an event E occurs if any one of k of these outcomes occurs as an outcome of the experiment. The classical definition of probability defines the probability of event E as $P(E) = k/n$.

EXAMPLE 2.10 Suppose a lot consisting of 100 items contains 5 defectives. If one item is randomly selected, the probability that it is defective is P(item is defective) = 5/100 or .05. There are $k = 5$ outcomes out of $n = 100$ that satisfy the event that the item is defective. It is common terminology to say that there is a 5% chance that a defective item will be chosen. Most people would regard this as a low probability.

EXAMPLE 2.11 Suppose 3 items are inspected and if at least one defective is found, the lot will be 100% inspected. Otherwise, the lot will be passed on. How likely is it that a lot containing 5 defectives will be passed on? The lot will be passed on if no defectives are found in the sample or if all 3 come from the 95 non-defectives. There are $k = \binom{95}{3}$ such samples possible. There are $n = \binom{100}{3}$ samples that could be drawn. The probability that the lot will be passed on, using EXCEL to do the calculations, is =COMBIN(95,3)/COMBIN(100,3) = .855999.

The **frequency probability definition** states that if an experiment were repeated many times, the probability of an event is the proportion of times the event occurs in the n repetitions.

EXAMPLE 2.12 If a jet from Chicago to Omaha is on time 80% of the time, the probability 0.80 is assigned to the event that the jet will arrive on time when flying from Chicago to Omaha. If an engineer communicates 50% of the time by e-mail with a company for which he consults, then the probability is 0.50 that he will communicate by e-mail with the company.

The third definition is of **personal or subjective probability evaluations**. Aeronautical engineers are assigning subjective probability when they assign 0.80 as the probability that a manned spaceflight to Mars in 2016 will be successful. A civil engineer states that the probability that the dikes will hold in New Orleans is 0.75. This is a subjective probability assignment. An army engineer estimating the probability that a terrorist organization will get hold of weapons of mass destruction to be 0.10 is another example of personal probability. Sometimes personal or subjective probability is the only kind available. The events mentioned in this paragraph do not lend themselves to repeatability.

Regardless of the source of the probability assignment, there are probability laws that the probability of events must satisfy. These will be considered in the following sections.

The Axioms of Probability

A probability function is actually a set function. The domain is composed of sets and the range is real numbers.

The usual mathematical function maps a real number into a real number. For example, $y = x^2$ is a real valued function. If the domain is 1, 2, and 3, then the range is 1, 4, and 9. The domain and the range are real numbers. See Figure 2.9.

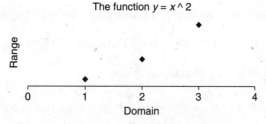

Figure 2.9   A function is a mapping from the real numbers to the real numbers.

Contrast the following example with the above example. Toss a coin twice. The sample space is $S = \{$HH, HT, TH, TT$\}$. Let E_1 be the event that the outcome has no heads, E_2 be the event that the outcome has 1 head, and E_3 be the event that the outcome has 2 heads. $P(E_1) = 1/4$, $P(E_2) = 2/4$, and $P(E_3) = 1/4$. Note that the probability function has domain E_1, E_2, E_3 and range 1/4, 2/4, 1/4. In other words, the domain is composed of sets and the range is real numbers.

The axioms of probability for a finite sample space are as follows. Suppose you have a finite sample space S and an event A in S. Define $P(A)$, the probability of A, to be a value of a set function that satisfies the following three conditions:

$$\text{Axiom 1: } 0 \le P(A) \le 1 \text{ for each event } A \text{ in } S.$$

$$\text{Axiom 2: } P(S) = 1.$$

$$\text{Axiom 3: If } A \text{ and } B \text{ are mutually exclusive events in } S, \text{ then } P(A \cup B) = P(A) + P(B).$$

Mutually exclusive events have empty intersections. The three axioms are consistent with the classical and the frequency definition of probability.

EXAMPLE 2.13   Suppose an experiment has a sample space that is made up of four mutually exclusive events, that is $S = \{A, B, C, D\}$. Check whether the following assignments of probabilities are permissible. (See if the axioms allow the assignments.)

(a) $P(A) = 0.25$, $P(B) = 0.25$, $P(C) = 0.25$, and $P(D) = 0.25$
(b) $P(A) = 0.25$, $P(B) = 0.35$, $P(C) = 0.45$, and $P(D) = -0.05$
(c) $P(A) = 0.25$, $P(B) = 0.25$, $P(C) = 0.35$, and $P(D) = 0.25$
(d) $P(A) = 0.20$, $P(B) = 0.25$, $P(C) = 0.25$, and $P(D) = 0.25$

(a) All the probabilities are between 0 and 1 and their sum is 1. This assignment is allowed.
(b) The assignment is not allowed because $P(D) < 0$.
(c) The sum of the probabilities is not valid because it exceeds 1.
(d) The sum of the probabilities is not valid because it is less than 1.

Some Elementary Theorems (Rules) of Probability

The third axiom of the three axioms in the last section can be extended from two mutually exclusive events to n mutually exclusive events by the use of mathematical induction. It says that if $A_1, A_2, \ldots, A_n$ are mutually exclusive events within a sample space, then $P(A_1 \cup A_2 \cup \ldots \cup A_n)$ is the sum of the probabilities of the n events.

EXAMPLE 2.14   An industrial engineer has found that when she takes a sample of size 5 of a product, 90% of the time there are no defectives in the sample, 3% of the time there is 1 defective, 2% of the time there are 2 defectives, 2% of the time there are 3 defectives, 2% of the time there are 4 defectives, and 1% of the time there are 5 defectives. What is the probability of taking a sample that has at least 3 defectives in the 5? The event "at least 3 defectives in the 5" is composed of 3 mutually exclusive events. The mutually exclusive events are 3 defectives, 4 defectives, and 5 defectives. The probability of at least 3 defectives is $.02 + .02 + .01 = .05$.

Suppose there are two events that are not mutually exclusive. What is $P(A \cup B)$? $(A \cup B)$ can be expressed as the union of three mutually exclusive events as follows: $(A \cup B) = (A \cap B^c) \cup (A \cap B) \cup (B \cap A^c)$. This relationship is shown in Figure 2.10.

From this mutually exclusive decomposition, we have the following from Axiom 3:

$$P(A \cup B) = P(A \cap B^c) + P(A \cap B) + P(B \cap A^c)$$

Now add and subtract $P(A \cap B)$ to the right side to obtain

$$P(A \cup B) = P(A \cap B^c) + P(A \cap B) + P(B \cap A^c) + P(A \cap B) - P(A \cap B)$$
$$P(A \cup B) = P(A) + P(B) - P(A \cap B)$$

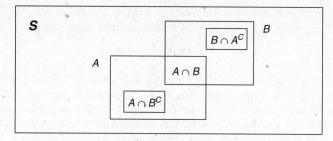

Figure 2.10 $(A \cup B) = (A \cap B^c) \cup (A \cap B) \cup (B \cap A^c)$.

This relationship is called the **general probability addition rule** for two events and if A and B are mutually exclusive $P(A \cup B) = P(A) + P(B)$ is called the **special addition rule**.

EXAMPLE 2.15   It is known that 60% of a class is female, 50% is majoring in chemical or electrical engineering, and 25% is female and majoring in chemical or electrical engineering. What percentage of class is female or majoring in chemical or electrical engineering? A is the event that a member of the class is female and B is the event that a member of the class is majoring in chemical or electrical engineering. The general probability addition rule gives $P(A \cup B) = P(A) + P(B) - P(A \cap B) = .60 + .50 - .25 = .85$. Thus 85% of the class female are majoring in chemical or electrical engineering. See the example at the beginning of Section 2.1 to check this out.

When three or more non-mutually exclusive events are involved, Venn diagrams become more complicated. The following Venn diagram shows three such events.

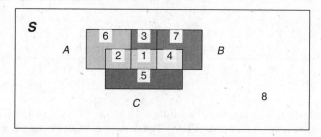

Event A is composed of regions 1, 2, 3, and 6. Event B is composed of regions 1, 3, 4, and 7. Event C is composed of regions 1, 2, 4, and 5. Region 8 is $(A \cup B \cup C)^c$ or everything that is in S, but outside the union of A, B, and C.

EXAMPLE 2.16   Suppose event A is that an engineering student owns a cell phone, event B is that an engineering student has a personal digital assistant (PDA), and event C is that an engineering student lives at home. Region 1 consists of all those students who own a cell phone, own a PDA, and live at home. Region 2 consists of those students who own a cell phone, live at home, but do not own a PDA. Region 3 consists of those students who own a cell phone, have a PDA, but do not live at home. Region 4 consists of those students who own a PDA, live at home, but do not have a cell phone. Region 5 consists of those students who live at home, do not have a cell phone, and do not own a PDA. Region 6 consists of those students who own a cell phone, do not own a PDA, and do not live at home. Region 7 consists of those students who have a PDA, do not own a cell phone, and do not live at home. Region 8 consists of all those students who do not own a cell phone, do not own a PDA, and do not live at home. It can be shown that the following relationship holds: (**This is called the general addition rule for three events.**)

$$P(A \cup B \cup C) = P(A) + P(B) + P(C) - P(A \cap B) - P(A \cap C) - P(B \cap C) + P(A \cap B \cap C)$$

Suppose the following probabilities are known.

Region	1	2	3	4	5	6	7	8
Probability	.02	.04	.03	.05	.16	.45	.20	.05

From Figure 2.10, $P(A \cup B \cup C)$ must equal 0.95. If the additions and subtractions on the right hand side of the equation for $P(A \cup B \cup C)$ are performed, the answer is $.54 + .30 + .27 - .05 - .06 - .07 + .02 = .95$. This confirms the equation for this example.

Suppose A and A^c are complementary events. Then $A \cap A^c$ is empty and $A \cup A^c = S$. $P(A \cap A^c) = P(A) + P(A^c)$ and since $A \cup A^c = S$, $P(A) + P(A^c) = 1$ or $P(A^c) = 1 - P(A)$. The rule of complements states that

$$P(A^c) = 1 - P(A)$$

EXAMPLE 2.17 An engineer selects a sample of 5 iPods from a shipment of 100 that contains 5 defectives. Find the probability that the sample contains at least one defective. At least one defective means 1 or 2 or 3 or 4 or 5 defectives. Remember that *or* and *union* mean the same. The probability $P(1$ or 2 or 3 or 4 or 5 defectives$) = P(1) + P(2) + P(3) + P(4) + P(5)$. The events 1 defective, 2 defectives, 3 defectives, 4 defectives, 5 defectives are mutually exclusive and we add the probabilities. However, the whole sample space consists of 0 defectives, 1 defective, 2 defectives, 3 defectives, 4 defectives, or 5 defectives. And since the probability of a certain event is 1, we know that

$$P(0) + P(1) + P(2) + P(3) + P(4) + P(5) = 1$$

The complement of at least one defective is no defectives and therefore the slick and easy solution is

$$P\{\text{at least one defective}\} = 1 - P\{0 \text{ defectives}\},$$

$$P\{\text{at least one defective}\} = 1 - \frac{\binom{95}{5}}{\binom{100}{5}} = 1 - \frac{57940519}{75287520} = 1 - .77 = .23.$$

EXCEL is used to evaluate the expression that is subtracted from 1. There are 57,940,519 ways that 5 can be chosen from the 95 non-defectives and 75,287,520 ways that 5 can be chosen from the 100. It is interesting to consider the work that the rule of complements saves in this problem. If the rule of complements is not used, the following must be evaluated:

$$\frac{\binom{5}{1}\binom{95}{4}}{\binom{100}{5}} + \frac{\binom{5}{2}\binom{95}{3}}{\binom{100}{5}} + \frac{\binom{5}{3}\binom{95}{2}}{\binom{100}{5}} + \frac{\binom{5}{4}\binom{95}{1}}{\binom{100}{5}} + \frac{\binom{5}{5}\binom{95}{0}}{\binom{100}{5}}$$

If EXCEL is used to evaluate this expression, the following is in cell A1: =COMBIN(5,1)*COMBIN(95,4)/COMBIN(100,5). Similar expressions are in cells A2:A5. The numerical values obtained are shown in the following table. The last number is the expression =SUM(A1:A5).

A	B
0.211425811	=COMBIN(5,1)*COMBIN(95,4)/COMBIN(100,5)
0.018384853	=COMBIN(5,2)*COMBIN(95,3)/COMBIN(100,5)
0.00059306	=COMBIN(5,3)*COMBIN(95,2)/COMBIN(100,5)
6.30915E-06	=COMBIN(5,4)*COMBIN(95,1)/COMBIN(100,5)
1.32824E-08	=COMBIN(5,5)*COMBIN(95,0)/COMBIN(100,5)
0.230410047	=SUM(A1:A5)

Note that the sum agrees with the answer found above. The amount of computation that can be saved using the rule of complements in problems is considerable.

Conditional Probability

The sample space that a probability is computed with respect to has a great bearing upon the probability. Consider the following simple examples.

EXAMPLE 2.18 If a die is tossed, the probability that the face 2 turns up is 1/6. If a die is tossed and it is known that the face was an even number, then the probability that the face 2 turned up is 1/3. This is twice what the probability is if nothing is known. If it is known that the face that turned up was an odd number, then the probability that the face 2 turned up is 0. We see that the probability of an event is conditioned by what other events we know have occurred.

EXAMPLE 2.19 Reconsider the example in the section on 'Sample Space, Events, and Operations on Events.' A class consisted of 20 engineering students. One student was to be selected to discuss the engineering program at Midwestern University. Figure 2.11 shows two events. Event A is that a female is selected and event B is that the student

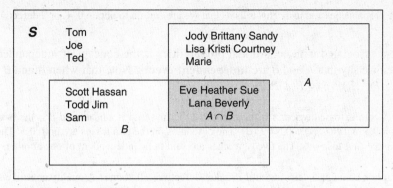

Figure 2.11   Events *A, B* and A and B.

is majoring in chemical or electrical engineering. The names of those students composing the two events and those students not falling in either event are given. The probability of event *A* is $P(A) = 12/20 = .6$. The probability of event *B* is $P(B) = 10/20 = .5$. Suppose it is known that event *B* has occurred and we wish to know the probability of *A*. In other words, it is known that the student is majoring in chemical or electrical engineering. We want to know the probability that the student is a female. The notation for such a conditional probability is $P(A|B)$. Looking at Figure 2.11, we know that we are in rectangle *B* and we want to find the probability that the chosen student is a female. The condition that the student was an electrical or chemical engineer, limits us to rectangle *B*. Within rectangle *B* there are five females. The probability is $5/10 = .5$. The conditional probability was found by dividing the number in event $A \cap B$ by the number in *B*. If the number in $A \cap B$ is divided by the number in *S* and the number in *B* is also divided by the number in *S*, the following formula results:

$$P(A|B) = \frac{P(A \cap B)}{P(B)}$$

The formula $P(A|B) = \dfrac{P(A \cap B)}{P(B)}$ is called the formula for conditional probability.

EXAMPLE 2.20   Suppose 1% of parts have type 1 defect and 2% have type 2 defect and 0.5% have both types of defects. A part is known to have type 1 defect. What is the probability that it has type 2 defect? Let *A* represent the event that a part has a type 1 defect and *B* represent the event that a part has also a type 2 defect. We are asked to find $P(B|A)$. $P(B|A) = \dfrac{P(A \cap B)}{P(A)} = \dfrac{.005}{.01} = .5$. Now suppose a part is known to have a type 2 defect. What is the probability it also has a type 1 defect? We are now asked to find $P(A|B) = \dfrac{P(A \cap B)}{P(B)} = \dfrac{.005}{.02} = .25$.

Take the conditional probability formula and solve it for $P(A \cap B)$ by multiplying both sides by $P(B)$. The following form of the formula is obtained.

$$P(A \cap B) = P(A|B)P(B)$$

EXAMPLE 2.21   A box contains 5 defective and 195 non-defective cell phones. A quality control engineer selects 2 cell phones at random **without replacement**. What is the probability that (a) neither is defective? (b) exactly 1 is defective? (c) both are defective?

(a) *P*(first is non-defective ∩ second is non-defective) = *P*(first is non-defective)*P*(second is non-defective|first is non-defective) = $\dfrac{195}{200}\dfrac{194}{199} = .9505$

(b) *P*((first is non-defective ∩ second is defective) ∪ (first is defective ∩ second is non-defective)). The union of two mutually exclusive events tells us to add.

= *P*(first is non-defective ∩ second is defective) + *P*(first is defective ∩ second is non-defective)

= *P*(first is non-defective)*P*(second is defective|first is non-defective) + *P*(first is defective)*P*(second is non-defective|first is defective)

= $\dfrac{195}{200}\dfrac{5}{199} + \dfrac{5}{200}\dfrac{195}{199} = .0490$.

(c) *P*(first is defective ∩ second is defective) = *P*(first is defective)*P*(second is defective|first is defective) = $\dfrac{5}{200}\dfrac{4}{199} = .0005$.

Notice that the three probabilities add to 1. This tells us that we are certain to obtain 0, 1, or 2 defectives when we select 2 from the 200.

Another concept associated with conditional probabilities is the concept of independence of events. If $P(A|B) = P(A)$ then we say that A and B are **independent events**. Note that when A and B are independent events, $P(A \cap B) = P(A|B)P(B) = P(A)P(B)$.

EXAMPLE 2.22　A coin is tossed twice. A is the event that the first toss is a head, and B is the event that the second toss is a head. $P(A) = 0.5$ and $P(B|A) = 0.5 = P(B)$. The outcomes on the two tosses are both 0.5. The coin has no idea what the outcome on the first toss was. The two outcomes are said to be independent of one another.

EXAMPLE 2.23　A box contains 5 defective and 195 non-defective cell phones. A quality control engineer selects two cell phones at random **with replacement**. What is the probability that (a) neither is defective? (b) exactly 1 is defective? (c) both are defective? Note that this example is just like the one preceding the last example, except the sampling is done with replacement. This makes the second draw independent of the first draw.

(a)　$P(\text{first is non-defective} \cap \text{second is non-defective}) = \dfrac{195}{200}\dfrac{195}{200} = .950625$

(b)　$P(\text{exactly one defective}) = \dfrac{195}{200}\dfrac{5}{200} + \dfrac{5}{200}\dfrac{195}{200} = .04875$

(c)　$P(\text{both are defective}) = \dfrac{5}{200}\dfrac{5}{200} = .000625$

Note that in the case of sampling with replacement, the three probabilities also add to 1. ($.950625 + .04875 + .000625 = 1$).

If n events are independent, then $P(A_1 \cap A_2 \cap \ldots \cap A_n)$ equals the product of the probabilities of the events A_i, $i = 1, 2, \ldots, n$.

EXAMPLE 2.24　If a die is rolled 5 times, what is the probability of obtaining five 6s in a row? The rolls are independent of one another. The probability of a 6 on any one roll is 1/6. The probability of five 6s in a row is $\left(\dfrac{1}{6}\right)^5$.

EXAMPLE 2.25　A machine has a probability of producing a defective equal to .05 every time it produces a product. If three of its products are selected randomly and independently during a shift, what is the probability that a quality engineer will find: (a) no defectives in the three? (b) 1 defective in the three? (c) 2 defectives in the three? (d) 3 defectives in the three?

(a)　$P(\text{non-defective }\textbf{and}\text{ non-defective }\textbf{and}\text{ non-defective}) = (.95)(.95)(.95) = 0.857375$

(b)　$P(\text{defective }\textbf{and}\text{ non-defective }\textbf{and}\text{ non-defective }\textbf{or}\text{ non-defective }\textbf{and}\text{ defective }\textbf{and}\text{ non-defective }\textbf{or}\text{ non-}$
　　defective **and** non-defective **and** defective) $= .05(.95)(.95) + .95(.05)(.95) + .95(.95)(.05) = .135375$

(c)　$P(\text{defective }\textbf{and}\text{ defective }\textbf{and}\text{ non-defective }\textbf{or}\text{ defective }\textbf{and}\text{ non-defective }\textbf{and}\text{ defective }\textbf{or}\text{ non-defective }\textbf{and}$
　　defective **and** defective) $= .05(.05)(.95) + .05(.95)(.05) + .95(.05)(.05) = .007125$

(d)　$P(\text{defective }\textbf{and}\text{ defective }\textbf{and}\text{ defective}) = .05(.05)(.05) = .000125$

Note: When you see and ($\cap$), multiply and when you see or ($\cup$), add. Also note that the probabilities add up to 1.

Bayes' Theorem

We now combine many of the ideas in the previous sections to come up with two very important results. One is known as the rule of total probability and the other is known as Bayes' theorem. They are developed with the help of an example.

EXAMPLE 2.26　A plant has three suppliers. S_1 supplies 30% of the parts to the plant, S_2 supplies 50% of the parts, and S_3 supplies the remaining 20%. One percent of the parts supplied by S_1 are defective, two percent of the parts supplied by S_2 are defective, and three percent of the parts supplied by S_3 are defective. The event of interest is parts received by the plant that are defective. The plant has defective parts in inventory if they came from supplier S_1 and are defective, or if they came from supplier S_2 and are defective, or if they came from supplier S_3 and are defective. Let D be the event that parts received by the plant are defective. D may be expressed as

$$D = (D \cap S_1) \cup (D \cap S_2) \cup (D \cap S_3)$$

This represents a disjoint union, and the probability of the disjoint union is the sum of the probabilities.

$$P(D) = P(D \cap S_1) + P(D \cap S_2) + P(D \cap S_3)$$
$$P(D) = P(S_1)P(D \mid S_1) + P(S_2)P(D \mid S_2) + P(S_3)P(D \mid S_3)$$
$$P(D) = (.3)(.01) + (.5)(.02) + (.2)(.03) = .019$$

The percentage of parts at the plant that are defective is 1.9%. Expanding this idea, we have the following **rule of total probability**.

If $S_1, S_2, \cdots, S_n$ are mutually exclusive events and one of the n events must occur, then

$$P(D) = P(S_1)P(D|S_1) + P(S_2)P(D|S_2) + \cdots + P(S_n)P(D|S_n)$$

EXAMPLE 2.27   To carry this example one step further, suppose we choose a defective item at the plant and ask the question "What is the probability that it came from supplier S_3?" This may be formulated as follows: Find $P(S_3 \mid D)$. Consider the following derivation:

$$P(S_3 \mid D) = \frac{P(D \cap S_3)}{P(D)} = \frac{P(S_3)P(D|S_3)}{P(S_1)P(D|S_1) + P(S_2)P(D|S_2) + P(S_3)P(D|S_3)}$$

$$P(S_3 \mid D) = \frac{(.20)(.03)}{(.30)(.01) + (.50)(.02) + (.20)(.03)} = \frac{.006}{.019} = .3158$$

Given that the part was defective, the probability it came from supplier S_1 is $P(S_1|D) = (0.30)(0.01)/0.019 = 0.1579$. Given that the part was defective, the probability it came from supplier S_2 is $P(S_2|D) = (0.50)(0.02)/0.019 = 0.5263$.

Note that $(0.3158 + 0.1579 + 0.5263 = 1)$.

If this last result is extended from the three suppliers to n, it is called **Bayes' Theorem**.

Mathematical Expectation

Now that we have established the concepts of probability and its properties, we turn our attention to one of the many uses of probability. What do we expect to happen if we know the probabilities concerning an event and what action do we take as a result of that expectation?

EXAMPLE 2.28   The following game is played. A coin is flipped. If it lands heads, you are paid $5. If it lands tails, you pay $5. If you perform this experiment many times, what do you expect your winnings to be? The answer is simple. Your expectation is 0. We arrive at this in the following way. Consider the following table:

Winning	$5, if heads	−$5, if tails
Probability	1/2	1/2

The expected winning is $5(1/2) – 5(1/2) = 0$. Such a game is called fair. If your expected winnings are zero, the game is said to be **fair**.

The definition of **mathematical expectation** is as follows. If you win amount a_1 with probability p_1, amount a_2 with probability p_2, . . . , amount a_n with probability p_n, your mathematical expectation is $\text{Exp} = \Sigma a_i p_i$. The amounts are any real number and the probabilities are between 0 and 1 and sum up to 1.

EXAMPLE 2.29 The game of **chuck-a-luck** is played as follows. Three dice are thrown. You pay $5 to play. You bet on the number of times the number 6 appears in the toss of the three dice. Let X represent the number of times 6 occurs when the three dice are tossed. The following table gives the probabilities associated with different values that X may assume.

X	0	1	2	3
$P(X)$	125/216	75/216	15/216	1/216

The rules for this game are summarized in the following table:

Winning	−$5	$1	$15	$25
Probability	125/216	75/216	15/216	1/216

What the table says is: If no 6s occur, you loose the $5 you paid to play. If one 6 occurs, you get your $5 back plus $1. If two 6s occur, you get your $5 back plus $15. If three 6s occur, you get your $5 back plus $25. Your mathematical expectation of the game is Exp = $(-5)(125/216) + (1)(75/216) + (15)(15/216) + (25)(1/216) = -300/216 = -1.38$. If you play the game 100 times, you expect to loose about $138. You will loose $1.38 per game on the average. This is not a fair game. It favors the "house."

Notes concerning the chuck-a-luck example:

1. How can the distribution of X be derived? By using EXCEL as well as copy and paste techniques, the 216 possible outcomes associated with rolling the three dice may easily be obtained. The outcomes are placed in the range A1:C216. Then the function =COUNTIF(A1:C1,"6") is entered in D1 and a click-and-drag is performed across the 216 observations. Then a count shows that 6 did not occur 125 times, 6 occurred once 75 times, 6 occurred twice 15 times and 6 occurred three times once. The table giving X and $P(X)$ are easily obtained from knowing the number of times 0, 1, 2, and 3 occurred. The reader is encouraged to use EXCEL to verify this.

2. The game could be made a fair game in many different ways. One way is to leave it alone except for one change. Give the $5 back as well as a grand prize of $325 if 6 occurs three times. This gives the following table:

Winning	−$5	$1	$15	$325
Probability	125/216	75/216	15/216	1/216

The mathematical expectation is Exp = $(-5)(125/216) + (1)(75/216) + (15)(15/216) + (325)(1/216) = 0$. The reader is encouraged to give another version of the game that will result in a fair game.

Note that the mathematical expectation does not tell you what will happen on one performance of the experiment, but rather what to expect from the experiment in the long run.

EXAMPLE 2.30 A game is played as follows. You pay $1 to play. A coin is flipped four times. If four tails or four heads are obtained, you get your $1 back plus $5 more. Otherwise you forfeit your $1. What is the mathematical expectation? You win $5 with probability 1/16 + 1/16. You win −$1 with probability 14/16. The mathematical expectation is 5(2/16) − 1(14/16) = −4/16 or −0.25. There is an average of a quarter lost per player per game. The probabilities are arrived at as follows. You win $5 if you toss all heads or all tails. Because of independence, $P(HHHH) = (1/2)^4 = 1/16$. Similarly, $P(TTTT) = (1/2)^4 = 1/16$. You add the two because they are mutually exclusive. You lose $1 if you get the complement of $HHHH$ or $TTTT$. The probability of the complement of all heads or all tails is $1 - 2/16 = 14/16$.

EXAMPLE 2.31 An engineering company prepares an estimate for a job. The cost of preparing the estimate is $10,000. The amount of profit over and above the $10,000 is $25,000 if their estimate is accepted. The probability that their estimate will be accepted is 0.7 and the probability that their estimate will not be accepted is 0.3. What is the expected profit? The expected profit is Exp = $25,000(.7) − $10,000(.3) = $14,500.

EXAMPLE 2.32 An engineering consulting company must decide between two jobs. The decision is based on the following information. Which job has the greater expected profit?

Job 1

Probability	.2	.8
Outcome	loss of $30,000	profit of $100,000

Job 2

Probability	.4	.6
Outcome	loss of $20,000	profit of $125,000

For Job 1, Exp = –30,000(.2) + 100,000(.8) = $74,000 and for Job 2, Exp = –20,000(.4) +125,000(.6) = $67,000. The engineering consulting company should take Job 1 because it has the greater expected profit. (This, of course, assumes all other factors are the same.)

What is the difference between the first three examples and the last two examples of this section? The difference is in the computation of the probabilities. In the games of chance, the probabilities are computed using the classical definition of equal outcomes. In the engineering examples, the probabilities are computed from limited observations or from personal or subjective sources. The probabilities in the engineering examples are not as accurate as those in the examples that use the classical method of computing probabilities. As long as the dice are balanced and the coins are fair, the mathematical expectation will accurately reflect what occurs on the average. There is also some question about the monetary values that are assigned in the real world examples.

SOLVED PROBLEMS

Sample space, events, and operations on events

2.1. An experiment has a continuous sample space equal to $15 \leq X \leq 20$. If events A and B are $A = \{X : 15 \leq X \leq 17\}$, $B = \{X : 16 \leq X \leq 19\}$, find the following: (a) $A \cap B$ (b) $A \cup B$ (c) A^c.

SOLUTION

(a) $A \cap B = \{X : 16 \leq X \leq 17\}$ (b) $A \cup B = \{X : 15 \leq X \leq 19\}$ (c) $A^c = \{17 < X \leq 20\}$

2.2. An experiment consists of selecting three items from a box containing several items. The three items are classified as defective (D) or non-defective (N) as they are selected. Give the sample space for this experiment. Which outcomes are in the events A, B, and C? (a) A, Exactly one item defective. (b) B, At least two defectives in the three. (c) C, At most one defective in the three.

SOLUTION

The sample space is

$$S = \{DDD, DND, DDN, DNN, NDD, NND, NDN, NNN\}.$$

(a) $A = \{DNN, NDN, NND\}$ (b) $B = \{NDD, DND, DDN, DDD\}$ (c) $C = \{NNN, DNN, NDN, NND\}$

2.3. An experiment consists of inspecting items from a production line until a defective (D) is found. Give the sample space for this experiment. Which outcomes are in the events A, B, and C? (a) A, Exactly one item is inspected. (b) B, At least two are inspected. (c) C, At most five items are inspected.

SOLUTION

The sample space is

$$S = \{D, ND, NND, NNND, \ldots\}.$$

(a) $A = \{D\}$ (b) $B = \{ND, NND, NNND, \ldots\}$ (c) $C = \{D, ND, NND, NNND, NNNND\}$

The multiplicative rule, permutations, and combinations

2.4. In how many different ways can one make a first, second, third, and fourth choice among 10 firms leasing construction equipment?

SOLUTION

The number of permutations are $P_4^{10} = \dfrac{10!}{4!} = 5040$. Using EXCEL, =PERMUT(10,4) = 5040.

2.5. An electronic controlling device requires six identical memory chips. In how many ways can this mechanism be assembled using six given chips?

SOLUTION

The number of permutations is $P_6^6 = \dfrac{6!}{0!} = 720$. Using EXCEL, =PERMUT(6,6) = 720.

2.6. In how many different ways can the director of a research laboratory choose two chemical engineers from among five and two industrial engineers from among four applicants?

SOLUTION

The chemical engineers can be chosen =COMBIN(5,2) = 10 and the industrial engineers can be chosen =COMBIN(4,2) = 6. The multiplication principle gives 6(10) = 60.

Probability

2.7. It is common in many industrial areas to fill boxes full of product. This occurs in an industry where the product is used in the home; for example, detergent. These machines fill to specification (A), under-fill (B), or over-fill (C). The practice of under-filling is the one that we try to avoid. Suppose it is known that $P(B) = 0.001$ and $P(A) = 0.990$. (a) Find $P(C)$. (b) What is the probability that the machine does not under-fill? (c) What is the probability that the machine over-fills or under-fills?

SOLUTION

(a) The events A, B, and C are mutually exclusive and make up the whole sample space. We have $P(A \text{ or } B \text{ or } C)$
= $P(A) + P(B) + P(C)$ =1 and solving for $P(C)$ we find $P(C) = 0.009$.
(b) $P(B^c) = 1 - P(B) = 1 - .001 = .999$.
(c) $P(B \text{ or } C) = P(B) + P(C) = .001 + .009 = .010$.

2.8. Specifications are given for the weight of a packaged product. The package is rejected if it is too heavy or too light. Historical data suggests that 0.95 is the probability that the product meets weight specifications and 0.002 is the probability that the product is too light. For each single packaged product the manufacturer invests $20 in production and the purchase price is $25.
(a) What is the probability that a package chosen at random from the production line is too heavy?
(b) For each 10,000 packages sold, what profit is received by the manufacturer if all packages meet weight specifications?
(c) Assuming that all defective packages are rejected and rendered worthless, how much is the profit reduced on 10,000 packages due to failure to meet weight specifications?

SOLUTION

(a) $1 - .95 - .002 = 0.048$, since the three events "too heavy," "too light," and "meets specifications" are mutually exclusive.
(b) Profit per package = $25 - $20 = $5. If all 10,000 meet specifications, profit is $5(10,000) = $50,000.
(c) Profit is reduced by $25 per package for 0.05(10,000) = 500 packages or $25(500) = $12,500.

2.9. A component fails a particular test (A) 15% of the time or a component displays strain but does not fail the test (B) 25% of the time.
(a) What is the probability that the component does not fail the test?
(b) What is the probability that a component works perfectly well (neither displays strain nor fails the test)?
(c) What is the probability that the component either fails or shows strain in the test?

SOLUTION

(a) $P(A^c) = 1 - P(A) = 1 - .15 = .85$
(b) $1 - P(A \text{ or } B) = 1 - .4 = .6$
(c) $P(A \text{ or } B) = .4$

The axioms of probability

2.10. A set of events are **collectively exhaustive** if at least one of the events occur when the experiment is performed. (True or False) (a) A die is rolled and the events A_1, the face 1 turns up, . . . , A_6, the face 6 turns up are collectively exhaustive events. (b) The events A, a face that is an even number turns up and B, a face that is on odd number turns up are collectively exhaustive. (c) A_1 is the event that a 1 or 2 turns up, A_2 is the event that a 3 or 4 turns up, A_3 is the event that a 5 or 6 turns up. The events A_1, A_2, and A_3 are collectively exhaustive.

SOLUTION

a, b, and c are all true.

2.11. (True or False) The events in parts a, b, and c of Problem 2.10 are collectively exhaustive and mutually exclusive.

SOLUTION

True

2.12. In problem 2.10, find (a) $P(A_1$ or A_2 or A_3 or A_4 or A_5 or $A_6)$ (b) $P(A$ or $B)$ (c) $P(A_1$ or A_2 or $A_3)$.

SOLUTION

(a) $1/6 + 1/6 + 1/6 + 1/6 + 1/6 + 1/6 = 1$
(b) $1/2 + 1/2 = 1$
(c) $1/3 + 1/3 + 1/3 = 1$

Some elementary theorems (rules) of probability

2.13. If A, B, and C are dependent events, find an expression for $P(A$ and B and $C)$. If A, B, and C are independent events, find an expression for $P(A$ and B and $C)$.

SOLUTION

For dependent events, $P(A$ and B and $C) = P(A)P(B|A)P(C|A, B)$ and for independent events $P(A$ and B and $C) = P(A)P(B)P(C)$.

2.14. A quality engineer selects 3 iPods from a box that contains 20 iPods of which 3 are defective. What is the probability that the first is defective and the others are not defective?

SOLUTION

Let A be the event that the first one selected is defective, B be the event that the second one is non-defective, and C be the event that the third one is non-defective. Refer to Problem 2.13. $P(A$ and B and $C) = P(A)P(B|A)$ $P(C|A, B) = 3/20(17/19)(16/18) = .119$.

2.15. A production line produces flash drives of which 1% are defective. A quality engineer selects a sample of three. What is the probability that the sample contains three defectives?

SOLUTION

The production line produces a large number of flash drives. When the sample is selected the items selected are independent of one another. The probability that any one is defective is 0.01. Refer to Problem 2.13. $P(A$ and B and $C) = P(A)P(B)P(C)$. A is the event that the first selected is defective, B is the event that the second selected is defective, and C is the event that the third selected is defective. $P(A$ and B and $C) = (.01)(.01)(.01) = .000001$.

Conditional probability

2.16. One thousand copper rods have the properties shown in the following table.

	Diameter		
Length	Too thin	OK	Too thick
Too short	10	5	5
OK	40	902	7
Too long	4	20	7

If the rod meets the length specification, find the probability the rod meets the diameter specifications.

SOLUTION

P(the rod meets the diameter specifications | the rod meets the length specifications) =

$$\frac{\frac{902}{1000}}{\frac{949}{1000}} = \frac{902}{949} = .95$$

2.17. In Problem 2.16, if the rod meets the diameter specifications, find the probability the rod meets the length specifications.

SOLUTION

P(the rod meets the length specifications | the rod meets the diameter specifications) =

$$\frac{\frac{902}{1000}}{\frac{927}{1000}} = \frac{902}{927} = .97$$

2.18. In Problem 2.16, find the unconditional probability that the rod meets the length specifications. Are the events "the rod meets the length specifications" and "the rod meets the diameter specifications" dependent or independent?

SOLUTION

P(the rod meets the length specifications) = 949/1000 = .949 and

P(the rod meets the length specifications | the rod meets the diameter specifications) = .97 and since these two probabilities are not equal, the two events are dependent.

Bayes' theorem

2.19. On the East Coast, it is known from health records that the probability of selecting an adult over 40 years of age with cancer is 0.05. The probability of diagnosing a person with cancer as having the disease is 0.78 and the probability of incorrectly diagnosing a person without cancer as having the disease is 0.06. Find the probability that a person is diagnosed as having cancer.

SOLUTION

Let A be the event that a person over 40 has cancer. A^c is the event that a person over 40 does not have cancer. Let D be the event that a person is diagnosed as having cancer and D^c is the event that a person is diagnosed as not having cancer. We have $P(A) = .05$, $P(A^c) = .95$, $P(D|A) = .78$, and $P(D|A^c) = .06$. $P(D) = P(A) P(D|A) + P(A^c) P(D|A^c)$.

$P(D) = (.05)(.78) + (.95)(.06) = .096$.

2.20. Referring to Problem 2.19, what is the probability that a person diagnosed as having cancer actually has the disease?

SOLUTION

$$P(A|D) = P(A \text{ and } D)/P(D) = P(A)P(D|A)/P(D) = (.05)(.78)/(.096) = .40625$$

2.21. Referring to Problem 2.19, what is the probability that a person diagnosed as having cancer does not have the disease?

SOLUTION

$$P(A^c|D) = P(A^c \text{ and } D)/P(D) = P(A^c)P(D|A^c)/P(D) = (.95)(.06)/(.096) = .59375$$

Mathematical expectation

2.22. A group of engineers decide to play the game of craps. A pair of dice is rolled in this game and the sum to appear on the dice is of interest. What is the mathematical expectation of the sum to appear when the dice are rolled?

SOLUTION

There are 6 possible outcomes on each die. The multiplication principle gives 6(6) = 36 outcomes possible. The sum X of the two dice varies from 2 to 12. One of the outcomes is $X = 2$ which occurs if a 1 appears on each die. The probability that $X = 2$ is 1/36 or .0278. The set of all outcomes and their probabilities are as follows:

X	2	3	4	5	6	7	8	9	10	11	12
P(X)	1/36	2/36	3/36	4/36	5/36	6/36	5/36	4/36	3/36	2/36	1/36

The expected sum is $\text{Exp} = \Sigma a_i p_i = 7$.

2.23. The number of defective welds in a length of pipe is 0 through 6 with the following probabilities.

X	0	1	2	3	4	5	6
P(X)	.60	.30	.05	.02	.01	.01	.01

Find the expected value of the number of defective welds.

SOLUTION

$$E(X) = (0)(.60) + (1)(.30) + (2)(.05) + (3)(.02) + (4)(.01) + (5)(.01) + (6)(.01) = 0.61$$

The average number of defective welds is expected to be 0.61.

2.24. The probability that a relay remains open is 0.5. An electrical circuit consists of three such relays. The number of relays that remain open in the three are 0, 1, 2, or 3. The number of relays that remain open is represented by X. The number of relays that remain open has the following values with the probabilities.

X	0	1	2	3
P(X)	.125	.375	.375	.125

Find the expected value of the number of open relays.

SOLUTION

$$E(X) = (0)(.125) + (1)(.375) + (2)(.375) + (3)(.125) = 1.5$$

In the long run, if we perform this circuit-opening experiment many times, the average number of open relays we can expect is 1.5.

SUPPLEMENTARY PROBLEMS

2.25. An electrical engineer has two boxes of resistors. The resistors in the first box are 9, 10, 11, and 12 ohms and the resistors in the second box are 18, 19, 20, and 21 ohms. The engineer chooses one resistor from each box and determines the resistance of the resistor. Looked at as an experiment, what is the sample space?

2.26. Refer to Problem 2.25. Three events are defined as: A is the event that the resistor from the first box has a resistance greater than 10, B is the event that the resistor from the second box has a resistance less than 19, and C is the event that the sum of the resistances is equal to 28. Give the outcomes in each of the three events.

2.27. Refer to Problems 2.25 and 2.26. Give the outcomes in the following events:

$$A \cap B \quad A \cup B \quad B \cap C^c$$

2.28. When ordering a computer, there are 3 choices of hard drive, 2 choices of the amount of memory, 3 choices of video card, and 2 choices of monitor. How many different computers could be ordered?

2.29. Twelve engineers have applied for a management position at a large firm. Three of them will be selected as finalists for the position. In how many ways can this selection be made?

2.30. A computer password consists of eight characters. A character is any lower case letter or digit. How many passwords are available?

2.31. As a project, a freshmen communications major is assigned the following project: Randomly select 125 students and count how many are talking on a cell phone at a randomly chosen time. The major counts 83 students that are talking on a cell phone. What is the relative frequency probability that a student randomly selected at a random time will be talking on a cell phone?

2.32. A professor of international studies is asked to estimate the probability that a terrorist act will be performed in a large European city within the next year. The professor states that there is a 10% chance. What is the subjective probability that this event occurs?

2.33. A box of bolts contains 4 thick bolts, 3 medium bolts and 3 thin bolts. A box of nuts contains 5 that fit thick bolts, 5 that fit medium bolts, and 5 that fit thin bolts. One bolt and one nut are chosen at random. What is the classical probability that the nut fits the bolt?

2.34. Answer as true or false. $P(A) - P(A \cap B) \leq P(A)$.

2.35. Answer as true or false. $P(A \text{ or } B) \geq P(A)$.

2.36. Answer as true or false. If A and B are mutually exclusive and $P(A) > 0$ and $P(B) > 0$, then $P(A \text{ or } B) > P(A)$.

2.37. The probabilities that a consumer testing service will rate a new antipollution device for cars very poor, poor, fair, good, very good, or excellent are 0.06, 0.14, 0.20, 0.3, 0.22, and 0.08. What are the probabilities that it will rate the device

   (a) very poor, poor, or fair?   (b) very good and excellent?

2.38. The probabilities are 0.80, 0.70, and 0.65 that a family, randomly chosen as part of a sample survey in a large city, owns a flat screen television or a computer or both. What is the probability that a family in this area own one or the other or both?

2.39. In a group of 150 graduate engineering students, 90 are enrolled in an advanced course in statistics, 60 are enrolled in a course in operations research, and 40 are enrolled in both. How many of these students are not enrolled in either course?

2.40. A process manufactures aluminum cans. The probability that a can has a flaw on the side is 0.02, the probability that it has a flaw on the top is 0.03, and the probability that it has a flaw on the top and the side is 0.01. What is the probability that a can will have flaw on its side, given that it has a flaw on the top?

2.41. In Problem 2.40, what is the probability that a can will have a flaw on the top given it has a flaw on the side?

2.42. Of the microcomputers manufactured by a certain process, 5% are defective. Four of the microcomputers are chosen at random. Assume they function independently. What is the probability that they all work?

2.43. A Ford has engines in three sizes. Of the Ford cars sold, 50% have the smallest engine, 40% have the medium engine, and 10% have the largest engine. Of the cars with the smallest engine, 15% fail an emissions test within two years of purchase. The failure figure for medium size engines is 10%, and the failure figure for the largest engines is 5%. What is the probability that this Ford will fail the emissions test within two years?

2.44. In Problem 2.43, a record for a failed emissions test is randomly chosen. What is the probability that it is for a Ford with the smallest engine?

2.45. In Problem 2.43, a record for a failed emissions test is randomly chosen. What is the probability that it is for a Ford with the largest engine?

2.46. A resistor in a certain circuit is specified to have a resistance in the range of 99 ohms to 101 ohms. An engineer obtains two resistors. The probability that both of them meet the specification is 0.30, the probability that exactly one of them meets the specification is 0.60, and the probability that neither of them meets the specifications is 0.10. Let X represent the number of resistors that meet specifications. Find the mathematical expectation of X.

2.47. The following table gives the distribution of the number of defects X in a randomly chosen printed-circuit board.

X	0	1	2	3
$P(X)$	.6	.2	.15	.05

Find the mathematical expectation of X.

2.48. A chemical supply company ships a chemical in 5 gallon drums. X represents the number of drums ordered by a randomly chosen customer. X has the following distribution.

X	1	2	3	4	5
$P(X)$	.1	.1	.2	.2	.4

Find the mathematical expectation of X.

CHAPTER 3

Probability Distributions for Discrete Variables

Random Variables

The concept of a random variable may be new to the reader of this book. A **random variable** assigns values to outcomes in the sample space. A random variable is best described by its probability distribution. The **probability distribution function** gives the values that X may assume and their probabilities. Random variables are best introduced by examples.

EXAMPLE 3.1 An experiment consists of rolling a pair of dice and letting the random variable X be the number of times the number 1 comes up when the pair of dice come to rest. X can take on the values 0, 1, and 2 since when a pair of dice is rolled the number 1 can turn up on none, or one, or both of the dice.

Figure 3.1 gives the EXCEL output for this experiment. The 36 outcomes are shown in the figure and in cell C2, the expression =COUNTIF(A2:B2, "1") is entered and a click-and-drag is performed from C2 to C19. This is also performed in G2:G19. The number 0 occurs 25 times, the number 1 occurs 10 times, and 2 occurs once. The probability of 0 is 25/36, the probability of 1 is 10/36, and the probability of 2 is 1/36. The probability distribution for X is shown in the upper right corner of Figure 3.1. **Note that the probability distribution function for X has the properties $p(x) \geq 0$ and $\Sigma\, p(x) = 1$, where the sum is over the x values.** The interpretation of the probability distribution is as follows: When the dice are tossed a large number of times, the number 1 will not turn up 69.4% of the time, the number 1 will turn up once about 27.8% of the time, and the number 1 will turn up on both dice 2.8% of the time. A plot of the probability

A	B	C	D	E	F	G	H	I
Die1	Die2	x		Die1	Die2	x		
1	1	2		4	1	1	x	p(x)
1	2	1		4	2	0	0	0.694
1	3	1		4	3	0	1	0.278
1	4	1		4	4	0	2	0.028
1	5	1		4	5	0		
1	6	1		4	6	0		
2	1	1		5	1	1		
2	2	0		5	2	0		
2	3	0		5	3	0		
2	4	0		5	4	0		
2	5	0		5	5	0		
2	6	0		5	6	0		
3	1	1		6	1	1		
3	2	0		6	2	0		
3	3	0		6	3	0		
3	4	0		6	4	0		
3	5	0		6	5	0		
3	6	0		6	6	0		

Figure 3.1 Sample space for rolling a pair of dice. $X =$ number of times 1 occurs on a roll of the dice, and the probability distribution of X in columns H and I.

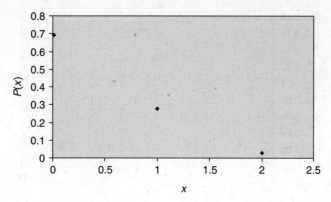

Figure 3.2   Plot of $P(x)$.

distribution function is given in Figure 3.2. It was created using the chart wizard in EXCEL. It is said to be **skewed to the right** or it has a tail to the right.

Many random variables may be defined on the sample space of an experiment. They will have different probability distributions.

EXAMPLE 3.2   Consider the same experiment of rolling a pair of dice. Let random variable Y equal the sum on the faces that are turned up.

Figure 3.3 gives the EXCEL output for this experiment. The 36 outcomes are shown and in cell C2, the expression =SUM(A2:B2) is entered in C2 and a click-and-drag is performed from C2 to C19. This is also performed in G2:G19. The probability distribution for Y is shown in the upper right corner of Figure 3.3. Note that the probability distribution for Y has the property $p(y) \geq 0$ and $\Sigma\, p(y) = 1$, where the sum is over the y values. The interpretation of the probability distribution is as follows. When the dice are tossed a large number of times, the sum 2 occurs 2.8% of the time, the sum 3 occurs 5.6 % of the time, etc. Figure 3.4 is a plot of $p(y)$ created by the chart wizard of EXCEL. It is said to be **symmetrical about 7**.

A	B	C	D	E	F	G	H	I
Die1	Die2	y		Die1	Die2	y		
1	1	2		4	1	5	y	p(y)
1	2	3		4	2	6	2	0.028
1	3	4		4	3	7	3	0.056
1	4	5		4	4	8	4	0.083
1	5	6		4	5	9	5	0.111
1	6	7		4	6	10	6	0.139
2	1	3		5	1	6	7	0.167
2	2	4		5	2	7	8	0.139
2	3	5		5	3	8	9	0.111
2	4	6		5	4	9	10	0.083
2	5	7		5	5	10	11	0.056
2	6	8		5	6	11	12	0.028
3	1	4		6	1	7		
3	2	5		6	2	8		
3	3	6		6	3	9		
3	4	7		6	4	10		
3	5	8		6	5	11		
3	6	9		6	6	12		

Figure 3.3   EXCEL output for the sample space of rolling a pair of dice, where Y = sum on the dice, and the probability distribution of Y is given.

Random variables are classified according to the number of values they can assume. The random variables encountered in this chapter can assume a finite or countable infinite number of values and are called **discrete random variables**. **Continuous random variables** are considered in the next chapter. Their probabilities are areas under curves. Another important function associated with discrete random variables is the **cumulative distribution function**. The **cumulative distribution function** is defined as

$$F(x) = P(X \leq x)$$

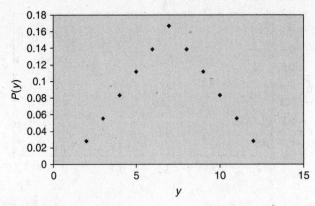

Figure 3.4   Plot of $P(y)$.

EXAMPLE 3.3   The cumulative distribution function associated with X in the first example of this section is

x	0	1	2
$F(x)$	0.694	0.972	1

and the cumulative distribution function associated with Y in the second example of this section is

Y	2	3	4	5	6	7	8	9	10	11	12
$F(y)$	0.028	0.084	0.167	0.278	0.417	0.584	0.723	0.834	0.917	0.973	1

The graph of $F(x)$ is given in Figure 3.5 and the graph of $F(y)$ is given in Figure 3.6.

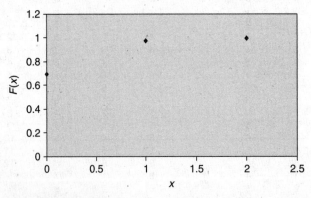

Figure 3.5   Plot of $F(x)$.

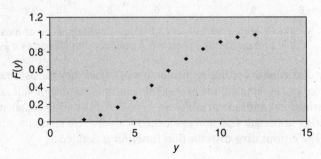

Figure 3.6   Plot of $F(y)$.

EXAMPLE 3.4 A cell phone manufacturer has been notified by an industrial engineer that a cell phone may have up to four surface flaws. If $X =$ the number of surface flaws per cell phone, then the probability distribution function has been found to have the following distribution.

x	0	1	2	3	4
$p(x)$	0.75	0.15	0.05	0.04	0.01

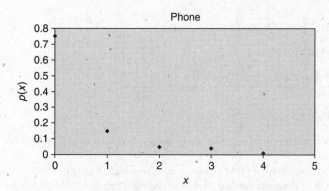

Figure 3.7 Probability distribution function of the number of surface flaws per cell phone.

The graph of $p(x)$ is shown in Figure 3.7.
The cumulative distribution function is shown in Figure 3.8.

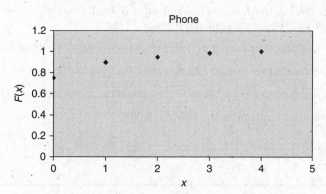

Figure 3.8 Cumulative distribution function of number of surface flaws per cell phone.

x	0	1	2	3	4
$F(x)$	0.75	0.90	0.95	0.99	1

The probability of at most 2 defects is $F(2) = 0.95$.

The probability of at least 3 defects is $1 - F(2) = 0.05$.

The probability of exactly 2 defects is $F(2) - F(1) = 0.95 - 0.90 = 0.05$ or $p(2) = 0.05$.

Remember that a probability distribution function has two basic properties: (1) $p(x) \geq 0$ and (2) $\Sigma\, p(x) = 1$.

EXAMPLE 3.5   Check to see if the following can serve as probability distribution functions.

(a) $p(x) = \dfrac{24}{50}\dfrac{1}{x}$ for $x = 1, 2, 3, 4$ (b) $p(x) = x/55$ for $x = 1, 2, 3, 4, 5, 6, 7, 8, 9, 10$

(c) $p(x) = \dfrac{x-4}{2}$ for $x = 3, 4, 5, 6$

(a) 1. all the probabilities are positive.

    2. $\sum p(x) = \dfrac{24}{50}\left\{1 + \dfrac{1}{2} + \dfrac{1}{3} + \dfrac{1}{4}\right\} = 0.48 + 0.24 + 0.16 + 0.12 = 1$

This function is a probability distribution function.

(b) 1. all the probabilities are positive.

    2. $\sum p(x) = \dfrac{1+2+3+4+5+6+7+8+9+10}{55} = \dfrac{55}{55} = 1$

This function is a probability distribution function.

(c) $P(3) = -0.5$. Since this probability is negative, this function is not a probability distribution function.

EXAMPLE 3.6   A quality control engineer is in charge of receiving parts by the truck load. Each truck contains several thousand parts. Generally 1% of the parts are defective. The engineer can live with 1% but not 3% or 4%. She uses the following plan. Randomly sample 5 parts from the truck. If there are no defectives in the sample, accept the truck load as suitable. If 1% the parts are defective, there is a 95% chance that she will accept the truck load, since $P(5$ out of 5 are non-defective) is $(0.99)^5 = 0.951$. Now if 4% of the parts are defective rather than 1%, find the probability that the sample of 5 will contain 0, 1, 2, 3, 4, or 5 defectives. That is, find the probability distribution of X, the number of defectives in a 4% defective truck load.

$P(0$ defectives in a 4% defective truck load$) = (0.96)^5 = 0.8154$

$P(1$ defective in a 4% defective truck load$) = (0.04)(0.96)^4 + (0.96)(0.04)(0.96)^3 + (0.96)^2(0.04)(0.96)^2$
    $+ (0.96)^3(0.04)(0.96) + (0.96)^4(0.04) = 5(0.04)(0.96)^4 = 0.1699$

$P(2$ defectives and 3 non-defectives in a 4% defective truck load$) = 10(0.04)^2(0.96)^3 = 0.0142$

$P(3$ defectives and 2 non-defectives in a 4% defective truck load$) = 10(0.04)^3(0.96)^2 = 0.0006$

$P(4$ defectives and 1 non-defective in a 4% defective truck load$) = 5(0.04)^4(0.96) = 0.0000$

$P(5$ defectives in a 4% defective truck load$) = (0.04)^5 = 0.0000$

The probability distribution of X, where X is the number of defectives in a sample of 5 from a 4% defective truck load is as follows:

x	0	1	2	3	4	5
$p(x)$	0.8154	0.1699	0.0142	0.0006	0.0000	0.0000

EXAMPLE 3.7   Consider the following sampling plan. One item is inspected and if it passes inspection, the industrial process continues. If it does not pass inspection, the industrial process is stopped and inspected. The industrial process continues until some item does not pass inspection. An item is selected from the line every hour. Let X be the hour in which the first item does not pass inspection. If the process is producing 90% acceptable items, give the probability distribution for X. The possible values of X are 1, 2, 3, $\cdots$. The probabilities are $p(1) = 0.1$, $p(2) = 0.9(0.1)$, $p(3) = (0.9)^2(0.1)$, and so forth. Summarizing we find that $p(x) = (0.9)^{x-1}(0.1)$ for $x = 1, 2, 3, \cdots$. Note that this is the first infinite discrete random variable we have discussed. Its possible outcomes are in a 1-to-1 correspondence with the positive integers.

First of all, note that the probability distribution function satisfies the two properties $p(x) \geq 0$ and $\sum p(x) = 1$. Certainly all the probabilities are positive. If we add all the probabilities, we obtain $0.1 + (0.9)(0.1) + (0.9)^2(0.1) + (0.9)^3(0.1) + \cdots$. This is an infinite geometric series with first term $a = 0.1$ and common ratio $r = 0.9$. In algebra, it is shown that such a series has sum equal to $\dfrac{a}{1-r} = \dfrac{0.1}{1-0.9} = 1$. When Figure 3.9 is consulted, it is seen that the above sum through the first 60 terms gives 0.9982. Furthermore, note that $F(29) = 0.9529$. This can be interpreted as follows: The probability that the

X	p(x)	F(x)	x	p(x)	F(x)	x	p(x)	F(x)
1	0.1	0.1	21	0.0122	0.8906	41	0.0015	0.9867
2	0.09	0.19	22	0.0109	0.9015	42	0.0013	0.988
3	0.081	0.271	23	0.0098	0.9114	43	0.0012	0.9892
4	0.0729	0.3439	24	0.0089	0.9202	44	0.0011	0.9903
5	0.0656	0.4095	25	0.008	0.9282	45	0.001	0.9913
6	0.059	0.4686	26	0.0072	0.9354	46	0.0009	0.9921
7	0.0531	0.5217	27	0.0065	0.9419	47	0.0008	0.9929
8	0.0478	0.5695	28	0.0058	0.9477	48	0.0007	0.9936
9	0.043	0.6126	29	0.0052	0.9529	49	0.0006	0.9943
10	0.0387	0.6513	30	0.0047	0.9576	50	0.0006	0.9948
11	0.0349	0.6862	31	0.0042	0.9618	51	0.0005	0.9954
12	0.0314	0.7176	32	0.0038	0.9657	52	0.0005	0.9958
13	0.0282	0.7458	33	0.0034	0.9691	53	0.0004	0.9962
14	0.0254	0.7712	34	0.0031	0.9722	54	0.0004	0.9966
15	0.0229	0.7941	35	0.0028	0.975	55	0.0003	0.997
16	0.0206	0.8147	36	0.0025	0.9775	56	0.0003	0.9973
17	0.0185	0.8332	37	0.0023	0.9797	57	0.0003	0.9975
18	0.0167	0.8499	38	0.002	0.9818	58	0.0002	0.9978
19	0.015	0.8649	39	0.0018	0.9836	59	0.0002	0.998
20	0.0135	0.8784	40	0.0016	0.9852	60	0.0002	0.9982

Figure 3.9   EXCEL output for $p(x) = (0.9)^{x-1}(0.1)$ and $F(x)$ for $x = 1$ to 60.

first defective will occur on the first inspection or on the second or on the third or $\cdots$ or on the twenty-ninth is 0.9529. The cumulative distribution function evaluated at 29 is equal to 0.9529.

The graph in Figure 3.10 continues on from 60 to infinity and the $p(x)$ values get closer and closer to 0, but never touch 0. The graph in Figure 3.11 continues from 60 to infinity and the $F(x)$ values get closer and closer to 1 but never equal 1. EXCEL is used to construct Figure 3.9 in the following manner. The numbers 1 through 60 are entered into A2:A61. The expression =0.9^(A2-1)*0.1 is entered into B2 and a click-and-drag is performed from B2 to B61. $F(x)$ is formed in the following manner. B2 is entered into C2. Then =C2+B3 is entered into C3 and a click-and-drag is performed from C3 to C61.

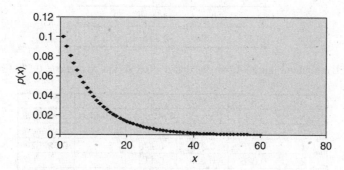

Figure 3.10   Plot of $p(x) = (0.9)^{x-1}(0.1)$ from Figure 3.9.

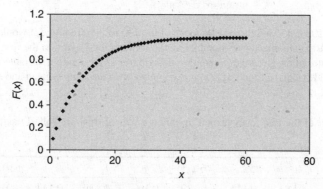

Figure 3.11   Plot of $F(x)$ from Figure 3.9.

Mean and Standard Deviation of a Probability Distribution

The mean of a random variable is its mathematical expectation. If a discrete random variable can take on the values $x_1, x_2, \cdots, x_n$ with probabilities $p(x_1), p(x_2), \cdots, p(x_n)$ the **mean** is defined to be

$$\mu = \sum x_i\, p(x_i), \text{ where } i \text{ goes from 1 to } n.$$

The sum may contain an infinite number of terms. The mean locates the center or balance point of a distribution. Another important measure associated with random variables is the variance. The **variance** of a random variable is defined to be

$$\sigma^2 = \sum (x_i - \mu)^2 p(x_i), \text{ where } i \text{ goes from 1 to } n.$$

The square root of the variance is the **standard deviation**.

$$\sigma = \sqrt{\sum (x_i - \mu)^2 p(x_i)}$$

The standard deviation is used more widely than the variance. However, it is necessary to find the variance before finding the standard deviation. Also, the variance is in square units and the standard deviation is in the units of measurement. There is an equivalent formula that is usually used to compute the variance and standard deviation. The variance may be found using the algebraically equivalent formula

$$\sigma^2 = \sum x_i^2 p(x_i) - \mu^2$$

The standard deviation gives a measure of the spread of the distribution. The parameters μ and σ are important measures to know for any distribution. Let us consider some of the distributions we studied in the section on random variables. We shall find their means and standard deviations and comment on their meanings.

EXAMPLE 3.8   The number of times a 1 turns up when rolling a pair of dice has the following probability distribution.

x	0	1	2
$p(x)$	0.694	0.278	0.028

The following EXCEL output in Figure 3.12 shows that $\mu = 0.334$ (in C5) and $\sigma = 0.528$ (in F3).

A	B	C	D	E	F
x	$p(x)$	$xp(x)$	$x\text{^}2p(x)$	Variance	Standard
0	0.694	0	0	0.278444	Deviation
1	0.278	0.278	0.278		0.527678
2	0.028	0.056	0.112		
		0.334	0.39		

Figure 3.12   EXCEL worksheet for computing mean and standard deviation.

If the dice were rolled 100 times and the sample mean of the 100 rolls were found it would be near 0.334, and if the standard deviation of the 100 outcomes were found using the formula for S given in the first chapter, it would be near 0.528. The probability distribution gives the population description of the random variable and when the probability distribution is known, μ and σ may be found. If $p(x)$ is not known, then the sample mean and sample standard deviation provide estimates of μ and σ.

EXAMPLE 3.9   The sum of the two faces turned up when a pair of dice are rolled has the following probability distribution.

x	2	3	4	5	6	7	8	9	10	11	12
$p(x)$	0.028	0.056	0.083	0.111	0.139	0.167	0.139	0.111	0.083	0.056	0.028

y	p(y)	yp(y)	y^2p(y)	Variance	Standard
2	0.028	0.056	0.112		Deviation
3	0.056	0.168	0.504	5.802951	2.408932
4	0.083	0.332	1.328		
5	0.111	0.555	2.775		
6	0.13	0.834	5.004		
7	0.167	1.169	8.183		
8	0.139	1.112	8.896		
9	0.111	0.999	8.991		
10	0.083	0.83	8.3		
11	0.056	0.616	6.776		
12	0.028	0.336	4.032		
		7.007	54.901		

Figure 3.13 EXCEL worksheet for computing mean and standard deviation.

The following EXCEL computation in Figure 3.13 shows that $\mu = 7$ and $\sigma = 2.4$.

The **empirical rule** applied to probability distributions states that for mound-shaped distributions such as the one in Figure 3.4, approximately 68% of the probability will be between $\mu - \sigma$ and $\mu + \sigma$, approximately 95% of the probability will be between $\mu - 2\sigma$ and $\mu + 2\sigma$, and 99.7% will be between $\mu - 3\sigma$ and $\mu + 3\sigma$. Between $\mu - \sigma = 4.6$ and $\mu + \sigma = 9.4$, there is $0.111 + 0.139 + 0.167 + 0.139 + 0.111 = 0.67$ or 67%. Between $\mu - 2\sigma = 2.2$ and $\mu + 2\sigma = 11.8$, there is 0.944 or 94.4%. All of the probability is contained between $\mu - 3\sigma$ and $\mu + 3\sigma$. We see that the empirical rule very closely predicts the probability distribution for this mound-shaped distribution.

EXAMPLE 3.10 Consider the random variable X, where X has the probability distribution $0.9^{(x-1)}(0.1)$, for $x = 1, 2, 3, \cdots$. This is a discrete random variable with an infinite domain. The mean is given by $\mu = \sigma x\, 0.9^{(x-1)}(0.1)$, where the sum goes from 1 to infinity. The standard deviation is given by $\sigma = \sqrt{\sum x^2 0.9^{x-1}(0.1) - \mu^2}$. EXCEL is used to evaluate the mean and standard deviation. Suppose only 10 terms in the infinite sum that represents the mean are used to start. The expression =A2*0.9^(A2-1)*0.1 is entered into cell B2 and a click-and-drag produces the output shown in cells B2:B11 in Figure 3.14. The sum is obtained in cell B12 and is 3.0264. Now Figure 3.15 shows the mean obtained using 50, 100, 150, and 200 terms in the sum.

A	B
X	xp(x)
1	0.1
2	0.18
3	0.243
4	0.2916
5	0.32805
6	0.354294
7	0.372009
8	0.382638
9	0.38742
10	0.38742
	3.026431

Figure 3.14 First 10 terms of the sum of the terms for the mean.

X	Sum
50	9.690733
100	9.997078
150	9.999978
200	10

Figure 3.15 Sum of the first 50, 100, 150, and 200 terms for the mean.

It is clear from Figure 3.15 that the sum is converging to 10.

This is a numerical technique that may be used for discrete random variables with infinite domains. We don't really need to add an infinite number of terms to see that $\mu = 10$. Once the mean is known the standard deviation can be found as follows.

Figure 3.16 shows the sum of $x^2 p(x)$ for 50, 100, 150, and 200 terms in the sum for the standard deviation. The variance is $190 - 10^2 = 90$ and the standard deviation is 9.4868.

X	Sum
50	170.9826
100	189.6762
150	189.9965
200	190

Figure 3.16   Sums helpful in evaluating the variance of X.

Chebyshev's Theorem

The empirical rule tells us what to expect concerning the probability distribution and its relationship to μ and σ for a probability distribution that is mound-shaped (also called normally distributed). But what about probability distributions that are not mound-shaped? A result proved by Chebyshev answers the question for probability distributions of any shape. **Chebyshev's Theorem** states that if we go at least k standard deviations from the mean, then we will find at most $1/k^2$ of the probability distribution. Thus, if we go at least 2 standard deviations from the mean, there will be at most $1/2^2 = 1/4 = 0.25$ or 25% of the distribution. If we go at least 3 standard deviations from the mean, there will be at most 1/9 or 11% of the distribution. We can also state the theorem as follows. Within 2 standard deviations of the mean there will be at least 75% of the distribution and within 3 standard deviations of the mean there will be at least 89% of the distribution.

$$\mu - 2\sigma \qquad\qquad \mu \qquad\qquad \mu + \sigma \qquad\qquad x$$

| At least 75% of the probability |

EXAMPLE 3.11   Let X be the sum on two dice that are rolled. In the section on random variables, we derived the probability distribution function and the cumulative probability distribution function. They are given in Figure 3.17.

x	p(x)	F(x)
2	0.028	0.028
3	0.056	0.084
4	0.083	0.167
5	0.111	0.278
6	0.139	0.417
7	0.167	0.584
8	0.139	0.723
9	0.111	0.834
10	0.083	0.917
11	0.056	0.973
12	0.028	1.000

Figure 3.17   Distributions of p(x) and F(x) for sum of dice.

We found that $\mu = 7$ and $\sigma = 2.4$. Show that there is at least 75% of the probability distribution within 2σ of μ and at least 89% within 3σ of μ. Now $\mu - 2\sigma = 2.2$ and $\mu + 2\sigma = 11.8$. The amount of probability within 2σ of μ is $F(11) - F(2) = 0.973 - 0.028 = 0.945$ and this is at least 75%. The amount of probability could also be found by adding the following: $p(3) + p(4) + p(5) + p(6) + p(7) + p(8) + p(9) + p(10) + p(11) = 0.945$. There is 100% between $\mu - 3\sigma = -0.2$ and $\mu + 3\sigma = 14.2$. The lower bound of 89% is seen to hold here.

EXAMPLE 3.12   Consider the random variable X, where X has the probability distribution $0.9^{(x-1)}(0.1)$, for $x = 1, 2, 3, \cdots$. We found the mean to be $\mu = 10$ and the standard deviation to be $\sigma = 9.49$. Figure 3.18 gives $p(x)$ and $F(x)$ for $x = 1$ through $x = 60$.

Find the percentage of probability within 2σ and 3σ of μ and compare with Chebyshev's theorem. Now, $\mu - 2\sigma = -8.98$ and $\mu + 2\sigma = 28.98$. The amount of probability between -8.98 and 28.98 is given by $F(28) = 0.9477$, which is at least 75%. Also $\mu - 3\sigma = -18.47$ and $\mu + 3\sigma = 38.47$. The amount of probability between -18.47 and 38.47 is given by $F(38) = 0.9818$, which is at least 89%.

x	p(x)	F(x)	x	p(x)	F(x)	x	p(x)	F(x)
1	0.1	0.1	21	0.0122	0.8906	41	0.0015	0.9867
2	0.09	0.19	22	0.0109	0.9015	42	0.0013	0.988
3	0.081	0.271	23	0.0098	0.9114	43	0.0012	0.9892
4	0.0729	0.3439	24	0.0089	0.9202	44	0.0011	0.9903
5	0.0656	0.4095	25	0.008	0.9282	45	0.001	0.9913
6	0.059	0.4686	26	0.0072	0.9354	46	0.0009	0.9921
7	0.0531	0.5217	27	0.0065	0.9419	47	0.0008	0.9929
8	0.0478	0.5695	28	0.0058	0.9477	48	0.0007	0.9936
9	0.043	0.6126	29	0.0052	0.9529	49	0.0006	0.9943
10	0.0387	0.6513	30	0.0047	0.9576	50	0.0006	0.9948
11	0.0349	0.6862	31	0.0042	0.9618	51	0.0005	0.9954
12	0.0314	0.7176	32	0.0038	0.9657	52	0.0005	0.9958
13	0.0282	0.7458	33	0.0034	0.9691	53	0.0004	0.9962
14	0.0254	0.7712	34	0.0031	0.9722	54	0.0004	0.9966
15	0.0229	0.7941	35	0.0028	0.975	55	0.0003	0.997
16	0.0206	0.8147	36	0.0025	0.9775	56	0.0003	0.9973
17	0.0185	0.8332	37	0.0023	0.9797	57	0.0003	0.9975
18	0.0167	0.8499	38	0.002	0.9818	58	0.0002	0.9978
19	0.015	0.8649	39	0.0018	0.9836	59	0.0002	0.998
20	0.0135	0.8784	40	0.0016	0.9852	60	0.0002	0.9982

Figure 3.18   EXCEL output for $p(x) = (0.9)^{x-1}(0.1)$ and $F(x)$ for $x = 1$ to 60.

EXAMPLE 3.13   The temperature associated with an industrial process has an unknown distribution. However, based on sampling, the approximate mean is known to be 25.3°C and the standard deviation is known to be 2.1°C. Applying Chebyshev's theorem, at least 75% of the time the temperature will be between 21.2°C and 29.5°C. At least 96% of the time the temperature will be between 14.8°C and 35.8°C. (Let $k = 5$ in Chebyshev's theorem.)

EXAMPLE 3.14   The pressure associated with an industrial process has a mean of 50.5 psi and a standard deviation of 0.2 psi. What can be said about the percentage of time that the pressure is between 49.5 and 51.5 psi? Applying Chebyshev's theorem, and noting that the two limits are 5 standard deviations on either side of the mean, the probability is at least $1 - \dfrac{1}{5^2} = \dfrac{24}{25}$ or 96% that the pressure will be between 49.5 and 51.5 psi.

Binomial Distribution

In the first three sections of this chapter, general properties of discrete random variables were discussed. Some types of discrete random variables occur so often that they have been singled out and studied in detail. Probably the most frequently applied discrete random variable and distribution is the binomial. We will now study this random variable, its distribution, and many of its properties.

Let us start by considering an example of a binomial random variable. Suppose X represents the number of heads to occur in 4 tosses of a fair coin. The random variable X can take on the five values 0, 1, 2, 3, or 4. The probability distribution of X has a binomial distribution. Each flip of the coin is called a **trial**. We say that X is based on 4 trials. Each trial has two outcomes possible, heads or tails. Since the coin is fair, the probability of a head is 0.5 and the probability of a tail is 0.5. We say that a head occurring is a **success** and a tail is a **failure**. The outcome of interest is the number of successes to occur in the 4 trials. The trials are **independent** of one another. What happens on any trial is not influenced by what happens on any other trial. The probabilities of success and failure remain the same from trial to trial. The letter p represents the success probability and the letter q represents the failure probability. The probability of 0 successes (heads) in the 4 trials is p^4 or $(0.5)^4$. We multiply the probabilities because of independence. No successes means 4 trials were obtained. The probability of a trial followed by a trial followed by a trial followed by another trial is $0.5(0.5)(0.5)(0.5)$ or 0.0625. The probability of 1 success (1 head) is the same as *HTTT* or *THTT* or *TTHT* or *TTTH*. In other words, the head could have occurred on the first trial or the second or the third or the fourth. We would add the four probabilities because the four events are mutually exclusive. The probability that $X = 1$ is $4(0.5)^4 = 0.25$. The event $X = 2$ is equivalent to the events *HHTT* or *HTHT* or *HTTH* or *TTHH* or *THTH* or *THHT* occurring. Multiplying and adding probabilities as the rules of probability say we should, we get $6(0.5)^4 = 0.375$. Similarly, the probability of 3 successes is 0.25 and the probability of 4 successes is 0.0625. An EXCEL plot of this distribution is shown in Figure 3.19.

x	0	1	2	3	4
p(x)	0.0625	0.25	0.375	0.25	0.0625

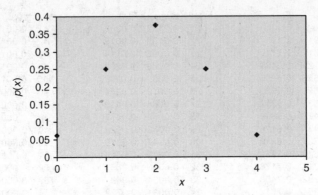

Figure 3.19   The binomial distribution with $n = 4$ and $p = 0.5$.

Having introduced the idea and terminology about a binomial distribution in the last paragraph, let us now generalize the ideas. Binomial random variables satisfy the following assumptions called the **binomial assumptions**.

1. There are only two possible outcomes for each trial called success and failure.

2. The probabilities of success and failure remain the same from trial to trial.

3. There are n trials and n is constant.

4. The n trials are independent of one another.

The **binomial random variable** is then defined as the number of successes in the n trials. A binomial random variable has a probability distribution given by the following

$$b(x) = \binom{n}{x} p^x q^{n-x} \text{ for } x = 0, 1, 2, \cdots, n$$

and

$$\binom{n}{x} = \frac{n!}{x!(n-x)!} \text{ are called } \textbf{binomial coefficients}.$$

Note that the binomial coefficients and the combination of n things taken x at a time are the same. The binomial coefficients count the number of outcomes that give x successes. In the above discussion concerning 4 flips of a coin, there were 6 outcomes that gave 2 heads. They were *HHTT* or *HTHT* or *HTTH* or *TTHH* or *THTH* or *THHT*. If we evaluate $\binom{4}{2} = \frac{4!}{2!(4-2)!} = 6$, we see that the binomial coefficient accounts for the 6 outcomes that give 2 heads.

EXAMPLE 3.15   Use EXCEL to find the binomial coefficients for $x = 0, 1, \cdots, 10$ and $n = 10$. The numbers 0 through 10 are entered into A3:A13. The function =COMBIN(10,A3) is entered into B3 and a click-and-drag is performed to give Figure 3.20.

A plot of the data in Figure 3.20 is shown in Figure 3.21.

A	B
X	Binomial coefficients
0	1
1	10
2	45
3	120
4	210
5	252
6	210
7	120
8	45
9	10
10	1

Figure 3.20   Binomial coefficients for $n = 10$.

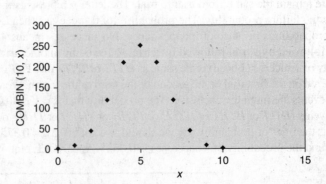

Figure 3.21   Plot of the binomial coefficients for $n = 10$.

EXAMPLE 3.16 A sampling plan works as follows: A sample of 10 items is selected from a large lot. X represents the number of defectives found in the sample. If $X = 0$, the lot is accepted. If X equals 1 or more, the complete lot is inspected. If the lot is 0.5% defective, find the probability that the lot is accepted and the probability that the lot will be completely inspected. First note that since the lot is large, taking a sample of 10 will not affect the probability of obtaining a defective item and the binomial assumptions may be assumed to hold. Now use EXCEL to calculate binomial probabilities as follows.

The labels x, $b(x)$, and $B(x)$ are entered in A1:C1. The numbers 0 through 10 are entered in A2:A12. To calculate the binomial distribution, enter =BINOMDIST(A2,10,0.005,0) into cell B2 and perform a click-and-drag. The first parameter in the function is the number of successes, the second is the number of trials, the third is the success probability, and if the fourth is a 0, the individual binomial probabilities are computed. If a 1 is entered in the fourth position, the cumulative binomial probabilities are computed.

A	B	C
x	$b(x)$	$B(x)$
0	0.9511101	0.95111
1	0.0477945	0.998905
2	0.0010808	0.999985
3	1.448E-05	1
4	1.274E-07	1
5	7.68E-10	1
6	3.216E-12	1
7	9.235E-15	1
8	1.74E-17	1
9	1.943E-20	1
10	9.766E-24	1

The probability that a 0.5% defective lot will be accepted is $B(0) = 0.95111$. The probability that a 0.5% defective lot will be rejected is $1 - B(0) = 0.04889$. If p is small, the binomial distribution is skewed to the right like this one.

EXAMPLE 3.17 Suppose the lot in the previous example is 5% defective rather than 0.5% defective. Compute the probabilities of accepting the lot as well as a complete inspection. If MINITAB is used to compute the probabilities, the pulldown **Calc $\Rightarrow$ Probability distributions $\Rightarrow$ Binomial** will give the Binomial Distribution dialog box which is filled as shown in Figure 3.22. The numbers 0 through 10 are entered into first column of the MINITAB worksheet. In the dialog box shown in Figure 3.22, probability is checked and cumulative probability is also checked. The below output will be the result.

```
Row   x    b(x)      cumulative
 1    0   0.598737    0.59874
 2    1   0.315125    0.91386
 3    2   0.074635    0.98850
 4    3   0.010475    0.99897
 5    4   0.000965    0.99994
 6    5   0.000061    1.00000
 7    6   0.000003    1.00000
 8    7   0.000000    1.00000
 9    8   0.000000    1.00000
10    9   0.000000    1.00000
11   10   0.000000    1.00000
```

The probability that a 5% defective lot will be accepted is 0.598, and the probability that a 5% defective lot will be completely inspected is $1 - 0.598 = 0.402$.

A shortcut formula for the mean and standard deviation may be derived from the formula for the mean and standard deviation and some algebraic manipulation. It may be shown that the mean is found by multiplying the number of trials times the success probability, i.e. $\mu = np$. The standard deviation, likewise, can be found by use of the shortcut formula $\sigma = \sqrt{npq}$.

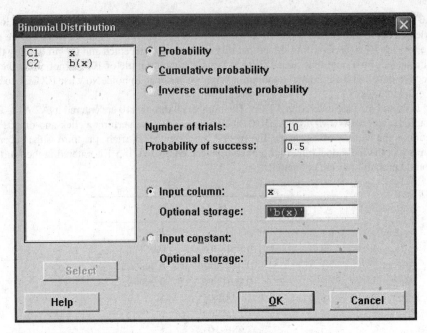

Figure 3.22 MINITAB binomial distribution dialog box.

EXAMPLE 3.18 Consider the following binomial distribution. Find the mean and standard deviation using the basic definition and the shortcut formula and show that you get the same answer.

x	0	1	2	3	4
$p(x)$	0.0625	0.25	0.375	0.25	0.0625

$$\mu = \sum xp(x) = 0(0.0625) + 1(0.25) + 2(0.375) + 3(0.25) + 4(0.0625) = 2$$

$$\mu = np = 4(0.5) = 2$$

$$\sigma^2 = \sum x_i^2 p(x_i) - \mu^2 = 0(0.0625) + 1(0.25) + 4(0.375) + 9(0.25) + 16(0.0625) - 4 = 1$$

$$\sigma^2 = npq = 4(0.5)(0.5) = 1$$

Finally, consider the effect of p on the binomial distribution. This may be illustrated by plotting three probability distributions. The three distributions are: (1) binomial with $n = 50$ and $p = 0.25$, (2) binomial with $n = 50$ and $p = 0.5$, and (3) binomial with $n = 50$ and $p = 0.75$. These three plots are shown in Figure 3.23.

When $p = 0.25$, the distribution is skewed to the right, when $p = 0.50$, the distribution is symmetrical, and when $p = 0.75$, the distribution is skewed to the left.

A note concerning binomial probabilities. Most statistics books contain tables of binomial probabilities for various values of n and p. The author believes that such tables are an anachronism. With the widespread availability of statistical software packages, many of which have the ability to compute binomial probabilities as well as many other probability distributions, the need for such tables no longer exists. In surveys of students in the author's classes, almost all students have EXCEL at home as well as being available on all university computers. In addition all engineers have EXCEL.

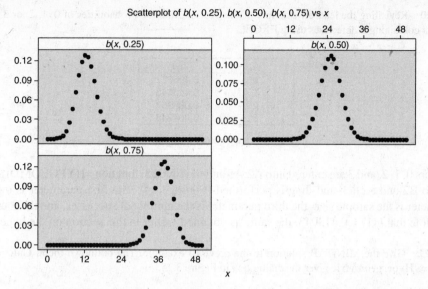

Figure 3.23   Binomial distributions with $n = 50$ and $p = 0.25$, $p = 0.50$, and $p = 0.75$.

Hypergeometric Distribution

EXAMPLE 3.19   Suppose personal digital assistants (PDAs) are packaged 25 per box and a box contains 5 defective PDAs. A sample of 3 is selected. It is of interest to compute the probability that the sample contains 1 defective. The 1 defective can come from the 5 defectives in $\binom{5}{1}$ ways. The other 2 can come from the 20 non-defectives in $\binom{20}{2}$ ways. The multiplicative rule states that the 1 defective and the 2 non-defectives can be selected in $\binom{5}{1}\binom{20}{2}$ ways. There are $5(190) = 950$ samples with 1 defective and 2 non-defectives. The total space is the number of ways that 3 may be selected from 25, which is $\binom{25}{3}$ or 2300.

The probability of selecting 3 that contain 1 defective is $950/2300 = 0.413$. This is called a hypergeometric probability.

Problems often arise in which the hypergeometric is applicable. There is a dichotomy of elements involved. That is, the elements may be classified into two types: defective/non-defective, male/female, hearts/non-hearts in a deck of cards, etc. A sample from the population that has been dichotomized into the two groups is selected. The generic terms success/failure is used to describe the two types of elements. The number of successes in the sample is the random variable of interest. With this introductory paragraph, we are ready to describe the random variable generally.

A population consists of N items of which s are classified as successes and $N-s$ are failures. A sample of size n is selected from the N. The hypergeometric random variable X is defined to be the number of successes in the sample. Paralleling the opening example of this section, the following probability distribution for a hypergeometric is obtained:

$$P(X = x) = \frac{\binom{s}{x}\binom{N-s}{n-x}}{\binom{N}{n}} \text{ for } x = 0, 1, \cdots, n \text{ and } x \leq s \text{ and } n - x \leq N - a$$

The term $\binom{s}{x}$ counts the number of ways x successes can be selected from s successes. The term $\binom{N-s}{n-x}$ counts the number of ways $(n-x)$ failures can be selected from $(N-s)$ failures. The term $\binom{N}{n}$ counts the number of ways n can be selected from N. The $x \leq s$ and $n - x \leq N - a$ parts are needed because you cannot select more items than are available. (For example you cannot select 5 items from 4.) The computational difficulty is made easier by computer software.

EXAMPLE 3.20   Revisiting the lead-off example of this section, find the probabilities of 0, 1, 2, or 3 defectives in a box of 25 PDAs containing 5 defectives using EXCEL.

x	h(x)
0	0.495652
1	0.413043
2	0.086957
3	0.004348

Figure 3.24   EXCEL output.

The numbers 0, 1, 2, and 3 are entered into A2:A5 and the EXCEL function =HYPGEOMDIST(A2,3,5,25) is entered into B2 and a click-and-drag is performed (Figure 3.24). The first parameter is the x value, the second parameter is the sample size, the third parameter is the number of successes, and the fourth is the total population. Note that $h(1) = 0.413043$, the same as obtained earlier in this section on the hypergeometric.

EXAMPLE 3.21   Give the MINITAB solution to the previous example. The pulldown menu **Calc ⇒ Probability Distributions ⇒ Hypergeometric** gives the dialog box in Figure 3.25.

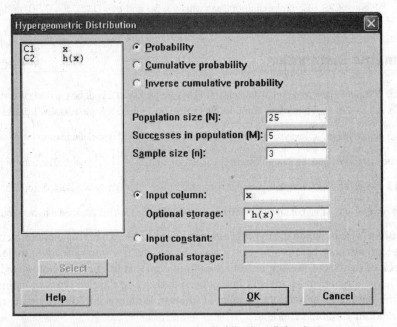

Figure 3.25   MINITAB hypergeometric distribution dialog box.

Row	x	h(x)	cumulative
1	0	0.495652	0.49565
2	1	0.413043	0.90870
3	2	0.086957	0.99565
4	3	0.004348	1.00000

The distribution given under $h(x)$ by MINITAB is the same as that obtained by EXCEL.

A shortcut formula for the mean and standard deviation may be derived from the formula for the mean and standard deviation and some algebraic manipulation. It may be shown that the mean of a hypergeometric distribution is $\mu = \dfrac{ns}{N}$. The standard deviation of a hypergeometric distribution, likewise, can be found by use of the shortcut formula $\sigma = \sqrt{\dfrac{ns(N-s)(N-n)}{N^2(N-1)}}$.

EXAMPLE 3.22   Consider the following hypergeometric distribution. Find the mean and standard deviation using the basic definition and the shortcut formula and show that you get the same answer. $N = 50$, $s = 10$, $n = 5$. The distribution of X = the number of successes in the sample of size 5 is as follows:

X	0	1	2	3	4	5
h(x)	0.310563	0.431337	0.20984	0.044177	0.003965	0.000119

$$\mu = \sum xp(x) = 0(0.310563) + 1(0.431337) + 2(0.20984) + 3(0.44177) + 4(0.003965) + 5(0.000119) = 1$$

Using the shortcut formula for the mean, $\mu = \dfrac{ns}{N} = \dfrac{5(10)}{50} = 1$.

$$\mu^2 = \sum x_i^2 p(x_i) - \mu^2 = 0(0.310563) + 1(0.431337) + 4(0.20984) + 9(0.44177)$$

$$+ 16(0.003965) + 25(0.000119) - 1 = 1.734694 - 1 = 0.7\,34694$$

$$\sigma = 0.857143$$

Using the shortcut formula for the standard deviation,

$$\sigma = \sqrt{\frac{ns(N-s)(N-n)}{N^2(N-1)}} = \sqrt{\frac{5(10)(40)(45)}{2500(49)}} = 0.857143$$

The hypergeometric and the binomial are very similar distributions when $n < 0.05N$. This is illustrated in the following example.

EXAMPLE 3.23   A box of mp3s are packaged for shipment. The box contains 500 mp3s of which 25 have a minor defect. A sample of 10 is selected. The number in the sample with a defect has a hypergeometric distribution with $N = 500$, $s = 25$, and $n = 10$. The number in the sample with a defect may also be approximated by a binomial with $n = 10$, $p = 0.05$ and $q = 0.95$. Since $n < 0.05N$, it may be assumed that as items are taken into the sample the 5% with defects do not change significantly. The hypergeometric probabilities in second row and binomial probabilities in third row in Figure 3.26 were generated using EXCEL.

Figure 3.27 gives a graphical comparison of the distributions in Figure 3.26.

x	0	1	2	3	4	5	6	7	8	9	10
h(x)	0.5959	0.3197	0.0739	0.0097	0.0008	4E-05	2E-06	3E-08	5E-10	4E-12	1E-14
b(i)	0.5987	0.3151	0.0746	0.0105	0.001	6E-05	3E-06	8E-08	2E-09	2E-11	1E-13

Figure 3.26   Comparison of hypergeometric and binomial probability distributions.

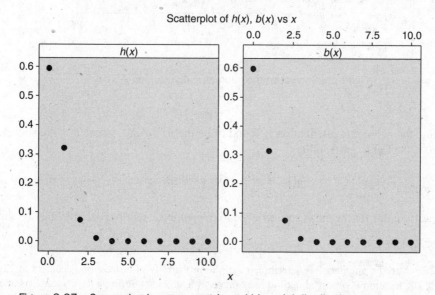

Figure 3.27   Comparing hypergeometric and binomial distributions.

Geometric Distribution

The **geometric random variable** has the same assumptions as the binomial, except the number of trials is not fixed. Random variable X is the trial number on which the first success occurs. The trials are independent of one another and the probability is p that a success occurs on a given trial and q that a failure occurs on a given trial.

EXAMPLE 3.24   Suppose a fair die is rolled until a 6 first occurs. The probability that a 6 occurs on the first roll is $\frac{1}{6}$. The probability that a 6 first occurs on the second roll is $P(2) = \frac{5}{6}\frac{1}{6} = \frac{5}{36}$. The trials are independent and so the probability of "not a 6" on the first roll followed by a 6 on the second roll is just the product of the two probabilities. The probability that a 6 occurs for the first time on the third roll is the same as "not a 6" on the first roll and "not a 6" on the second roll and a 6 on the third roll. The product of the probabilities is $P(3) = \frac{5}{6}\frac{5}{6}\frac{1}{6} = \frac{25}{216}$. If this is continued, the geometric probability distribution is found to be

$$P(x) = \left(\frac{5}{6}\right)^{x-1}\frac{1}{6} \quad \text{for } x = 1, 2, 3, \cdots.$$

EXCEL was used to generate Figures 3.28 and 3.29. It shows the probabilities associated with the first 20 values of the geometric variable X. The first 20 values account for over 97% of the probability associated with the distribution.

x	1	2	3	4	5	6	7	8	9	10
$p(x)$	0.167	0.139	0.116	0.096	0.08	0.067	0.056	0.047	0.039	0.032
x	11	12	13	14	15	16	17	18	19	20
$p(x)$	0.027	0.022	0.019	0.016	0.013	0.011	0.009	0.008	0.006	0.005

Figure 3.28   First 20 probabilities for the geometric distribution $p(x) = \left(\frac{5}{6}\right)^{x-1}\frac{1}{6}$.

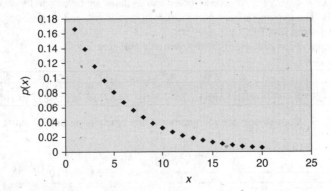

Figure 3.29   Plot of the geometric distribution $P(x) = \left(\frac{5}{6}\right)^{x-1}\frac{1}{6}$.

Generalizing, the geometric distribution is found to be the following, where p is the success probability and $q = (1 - p)$ is the failure probability,

$$g(x) = q^{x-1}p \quad \text{for } x = 1, 2, 3, \cdots.$$

The shortcut formulas for the mean and variance for a geometric random variable may be shown to be the following: $\mu = \frac{1}{p}$ and $\sigma^2 = \frac{1-p}{p^2}$.

EXAMPLE 3.25   A traffic engineer has determined that the probability of at least one accident at 72nd and Dodge per day is 0.25. The location is observed until at least one accident occurs during a day. Success is at least one accident occurs during a day. Failure is no accident occurs during a day. The success probability is $p = 0.25$ and the failure probability is $q = 0.75$. The probability distribution function is $g(x) = 0.75^{x-1}0.25$ for $x = 1, 2, 3, \cdots$. Find the mean and variance using both the fundamental definitions and the shortcut formulas.

Even though the sums involved in finding the mean and the variance go from 1 to infinity, if the sums are truncated at 30, the mean and variance will be very close. Refer to Figure 3.30 in the following discussion. First of all, note that the sum of the probabilities, $\Sigma g(x) = 0.944 + 0.053 + 0.003 = 1$ to three decimal places. The mean is approximated by $3.12 + 0.712 + 0.070 = 3.994$ is very close to the true shortcut value $\mu = \dfrac{1}{p} = \dfrac{1}{0.25} = 4$. The variance is $\sigma^2 = \Sigma x_i^2 p(x_i) - \mu^2$ and is approximated by $16.29 + 9.849 + 1.656 - 4^2 = 11.795$ which is very close to the true shortcut value $\sigma^2 = \dfrac{1-p}{p^2} = \dfrac{-0.25}{0.25^2} = 12$.

x	1	2	3	4	5	6	7	8	9	10	Sum
g(x)	0.25	0.188	0.141	0.105	0.079	0.059	0.044	0.033	0.025	0.019	0.944
xg(x)	0.25	0.375	0.422	0.422	0.396	0.356	0.311	0.267	0.225	0.188	3.212
x^2g(x)	0.25	0.75	1.266	1.688	1.978	2.136	2.18	2.136	2.027	1.877	16.29

x	11	12	13	14	15	16	17	18	19	20	Sum
g(x)	0.014	0.011	0.008	0.006	0.004	0.003	0.003	0.002	0.001	0.001	0.053
xg(x)	0.155	0.127	0.103	0.083	0.067	0.053	0.043	0.034	0.027	0.021	0.712
x^2g(x)	1.703	1.52	1.338	1.164	1.002	0.855	0.724	0.609	0.509	0.423	9.849

x	21	22	23	24	25	26	27	28	29	30	Sum
g(x)	8E-04	6E-04	4E-04	3E-04	3E-04	2E-04	1E-04	1E-04	8E-05	6E-05	0.003
xg(x)	0.017	0.013	0.01	0.008	0.006	0.005	0.004	0.003	0.002	0.002	0.07
x^2g(x)	0.35	0.288	0.236	0.193	0.157	0.127	0.103	0.083	0.067	0.054	1.656

Figure 3.30   Using EXCEL to find the mean and variance of $g(x) = 0.75^{(x-1)}0.25$.

The geometric distribution has a **countably infinite sample space**. However, if a package such as EXCEL is available, it is seen that a finite number of terms are sufficient to handle any computations involving the distribution.

Negative Binomial Distribution

The negative binomial distribution is the distribution of X, where X is the number of trials needed before the rth success. Consider an example before generalizing the formula for X.

EXAMPLE 3.26 Suppose we roll a die until a 6 occurs for the third time. Note that the third occurrence of the 6 might be on the third or fourth or fifth or $\cdots$ roll of the die. If the third occurrence of a 6 is on the xth roll, then there must have been two 6s in the proceeding $(x-1)$ rolls. The probability of two 6s in $x-1$ rolls of a die is a binomial probability. Therefore, the probability of two 6s in $(x-1)$ rolls and then a 6 on the xth roll is

$$P(X = x) = \binom{x-1}{2}\left(\frac{1}{6}\right)^2\left(\frac{5}{6}\right)^{(x-1)-2}\left(\frac{1}{6}\right) = \binom{x-1}{2}\left(\frac{1}{6}\right)^3\left(\frac{5}{6}\right)^{(x-3)}, \text{ for } x = 3, 4, \cdots.$$

Using EXCEL, the numbers 3 through 12 are entered into B1:K1 and the expression =COMBIN(B1-1,2)*(1/6)^3*(5/6)^(B1-3) is entered into B2 and a click-and-drag is performed from B2 through K2 to form the first two rows of Figure 3.31. This same procedure is used to form the remaining rows of the table in Figure 3.31.

The cumulative probability at 52 is $0.323 + 0.412 + 0.185 + 0.059 + 0.016 = 0.995$ in Figure 3.31, that is $F(52) = 0.995$. A plot of the approximate negative binomial distribution given in Figure 3.31 is shown in Figure 3.32. The plot corresponds to x values from 3 to 52. The y value at $x = 52$ is 0.0008. Even though this distribution continues forever, the amount of added probability from 53 forward is negligible (less than 0.005).

Generalizing from the last example, we have the following: Suppose the probability of a success is p and the probability of a failure is q on any given trial. The trials are independent of one another. The probability that the rth success occurs on the xth trial where $x = r, (r+1), \cdots$ is

$$NB(x) = \binom{x-1}{r-1}(p)^r(q)^{(x-r)}, \text{ for } X = r, (r+1), \cdots.$$

x	3	4	5	6	7	8	9	10	11	12	Sum
P(x)	0.005	0.012	0.019	0.027	0.033	0.039	0.043	0.047	0.048	0.049	0.323

x	13	14	15	16	17	18	19	20	21	22	Sum
P(x)	0.049	0.049	0.047	0.045	0.043	0.041	0.038	0.036	0.033	0.03	0.412

x	23	24	25	26	27	28	29	30	31	32	Sum
P(x)	0.028	0.025	0.023	0.021	0.019	0.017	0.015	0.014	0.012	0.011	0.185

x	33	34	35	36	37	38	39	40	41	42	Sum
P(x)	0.01	0.009	0.008	0.007	0.006	0.005	0.005	0.004	0.004	0.003	0.059

x	43	44	45	46	47	48	49	50	51	52	Sum
P(x)	0.003	0.002	0.002	0.002	0.002	0.001	0.001	0.001	9E-04	8E-04	0.016

Figure 3.31   Probabilities for $X = 3$ through $X = 52$ for the negative binomial distribution $P(X = x) = \binom{x-1}{2}\left(\frac{1}{6}\right)^3\left(\frac{5}{6}\right)^{(x-3)}$.

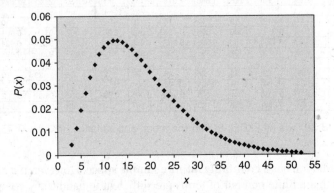

Figure 3.32   Plot of the negative binomial distribution

$$P(X = x) = \binom{x-1}{2}\left(\frac{1}{6}\right)^3\left(\frac{5}{6}\right)^{(x-3)}.$$

The shortcut formula for the mean of a negative binomial is $\mu = \dfrac{r}{p}$ and the variance is $\sigma^2 = \dfrac{r(1-p)}{p^2}$. Sometimes algebraic manipulations are used to derive the shortcut formulas for the mean and variance for the various distributions. Sometimes techniques utilizing moment generating functions are utilized in the proofs. (Moment generating functions are discussed in mathematical statistics courses.) As engineering students, we are mainly interested in the results, not the proofs. In a course in mathematical statistics, the derivations of the formulas would be the focus of the course and applications would be of secondary interest.

EXAMPLE 3.27   This example compares the mean found using the defining formula for the mean and the shortcut formula for the mean of the negative binomial distribution. Figure 3.33 is easily produced using EXCEL. The mean has an approximate value of $2.845 + 7.034 + 4.945 + 2.152 + 0.733 = 17.71$. This approximate value of the mean is obtained by using the terms corresponding to $X = 3$ through $X = 52$. Keep in mind that the sum defining the mean actually has an infinite number of terms in it. The true mean using the shortcut formula is $\mu = \dfrac{r}{p} = \dfrac{3}{1/6} = 18$.

EXAMPLE 3.28   A quality control engineer has decided to inspect play stations until 4 defectives are found. Suppose % of the play stations are defective. Let X stand for the number of inspections needed to find 4 defectives. (a) What is the mean number of inspections needed to find 4 defectives? (b) What is the standard deviation of X? (c) What is the probability that X will be greater than $\mu + \sigma$?

Note that X has a negative binomial distribution with $r = 4$, $p = 0.01$, and $q = 0.99$.

(a) $\mu = \dfrac{r}{p} = \dfrac{4}{0.01} = 400$ (b) $\sigma^2 = \dfrac{r(1-p)}{p^2} = \dfrac{4(.99)}{0.0001} = 39600$ or $\sigma = 199$

(c) Using EXCEL, the numbers 0 through 595 are placed in A1:A596 by using a click-and-drag and =NEGBINOMDIST(A1,4,0.01) is entered into B1 and a click-and-drag is performed from B1 to B596. The

x	3	4	5	6	7	8	9	10	11	12	Sum
P(x)	0.005	0.012	0.019	0.027	0.033	0.039	0.043	0.047	0.048	0.049	0.323
xP(x)	0.014	0.046	0.096	0.161	0.234	0.313	0.391	0.465	0.533	0.592	2.845

x	13	14	15	16	17	18	19	20	21	22	Sum
P(x)	0.049	0.049	0.047	0.045	0.043	0.041	0.038	0.036	0.033	0.03	0.412
xP(x)	0.642	0.68	0.709	0.727	0.736	0.736	0.728	0.714	0.694	0.669	7.034

x	23	24	25	26	27	28	29	30	31	32	Sum
P(x)	0.028	0.025	0.023	0.021	0.019	0.017	0.015	0.014	0.012	0.011	0.185
xP(x)	0.642	0.611	0.579	0.545	0.511	0.477	0.443	0.41	0.379	0.348	4.945

x	33	34	35	36	37	38	39	40	41	42	Sum
P(x)	0.01	0.009	0.008	0.007	0.006	0.005	0.005	0.004	0.004	0.003	0.059
xP(x)	0.319	0.292	0.266	0.242	0.219	0.198	0.179	0.161	0.145	0.13	2.152

x	43	44	45	46	47	48	49	50	51	52	Sum
P(x)	0.003	0.002	0.002	0.002	0.002	0.001	0.001	0.001	9E-04	8E-04	0.016
xP(x)	0.117	0.104	0.093	0.083	0.074	0.066	0.058	0.052	0.046	0.04	0.733
x	3	4	5	6	7	8	9	10	11	12	

Figure 3.33   Mean of the negative binomial is 18.

function =SUM(B1:B596) is entered in B597. This gives the probability $P(X \le \mu + \sigma)$, which is 0.84925. The probability we are seeking is $1 - 0.84925$ or 0.15075.

Poisson Distribution

A distribution that arises frequently in engineering is the Poisson distribution. In particular it arises when there are a large number of independent trials and a small success probability. The random variable X counts the number of successes to occur. It has a probability distribution represented by $Po(x)$ and given by the following

$$Po(x) = e^{-\lambda}\frac{\lambda^x}{x!}, x = 0, 1, 2, \cdots.$$

It can be shown that $\mu = \lambda$ and $\sigma^2 = \lambda$. The Poisson distribution and the binomial distribution give very similar probabilities when n is large and p is small as the next example will illustrate.

EXAMPLE 3.29   A traffic survey counts the number of accidents occur at 72nd and Dodge during a week. A large number of cars pass through the intersection and the probability that a given car is involved in an accident is $p = 0.0001$. During the week the traffic survey is conducted, 100,000 cars pass through the intersection. The mean number of accidents during such a week is $\mu = np = 100000(0.0001) = 10$. Figure 3.34 gives the binomial probabilities and Poisson probabilities that $X = 0$ up to 20. Note how close the binomial and Poisson probabilities are to one another.

The EXCEL expression =BINOMDIST(A2,100000,0.0001,0) is entered into B2 and a click-and-drag is executed from B2 to B21. The EXCEL expression =POISSON(A2,10,0) is entered into C2 and a click-and-drag is executed from C2 to C21. The 10 in the Poisson expression is the mean of the Poisson. Remember $\lambda = np = 100,000(0.0001) = 10$. The 0 in the Poisson function causes the individual Poisson probabilities rather than cumulative Poisson probabilities to be printed. Figure 3.35 gives a plot of the two probability distributions. The distributions are very similar to the normal for large n and small p. The normal distribution is continuous and will be discussed in the next chapter.

Most of the probability is associated with the values from 0 to 20. The sum of the Poisson probabilities in Figure 3.34 is 0.998412 and the sum of the binomial probabilities is 0.998413. Even though the variable takes on values from 0 to 100,000, there is very little probability beyond $X = 21$. Figure 3.35 is a MINITAB comparison of the graphs of the two probability functions.

EXAMPLE 3.30   The number of hits on a website during an 8-hour period follows a Poisson distribution. The mean number per 8-hour period is 10. (a) What is the probability of 15 or fewer hits during an 8-hour period using MINITAB? (b) What is the probability of 25 or more hits during a 16-hour period using EXCEL?

(a) Using MINITAB, the pulldown **Calc** $\Rightarrow$ **Probability Distributions** $\Rightarrow$ **Poisson Distribution** gives the dialog box shown in Figure 3.36. The output for the dialog box is as follows.

Cumulative Distribution Function

```
Poisson with mean = 10
 x P( X ≤ x )
15 0.951260
```

x	Binomial	Poisson
0	4.53772E-05	4.53999E-05
1	0.000453818	0.000453999
2	0.002269293	0.002269996
3	0.007564915	0.007566655
4	0.018913611	0.018916637
5	0.037829491	0.037833275
6	0.063052305	0.063055458
7	0.090078325	0.090079226
8	0.112601284	0.112599032
9	0.12511504	0.125110036
10	0.125116292	0.125110036
11	0.113742083	0.113736396
12	0.094784122	0.09478033
13	0.072909404	0.072907946
14	0.052076583	0.052077104
15	0.034716333	0.03471807
16	0.021696623	0.021698794
17	0.012761954	0.012763996
18	0.007089478	0.007091109
19	0.003731006	0.003732163
20	0.001865335	0.001866081
Sum	0.998413	0.998412

Figure 3.34   Comparing the binomial and
Poisson distributions ($n =$
100000 and $p = 0.0001$).

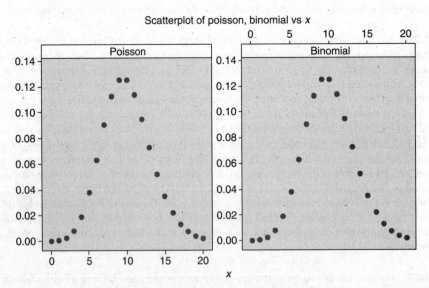

Figure 3.35   Comparing the binomial and Poisson distributions for $n = 100000$
and $p = 0.0001$.

The probability of 15 or fewer hits is 0.951260.

(b) The mean number of hits during a 16-hour period is Poisson with $\lambda = 20$. The probability of 24 or fewer hits during a 16-hour period is given by the EXCEL command =Poisson(24,20,1) = 0.843227. The probability of 25 or more hits during the 16-hour period is $1 - 0.843227 = 0.156773$. Note that 25 or more is the complement of 24 or less. The EXCEL command = Poisson(24,20,1) gives the cumulative probability of 24 or fewer for a Poisson distribution with mean 20.

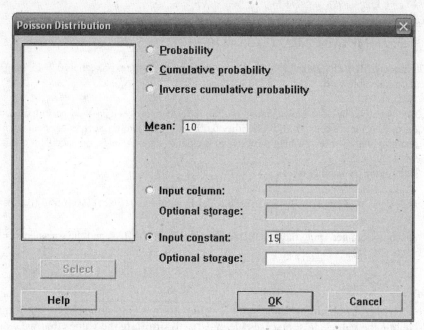

Figure 3.36   MINITAB dialog box for calculating $P(X \le 15)$ for the Poisson distribution.

Multinomial Distribution

The multinomial distribution is an extension of the binomial distribution. There are n independent trials with not 2 but k possible outcomes per trial. Define discrete random variables $X_1, \cdots, X_k$ and probabilities $p_1, \cdots, p_k$ as follows: X_i = the number of times that outcome i occurs in the n trials and p_i = the probability that outcome i occurs. The **joint probability mass function** that $X_1 = x_1, \cdots, X_k = x_k$ is

$$P(X_1 = x_x, ..., X_k = x_k) = \frac{n!}{x_1! \cdots x_k!} p_1^{x_1} p_2^{x_2} \cdots p_k^{x_k}, x_i = 0, 1, \cdots n \text{ and } \sum x_i = n$$

EXAMPLE 3.31   A fair die is thrown 5 times. On any one throw, **outcome 1** is that an even number appears, **outcome 2** is that a 1 or 3 appears, and **outcome 3** is that a 5 appears. Find the probability that in the 5 throws, outcome 1 occurs twice, outcome 2 occurs twice, and outcome 3 occurs once. The joint probability mass function is as follows.

$$P(X_1 = 2, X_2 = 2, X_3 = 1) = \frac{5!}{2!2!1!} 0.5^2 0.33^2 0.17^1$$

The expression may be evaluated using EXCEL. The EXCEL expression used to evaluate the probability is =MULTINOMIAL(2,2,1)*0.5^2*0.33^2*0.17^1 = 0.138848. The EXCEL multinomial function is used to evaluate $\frac{n!}{x_1! \cdots x_k!}, \Sigma x_i = n$.

EXAMPLE 3.32   Based on past experience, the makeup of Dr. Stephens' statistics for engineers class has consisted of 20% civil engineers, 30% chemical engineers, 40% industrial engineers, and 10% electrical engineers. Given that these probabilities still hold, what is the probability that her 2010 classes will consist of 25 civil engineers, 30 chemical engineers, 40 industrial engineers, and 10 electrical engineers?

The expression for the probability is

$$P(X_1 = 25, X_2 = 30, X_3 = 40, X_4 = 10) = \frac{105!}{25!30!40!10!} 0.2^{25} 0.3^{30} 0.4^{40} 0.1^{10}$$

The EXCEL expression yields 0.00074 and is given as follows.

$$=\text{MULTINOMIAL}(25,30,40,10)*0.2\text{^}25*0.3\text{^}30*0.4\text{^}40*0.1\text{^}10$$

The EXCEL function MULTINOMIAL(25, 30, 40, 10) combined with the exponentials 0.2^25*0.3^30*0.4^40*0.1^10 save the student considerable work.

EXAMPLE 3.33   A traffic engineer knows that at a certain intersection over a 24-hour period, no accidents occur with probability 0.25, one accident occurs with probability 0.60, and two or more accidents occur with probability 0.15. What is the probability that over ten 24-hour periods, no accidents occur 3 times, one accident occurs 6 times and two or more accidents occur once?

The expression for the probability is

$$P(X_1 = 3, X_2 = 6, X_3 = 1) = \frac{10!}{3!6!1!}0.25^3 0.6^6 0.15^1 = 840(0.015625)(0.046656)(0.15) = 0.0918$$

The same answer is obtained using the EXCEL function MULTINOMIAL and the exponents (=MULTINOMIAL (3,6,1)*0.25^3*0.6^6*0.15).

Simulation

Often we are interested in obtaining some sample values from a random variable, but it may be expensive or difficult to obtain the values. In such cases we may try to **simulate** the values that we would likely obtain without actually observing the variable.

EXAMPLE 3.34   Consider the sums that we would obtain from observing the outcomes when a pair of dice is rolled 50 times. A random number generator is used to simulate what we might obtain. We found earlier in the section on random variables that the probability distribution for the sum on the dice had the distribution given in first, second and third columns of Figure 3.37. Fourth column gives ranges of three-digit random numbers determined by the third column.

x	P(x)	Cumulative	Range
2	0.028	0.028	000-027
3	0.056	0.084	028-083
4	0.083	0.167	084-166
5	0.111	0.278	167-277
6	0.139	0.417	278-416
7	0.166	0.583	417-582
8	0.139	0.722	583-721
9	0.111	0.833	722-832
10	0.083	0.916	833-915
11	0.056	0.972	916-971
12	0.028	1.000	972-999

Figure 3.37   Simulating the sum on a pair of dice.

In fourth column are shown ranges of three-digit random numbers. EXCEL produces three-digit numbers as follows: The expression =RANDBETWEEN(0,999) is placed in a cell and a click-and-drag is performed over 5 columns and 10 rows. The random numbers generated by EXCEL are shown in Figure 3.38.

160	183	539	69	251
719	587	343	825	942
48	575	729	447	132
843	985	662	663	643
378	360	755	845	859
606	626	144	683	284
642	363	57	772	703
809	827	702	232	172
344	864	16	457	222
638	36	912	240	406

Figure 3.38   Fifty random numbers between 000 and 999 generated by EXCEL.

The probability that a three-digit random number falls in the ranges in Figure 3.37 is equal to the probabilities given in the column labeled $p(x)$ in Figure 3.37. The probabilities for the sums on the dice are, therefore, given by the $p(x)$ column of Figure 3.37.

When the random numbers in Figure 3.38 are translated to outcomes when rolling the dice, the results are shown in Figure 3.39.

4	5	7	3	5
8	8	6	9	11
3	7	9	6	4
10	12	8	8	7
6	6	9	10	10
8	8	4	8	6
8	6	3	9	8
9	9	8	5	5
6	10	2	7	5
8	3	10	5	6

Figure 3.39   Outcomes on the dice determined by the random numbers.

When a dot plot of the sample is generated by MINITAB, the results are shown in Figure 3.40. Suppose another simulated sample is desired. Figure 3.41 shows a random sample of 50 random numbers between 0 and 999 generated by MINITAB. These random numbers are converted to simulated outcomes when rolling the dice by using Figure 3.37. This gives a second simulated sample shown in Figure 3.42. A histogram of this second sample is

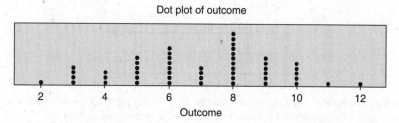

Dot plot of outcome

Outcome

Figure 3.40   Dot plot of simulated sample of outcomes from Figure 3.39.

85	389	174	700	609
215	351	604	622	824
601	202	292	52	82
110	938	574	555	67
131	935	111	448	986
836	631	448	570	90
3	727	597	78	800
410	347	111	979	123
92	415	531	131	963
146	615	770	179	682

Figure 3.41   Fifty random numbers between 000 and 999 generated by MINITAB.

4	6	5	8	8
5	6	8	8	9
8	5	6	3	3
4	11	7	7	3
4	11	4	7	12
10	8	7	7	4
2	9	8	3	9
6	6	4	12	4
4	6	7	4	11
4	8	9	5	8

Figure 3.42   Second sample of simulated outcomes on the dice determined by the random numbers in Figure 3.41.

shown in Figure 3.43. Random samples between 000 and 999 may be obtained using EXCEL, MINITAB, SAS, SPSS, or STATISTIX.

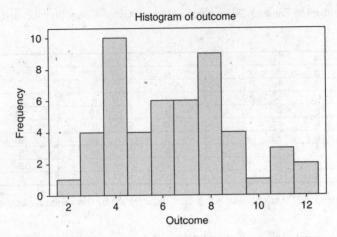

Figure 3.43   Histogram of second simulated sample.

EXAMPLE 3.35 In order to simulate the number of defective items in samples from a production line an engineer uses random numbers. The probability distribution for X, the number of defectives found in samples, is known to be the following.

x	0	1	2	3
$P(x)$	0.50	0.25	0.20	0.05

The engineer wishes to simulate what his results would be for a 30-day period. Give results that would simulate what he might experience over a 30-day period. First build a table like the one shown in Figure 3.44.

x	$p(x)$	Cumulative	Range
0	0.50	0.50	00–49
1	0.25	0.75	50–74
2	0.20	0.95	75–94
3	0.05	1.00	95–99

Figure 3.44   Simulation setup.

Then select 30 two-digit random numbers between 00 and 99.

45	50	64	18	91	98	89	83	78	81
99	96	44	92	23	49	91	30	29	80
15	34	46	46	72	4	4	47	31	65

These random numbers are replaced by simulated number of defectives using Figure 3.44 to obtain the following simulated results.

0	1	1	0	2	3	2	2	2	2
3	3	0	0	0	0	2	0	0	2
0	0	0	0	1	0	0	0	0	1

Figure 3.45 shows a time series plot of what the engineer might expect over a 30-day period. The index numbers represent days and the defective number represent the number of defectives found on a given day.

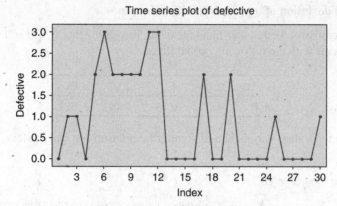

Figure 3.45 Simulated time series plot over a 30-day period for the number of defectives found.

SOLVED PROBLEMS

Random variables

3.1. A large engineering company employs a large number of individuals with engineering backgrounds. Of these, 2% have Ph.D degrees, 12% have masters degrees, 38% have bachelors degrees, and the remainder have a technical background, but no degree. The educational backgrounds are coded in the personnel department as 1, 2, 3, or 4 depending on whether the employee has a Ph.D, masters degree, bachelors degree, or no degree. Random variable X represents the coded educational background. Give the probability distribution of X, the cumulative probability distribution of X, and find $P(X < 3)$.

SOLUTION

x	1	2	3	4
$p(x)$	0.02	0.12	0.38	0.48

x	1	2	3	4
$F(x)$	0.02	0.14	0.52	1.00

$$P(X < 3) = F(2) = 0.14$$

3.2. Daily sales records for an engineering consulting firm shows that it will attract 0, 1, or 2 consulting jobs per day with probabilities as given below.

Consulting jobs	0	1	2
Probability	0.10	0.75	0.15

Find the probability distribution for X, the number of consulting jobs in a two-day period, assuming the number of consulting jobs are independent from day to day.

SOLUTION

$X = 0$ means no consulting jobs on day 1 and day 2. Due to independence the probability is $P(X = 0) = 0.10(0.10) = 0.01$. $X = 1$ means 0 consulting jobs on day 1 and 1 on day 2 or 1 on day 1 and 0 on day 2. The probability is $(0.10)(0.75) + (0.75)(0.10) = 0.075 + 0.075 = 0.15$. Similarly, $P(X = 2) = (0.10)(0.15) + (0.15)(0.10) + (0.75)(0.75) = 0.5925$, $P(X = 3) = (0.15)(0.75) + (0.75)(0.15) = 0.225$, and $P(X = 4) = (0.15)(0.15) = 0.0225$. The probability distribution for X is

x	0	1	2	3	4
$p(x)$	0.01	0.15	0.5925	0.225	0.0225

Mean and standard deviation of a probability distribution

3.3. Daily sales records for a computer manufacturing firm show that it will sell 0, 1, 2, or 3 mainframe computer systems with the following probabilities.

x	0	1	2	3
$p(x)$	0.6	0.15	0.15	0.10

Find the expected value, variance, and standard deviation for daily sales.

SOLUTION

The mean or expected value is

$$\mu = \sum x_i \, p(x_i) = 0(0.6) + 1(0.15) + 2(0.15) + 3(0.10) = 0.75$$

The variance is

$$\sigma^2 = \Sigma \, x_i^2 p(x_i) - \mu^2 = 0(0.6) + 1(0.15) + 4(0.15) + 9(0.10) - 0.5625 = 1.0875$$

And the standard deviation is the square root of 1.0875 or 1.0428.

3.4. Three dice are rolled and their sum is recorded. Use EXCEL to find the distribution of X = sum on the three dice. Use the distribution to find the mean and variance.

SOLUTION

The EXCEL worksheet is used to list all 216 outcomes and the sum of all the outcomes. The distribution of X is found to be as follows.

3	0.005
4	0.014
5	0.028
6	0.046
7	0.069
8	0.097
9	0.116
10	0.125
11	0.125
12	0.116
13	0.097
14	0.069
15	0.046
16	0.028
17	0.014
18	0.005

The mean is 10.5, the variance is 8.75, and the standard deviation is 2.96.

Chebyshev's theorem

3.5. The manager of a job shop does not know the probability distribution of the time required to complete an order. From past performance she has been able to estimate the mean and standard deviation to be 14 days and 2 days. Find an interval so that the probability is at least 75% that the order is finished in that time.

SOLUTION

Two standard deviations subtracted from the mean is $14 - 4 = 10$ days and two standard deviations added to the mean is $14 + 4 = 18$ days. The probability is at least 75% that the order is finished in 10 to 18 days.

3.6. The daily production of electric motors at a particular factory averages 120 with a standard deviation of 5. (a) What fraction of the days will have a production between 110 and 130? (b) Find the shortest interval certain to contain at least 96% of the daily production of electric motors.

SOLUTION

(a) 110 is 2 standard deviations below the mean daily production and 130 is 2 standard deviations above the mean daily production. The daily production will be between these two limits on at least $1 - 1/2^2 = 3/4$ days.

(b) $(1 - 1/k^2) = 0.96$ implies $1/k^2 = 0.04$ implies $k^2 = 25$ or $k = 5$.

The interval $(120 - 5(5) = 95, 120 + 5(5) = 145)$ will contain at least $24/25 = 96\%$ of the daily production.

Binomial distribution

3.7. Chemical engineers claim that 70% of all Americans have perfluorooctanoic acid (PFOA) in their blood stream. This chemical is a likely human carcinogen. Given that the 70% figure is correct, find the following for blood tests for PFOA in human blood. (Thirty Americans were randomly tested for PFOA.)

(a) What is the mean or expected number that you would find in a sample of 30 who test positive for PFOA?

(b) What is the standard deviation of the number found in samples of 30 who test positive for PFOA?

(c) What is the probability of finding 15 or fewer in a sample of 30 who test positive for PFOA?

(d) What is the probability of finding 25 or more in 30 who test positive for PFOA?

SOLUTION

(a) The mean or expected number for a binomial random variable is $np = 30(0.7) = 21$.

(b) The standard deviation for a binomial random variable is the square root of npq which is 6.3 or 2.5.

(c) The MINITAB output for the cumulative probability of 15 is

Cumulative Distribution Function
Binomial with $n = 30$ and $p = 0.7$

x	$P(X \le x)$	
15	0.0169373	The answer is 0.0169

(d) The probability of 25 or more is $P(X \ge 25) = 1 - P(X \le 24)$.

Cumulative Distribution Function
Binomial with $n = 30$ and $p = 0.7$

x	$P(X \le x)$	
24	0.923405	$P(X \ge 25) = 1 - 0.9234 = 0.0766$

3.8. An industrial firm supplies 10 manufacturing plants with a certain chemical. The probability that any one firm calls in an order on a given day is 0.3, and this is the same for all 10 plants. Let X represent the number of plants to call in an order on a given day. Find the mean of X, the standard deviation of X, the probability that the number of plants calling in orders is at most 3, and the probability that the number of plants calling in orders is at least 3.

SOLUTION

The mean of X is $10(0.3) = 3$, and the standard deviation is the square root of $10(0.3)(0.7)$ or 1.45. The cumulative binomial distribution with $n = 10$, and $p = 0.3$ is as follows:

x	$F(x)$
0	0.028248
1	0.149308
2	0.382783
3	0.649611
4	0.849732
5	0.952651
6	0.989408
7	0.99841
8	0.999856
9	0.999994
10	1

The probability that X is at most 3 is $F(3) = 0.650$ and the probability that X is at least 3 is $1 - P(X \leq 2) = 1 - F(2) = 1 - 0.383 = 0.617$.

The hypergeometric distribution

3.9. Of 50 buildings in an industrial park, 12 have electrical code violations. If 10 buildings are selected at random for inspection, (a) what is the expected number to have electrical code violations? (b) What is the standard deviation of the number to have electrical code violations? (c) What is the probability that 3 or fewer have electrical code violations?

SOLUTION

(a) The expected number to have electrical code violations is $\mu = \dfrac{ns}{N} = \dfrac{10(12)}{50} = 2.4$.

(b) The standard deviation of the number to have electrical code violations is

$$\sigma = \sqrt{\frac{ns(N-s)(N-n)}{N^2(N-1)}} = \sqrt{\frac{10(12)(50-12)(50-10)}{50^2(50-1)}} = \sqrt{1.4890} = 1.22$$

(c) Using MINITAB, the probability that 3 or fewer have electrical code violations is given by the following dialog box that results from the pulldown **Calc ⇒ Probability distribution ⇒ Hypergeometric** and is filled out as shown,

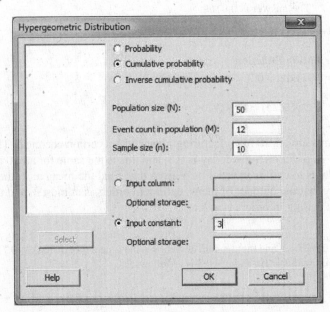

Cumulative Distribution Function

Hypergeometric with $N = 50$, $M = 12$, and $n = 10$

x $P(X \le x)$

3   0.820944 $P(X \le 3) = 0.821$

3.10. Under the condition that $S/N < 0.05$ where S is the number of successes and N is the population size, the binomial distribution and the hypergeometric distribution give probabilities that are very close to one another. Suppose a shipment of 120 burglar alarms contain 5 that are defective. If 3 of these alarms are selected and shipped to a customer, calculate the probability that 1 of the 3 will be defective using (a) the hypergeometric distribution and (b) the binomial distribution.

SOLUTION

(a) Using EXCEL, the probability of getting 1 defective in the sample is =HYPGEOMDIST(1,3,5,120) = 0.1167.

(b) Using EXCEL and the binomial distribution, the probability of getting 1 defective in the sample of 3 is =BINOMDIST(1,3,0.0417,0) where $p = 5/120 = 0.0417$. The probability using the binomial distribution is =BINOMDIST(1,3,0.0417,0) = 0.1149. Note how close 0.1167 and 0.1149 are. The approximation is close as long as $S/N < 0.05$.

Geometric distribution

3.11. A recruiting firm finds that 20% of the applicants have advanced training in computer programming. Applicants are interviewed sequentially and are selected at random from the pool of applicants. (a) Find the probability that the first applicant having advanced training is found on the fifth interview. (b) Find the mean number of applicants needed before finding the first with advanced training in computer programming.

SOLUTION

(a) The first applicant having advanced training occurs on the fifth trial if such an applicant is not found on the first four trials. $P(X = 5) = (0.8)^4(0.2) = 0.0819$.

(b) The mean of X is $1/p = 1/0.2 = 5$.

3.12. A structural engineer performs a test of weld strength that involves loading welded joints until a fracture occurs. For a certain type of weld, 95% of the fractures occur in the weld itself while the other 5% occur in the beam. A number of welds are tested. Let X be the number of tests up to and including the first test that results in a beam fracture. (a) What is the distribution of X? (b) What is the probability that X is less than 10? (c) What is the mean of X?

SOLUTION

(a) X has a geometric distribution with probability distribution $P(x) = 0.95^{x-1}(0.05)$.

(b)

A	B
1	0.05
2	0.0475
3	0.045125
4	0.042869
5	0.040725
6	0.038689
7	0.036755
8	0.034917
9	0.033171
	0.369751

The numbers 1 through 9 are entered into column A of EXCEL. The expression =0.95^(A1-1)*0.05 is entered into column B and a click-and-drag is performed through B9. The expression =SUM(B1:B9) is entered into B10. $P(X < 10) = 0.3698$.

(c) The mean of X is $1/0.05 = 20$.

Negative binomial distribution

3.13. A petroleum engineer knows that within a certain region, the probability that a drilling will lead to discovering oil is 0.45. Let X represent the drilling on which oil is discovered for the third time. The drillings are independent of one another.

(a) Give the probability distribution function for random variable X.
(b) Find $P(X \le 10)$.
(c) Find the mean and variance of the random variable X.

SOLUTION

(a) $NB(x) = \binom{x-1}{r-1}(p)^r (q)^{(x-r)}$, for $X = r, (r+1), \cdots$

$NB(x) = \binom{x-1}{3-1}(0.45)^3 (0.55)^{(x-3)}$, for $X = 3, 4, \cdots$

(b)

A	B
3	0.091125
4	0.150356
5	0.165392
6	0.151609
7	0.125078
8	0.09631
9	0.070627
10	0.049943
	0.90044

3 through 10 are entered into A1:A8 and =COMBIN(A1-1,2)*(0.45)^3*(0.55)^(A1-3) is entered into B1 and a click-and-drag is performed from B1 to B8 and then =SUM(B1:B8) is entered into B9 using EXCEL. $P(X \le 10) = 0.90044$.

(c) $\mu = \dfrac{r}{p} = \dfrac{3}{.45} = 6.67$ and $\sigma^2 = \dfrac{r(1-p)}{p^2} = \dfrac{3(0.55)}{0.45^2} = 8.148$ and $\sigma = 2.85$.

3.14. In a test of weld strength, 80% of tests result in a fracture in the weld, while the other 20% result in a fracture in the beam. Let X represent the number of tests up to and including the third beam fracture. (a) What is the distribution function of X? (b) What is the probability that X is equal to or less than 10? (c) Find the mean and variance of the random variable X.

SOLUTION

(a) $NB(x) = \binom{x-1}{3-1}(0.20)^3 (0.80)^{(x-3)}$, $x = 3, 4, \ldots$

(b)

3	0.008
4	0.0192
5	0.03072
6	0.04096
7	0.049152
8	0.05505
9	0.05872
10	0.060398
	0.3222

3 through 10 are entered into A1:A8 and =COMBIN(A1-1,2)*(0.20)^3*(0.80)^(A1-3) is entered into B1 and a click-and-drag is performed from B1 to B8 and then =SUM(B1:B8) is entered into B9 using EXCEL. $P(X \le 10) = 0.3222$.

(c) $\mu = \dfrac{r}{p} = \dfrac{3}{0.20} = 15$ and $\sigma^2 = \dfrac{r(1-p)}{p^2} = \dfrac{3(0.80)}{0.20^2} = 60$ and $\sigma = 7.75$.

Poisson distribution

3.15. An industrial engineer knows that 0.03% of plastic containers manufactured by a certain process have small holes that render them unfit for use. Let X represent the number of containers in a random sample of 10,000 that have this defect. Find the following: (a) $P(X = 3)$ (b) $P(X \le 2)$ (c) $P(1 \le X < 4)$ (d) mean and standard deviation of X.

SOLUTION

(a) The binomial mean is $np = 3$. The Poisson mean is $\lambda = 3$ since the Poisson and binomial are approximately equal for large n and small p. The EXCEL expression =POISSON(3,3,0) gives the Poisson probability at $X = 3$. This equals 0.224.

(b) The probability that $X \le 2$ is the cumulative distribution function evaluated at 2. The EXCEL expression =POISSON(2,3,1) = 0.4232 gives the cumulative probability of $X = 2$ for a Poisson having mean = 3.

(c) The probability that $1 \le X < 4$ is the probability $X = 1, 2$ or 3. The EXCEL expression =POISSON(3,3,1) − POISSON(0,3,1) gives the answer 0.5974.

(d) The mean = 3 and the standard deviation = $\sqrt{3} = 1.73$.

3.16. An industrial engineer knows that the number of industrial accidents averages 3 per week in the industry in which she is employed. X = the number of accidents per week, is known to have a Poisson distribution. Find the following probabilities: (a) the probability of no accidents in a given week, (b) the probability of at most 4 accidents in a given week, and (c) the probability of at least 2 accidents in a given week.

SOLUTION

Using MINITAB the answers to a, b, and c are:

(a) Probability Density Function

Poisson with mean = 3
x $P(X = x)$
0   0.0497871

(b) Cumulative Distribution Function

Poisson with mean = 3

x $P(X \leq x)$
4   0.815263

(c) The probability of at least 2 accidents in a given week is equal to 1 minus the probability of at most 1 accident in a given week.

Cumulative Distribution Function

Poisson with mean = 3

x $P(X \leq x)$

1   0.199148

$P(X \geq 2) = 1 - P(X \leq 1) = 1 - 0.199 = 0.801$

Multinomial distribution

3.17. The probabilities that the light bulb of the projector used in Dr. Stephens' statistical engineering software course will last less than 40 hours, 40 to 80 hours, or more than 80 hours are 0.3, 0.5, and 0.2. Find the probability that among 8 such bulbs, 2 will last less than 40 hours, 5 will last between 40 and 80 hours, and 1 will last more than 80 hours.

SOLUTION

$$P(X_1 = 2, X_2 = 5, X_3 = 1) = \frac{8!}{2!5!1!} 0.3^2 0.5^5 0.2^1$$

The EXCEL expression =MULTINOMIAL(2,5,1)*0.3^2*0.5^5*0.2^1 gives the answer 0.0945.

3.18. Items under inspection are subject to two types of defects. About 70% of the items in a large lot are defect-free, 20 % have a type A defect, and 10% have a type B defect. Six of the items are randomly selected. Find the probability that 3 have no defects, 1 has a type A defect, and 2 have a type B defect.

SOLUTION

$$P(X_1 = 3, X_2 = 1, X_3 = 2) = \frac{6!}{3!1!2!} 0.7^3 0.2^1 0.$$

The EXCEL expression =MULTINOMIAL(3,1,2)*0.7^3*0.2^1*0.1^2 gives 0.041.

Simulation

3.19. A fire insurance company has 4500 policy holders and the probability is 1/1200 that any one of these policy holders will file at least one claim during any given week. The probability that during a given week 0, 1, …, 4500 policy holders will file at least one claim is a binomial distribution with mean equal to $np = 4500(1/1200) = 3.75$. It also has an approximate Poisson distribution with mean $\mu = \lambda = 3.75$. Simulate the experience over 10 weeks for the policy holders by simulating a Poisson distribution with parameter $\lambda = 3.75$.

SOLUTION

We wish to simulate a Poisson random variable having mean $\lambda = 3.75$ over 10 weeks by giving the 10 weekly simulated outcomes. MINITAB has the capacity of simulating several discrete and continuous random variables as shown in the following MINITAB output. This output was produce by the pulldown **Calc ⇒ Random data.**

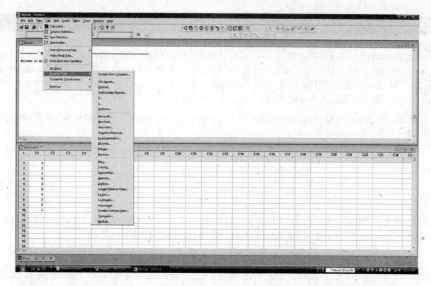

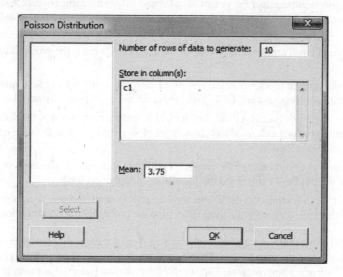

All the distributions that MINITAB will simulate are shown in the output snapshot. Choosing Poisson will give the following dialog box.

```
MTB > print c1
```

Data Display

```
C1
 7 3 4 1 5 2 1 5 2 5
```

This is our simulated sample for 10 weeks.

3.20. A petroleum engineer knows that within a certain region, the probability that a drilling will lead to discovering oil is 0.45. Let X represent the drilling on which oil is discovered for the third time. The drillings are independent of one another. Simulate 15 samples of X by using MINITAB.

SOLUTION

The variable X has a negative binomial distribution. MINITAB will be used to obtain a sample of size 15. See Problem 3.19 to find out how to use the random request of MINITAB.

Data Display

```
C1
 9 8 9 7 5 5 4 4 7 5 8 7 6 10 8
```

The simulated sample is shown.

SUPPLEMENTARY PROBLEMS

3.21. A regular tetrahedron is a solid that has four faces that are identical. The numbers 1, 2, 3, and 4 are stamped on the four faces. When the tetrahedron is tossed upon a table, one of the faces will be down and the other three will be facing up. The down face will be 1, 2, 3, or 4. Suppose two tetrahedra are tossed and that X represents the sum on the two faces that are facing down. X can take on the values 2, 3, ..., 8.

  (a) Use EXCEL or some other software package to generate the $4 \times 4 = 16$ possible outcomes when the above experiment is performed.

  (b) Find the probability distribution for X, the sum of the two down-turned faces when the two tetrahedra are tossed.

3.22. Find μ and σ^2 and σ for the random variable in Problem 3.21.

3.23. Refer to Problems 3.21 and 3.22. According to Chebyshev's theorem, how much probability is between $\mu - 2\sigma = 1.84$ and $\mu + 2\sigma = 8.16$? What is the actual probability between 1.84 and 8.16?

3.24. For a large shipment of integrated circuits the probability of failure for any one chip is 0.01. Assuming the assumptions underlying the binomial distribution are met, find the probability that at most two chips fail in a random sample of size 20.

3.25. A group of six engineers got together for a night of poker and before the games began, they posed some probability problems concerning a deck of cards to sharpen their thinking. See how many of the problems they posed you can workout.

  (a) If 5 cards are dealt (with replacement) to one of the engineers, what is the probability all 5 are from the same suit?

  (b) If 5 cards are dealt (with replacement) to one of the engineers until a heart first appears. Let X stand for the deal on which a heart first appears. Give the formula for the probability distribution.

  (c) Each of the six engineers is dealt a card, they look at it, and then replace it in the deck before the next card is dealt to the next engineer. Let X represent the number of face cards dealt to the six engineers. Give the formula for the probability distribution.

  (d) A hand of 4 cards is dealt to one of the engineers without replacement. The random variable X is defined as the number of aces in the 4 cards. Find the probability distribution for X.

  (e) A card is dealt, put back into the deck, and another card is dealt and this is continued. Let X represent the round on which a club occurs for the third time. Find the probability distribution for X.

3.26. An engineer has 50 of her favorite websites book-marked. Ten of them are technical in nature and she often uses them to get information pertinent to her job. The other 40 are non-technical sites such as Amazon.com from which she orders books for her reading pleasure. Suppose 5 of the sites are selected at random.

  (a) Use EXCEL to find the probabilities that in the 5 selected sites, 0, 1, 2, 3, 4, or 5 are technical sites.

  (b) Do part (a) using the built-in function in MINITAB.

3.27. The number of calls that an industrial engineer makes on her cell phone during a day has a Poisson distribution and the average is 15. Answer the following questions regarding the random variable X = the number of calls made per day on the cell phone.

  (a) Give the probability distribution function for random variable X.

  (b) Calculate the probabilities and the cumulative probabilities for $X = 0$ to $X = 20$.

  (c) What is the probability that she makes more than 30 calls?

  (d) What is the probability that she makes between 5 and 15 calls inclusive?

3.28. A traffic engineer is studying accidents and finds that 15% are 1-car accidents, 65% are 2-car accidents, 15% are 3-car accidents, and 5% are 4-car accidents in the area he is studying. He randomly selects 25 accidents. Find the following probabilities:

  (a) All 25 are 2-car accidents.

  (b) All 25 involve the same number of cars.

  (c) In the 25 selected, 6 are 1-car accidents, 6 are 2-car accidents, 6 are 3-car accidents, and 7 are 4-car accidents.

3.29. The number of wrecks on an interstate during a day is Poisson distributed with a mean equal to 4. Simulate the number of wrecks over a 30-day period using MINITAB.

3.30. A circuit board contains 500 diodes. Each diode has probability $p = 0.009$ of failing. A board works if none of its diodes fail. Find the probability the board works using the binomial distribution. Also find the probability the board works using the Poisson distribution.

CHAPTER 4

Probability Densities for Continuous Variables

Continuous Random Variables

Continuous random variables possibly take on an infinite number of values. Therefore, it is not possible to build a table having values and probabilities as was done in the discrete case. Characteristics such as pressure, temperature, density, and other engineering measurements take on values continuously in intervals. For example, adult female heights for a population take on values between 50 and 75 inches. We might represent their heights by the curve shown in Figure 4.1. This curve shows the probabilities associated with the variable height. The curve tells us that there are more females with heights near 62.5 inches than there are with heights close to 50 or 75 inches. There are a few very short and a few very tall, but most are in the middle near 62.5 inches.

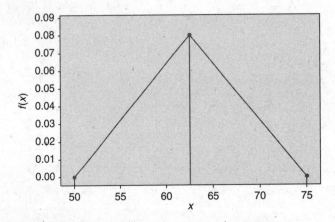

Figure 4.1   Density function for adult female heights.

A **probability density function (pdf)** shows how the characteristic is distributed. It has two properties: (1) **the curve is always positive,** $f(x) \geq 0$ and (2) **the total area under the curve is 1, that is** $\int_{50}^{75} f(x)dx = 1.$ In this case the equation of the pdf, shown in Figure 4.1, is as follows:

$$f(x) = \begin{cases} 0.0064x - 0.32, & 50 < x < 62.5 \\ -0.0064x + 0.48, & 62.5 \leq x \leq 75 \\ 0, & \text{otherwise} \end{cases}$$

If the graph is sketched, Figure 4.1 is obtained. The area under the curve is given by

$$\int_{50}^{62.5} (0.0064x - 0.32)dx + \int_{62.5}^{75} (-0.0064x + 0.48)dx.$$

MAPLE is a software package that performs many different mathematical functions, including integration. The software package MAPLE may be used to find the area under the graph as follows:

```
> evalf(int(0.0064*x-0.32,x=50..62.5)+
int(-.0064*x+0.48,x=62.5..75));    1.000000000
```

The student should study the command structure for performing the integration since it will be used quite often in this chapter. The number 1 is shown after the integration command, indicating that the area under the curve is equal to 1. Probabilities are interpreted as areas under the pdf curve. The probability that a female has a height over 67 inches is represented by $(X \geq 67)$ and is equal to the shaded area in the tail of Figure 4.2.

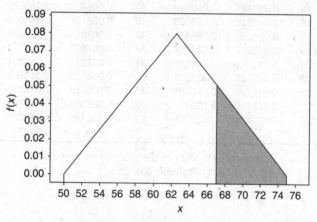

Figure 4.2 Females who are over 67 inches tall are represented by the gray area.

The probability or area is equal to $\int_{67}^{75} (-0.0064x + 0.48)dx$. The anti-derivative of this integrand is $-0.0032x^2 + 0.48x$ and is evaluated at the upper limit minus the evaluation at the lower limit or $18 - 17.7952 = 0.2048$. Or, if MAPLE is used to find the value of the integral, we find:

```
> evalf(int(-.0064*x+.48,x=67..75));    0.2048000000
```

As can be seen, we obtain the same answer using MAPLE as we do if we work the problem "by hand" as we would in a calculus course. We will use MAPLE to evaluate most of the integrals in this book. Using MAPLE to do the integration allows us to concentrate on the statistics and not to be constantly reviewing our calculus.

EXAMPLE 4.1   Engineers have developed a new tire with lifetime in thousands of miles, given by the following density curve.

$$f(x) = \begin{cases} 0.02e^{-0.02x}, & x > 0 \\ 0, & x < 0 \end{cases}$$

What percentage of the tires have mileages less than 60,000 miles? The answer is the integral from 0 to 60 of the density function or using MAPLE,

```
> evalf(int(0.02*exp(-.02*x),x=0..60));    0.6988057881.
```

About 70% of the tires will have lifetimes less than 60,000 miles.

Let's suppose the student is at home rather than at school and has EXCEL, but not MAPLE on a home computer. The interval from 0 to 60 is divided into 30 equal parts of length $60/30 = 2$ and the midpoints of the intervals are entered into column A. The expression =0.02*EXP(−0.02*A1) is entered into B1 and a click-and-drag is performed from B1 to B30. The expression =2*B1 is entered into C1 and a click-and-drag is performed from C1 to C30. Each cell from C1 to C30 contains the area of a rectangle and when the areas are added up, the sum (=SUM(C1:C30)) approximates the value of the integral. The value of the sum (0.69875) is very close to the MAPLE answer, 0.6988057881. The results are shown in Figure 4.3.

x	f(x)	f(x)Δx	x	f(x)	f(x)Δx
1	0.019604	0.039208	31	0.010759	0.021518
3	0.018835	0.037671	33	0.010337	0.020674
5	0.018097	0.036193	35	0.009932	0.019863
7	0.017387	0.034774	37	0.009542	0.019085
9	0.016705	0.033411	39	0.009168	0.018336
11	0.01605	0.032101	41	0.008809	0.017617
13	0.015421	0.030842	43	0.008463	0.016926
15	0.014816	0.029633	45	0.008131	0.016263
17	0.014235	0.028471	47	0.007813	0.015625
19	0.013677	0.027354	49	0.007506	0.015012
21	0.013141	0.026282	51	0.007212	0.014424
23	0.012626	0.025251	53	0.006929	0.013858
25	0.012131	0.024261	55	0.006657	0.013315
27	0.011655	0.02331	57	0.006396	0.012793
29	0.011198	0.022396	59	0.006146	0.012291
				Sum =	0.69875

Figure 4.3   Approximation of the integral as the sum of areas under the rectangles.

We are using the approximation

$$\int_{0}^{60} 0.02e^{-0.02x}\,dx \approx \sum_{1}^{30} f(x_i)\Delta x.$$

The term $\Delta x = 2$ and $f(x_i) = 0.02e^{-0.02x_i}$.

Figure 4.4 shows the area that is approximated above.

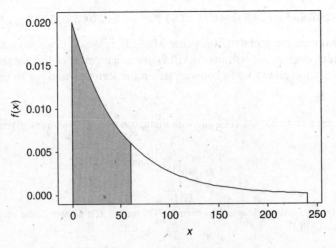

Figure 4.4 $P(X \leq 60)$.

Another property of continuous random variables is that the probability of a single point is zero, that is $P(X = a) = 0$. The area associated with a single point is zero. From this we have the following result.

$$P(a \leq X \leq b) = P(X = a) + P(a < X < b) + P(X = b) = 0 + P(a < X < b) + 0 = P(a < X < b)$$

For a discrete variable, we found that $\mu = \Sigma xp(x)$. The corresponding result for a continuous random variable is $\mu = \int_{-\infty}^{\infty} xf(x)dx$. And corresponding to the property for a discrete random variable that $\sigma^2 = \Sigma(x_i - \mu)^2 p(x_i)$ is the result $\sigma^2 = \int_{-\infty}^{\infty}(x - \mu)^2 f(x)dx$ for a continuous variable. A shortcut formula that is equivalent for the variance is $\sigma^2 = \int_{-\infty}^{\infty} x^2 f(x)dx - \mu^2$. Likewise, the cumulative distribution function for a continuous random variable is defined as follows $F(x) = P(X \leq x) = \int_{-\infty}^{x} f(x)dx$.

EXAMPLE 4.2 For the tire mileages pdf given above, find the mean, the variance, the standard deviation, and the cumulative distribution function.

$$f(x) = \begin{cases} 0.02e^{-0.02x}, x > 0 \\ 0, x < 0 \end{cases}$$

The mean mileage is given by $\mu = \int_{-\infty}^{\infty} xf(x)dx = \int_{0}^{\infty} 0.02e^{-0.02x} xdx$. The MAPLE solution is as follows:

```
> evalf(int(0.02*x*exp(-0.02*x),x=0..infinity));    50
```

We see that the mean mileage of these tires is 50,000 miles.

The variance, using the defining formula, is $\sigma^2 = \int_{-\infty}^{\infty}(x - 50)^2 f(x)dx = \int_{0}^{\infty}(x - 50)^2 0.02e^{-0.02x}dx$. The MAPLE solution is as follows:

```
> evalf(int((x-50)^2*0.02*exp(-0.02*x),x=0..infinity));    2500
```

Or, if we use the shortcut formula, $\sigma^2 = \int_{-\infty}^{\infty} x^2 f(x)dx - \mu^2 = \int_{0}^{\infty} x^2 0.02e^{-0.02x}dx - 50^2$. The MAPLE solution is as follows:

```
> evalf(int(x^2*0.02*exp(-0.02*x),x=0..infinity))-2500;    2500
```

The standard deviation is 50.
The cumulative distribution function is

$$F(x) = \begin{cases} 0, x < 0 \\ \int_{0}^{x} 0.02 * e^{-0.02t} dt, x > 0. \end{cases}$$

The MAPLE solution is as follows:

```
> evalf(int(0.02*exp(-0.02*t),t=0..x));    1-1.e^(-0.02000000000 x)
```

Note that the dummy integration variable is t and x is considered a fixed number. The function, $F(x)$ is

$$F(x) = \begin{cases} 0, x < 0 \\ 1 - e^{-0.02x}, x \geq 0. \end{cases}$$

Note the following values: $F(0) = 0$, $F(\infty) = 1$, and $F(60) = 1 - e^{-1.2} = 0.6988$. $F(60) = P(X < 60)$ is the same value calculated when we first encountered this pdf.

EXAMPLE 4.3   Civil engineers have determined that the daily water usage in a particular city in hundreds of thousands of gallons is a random variable having the probability density $f(x) = \begin{cases} (1/9)xe^{-x/3}, & x > 0 \\ 0, & x < 0 \end{cases}$. The graph for this water usage is shown in Figure 4.5.

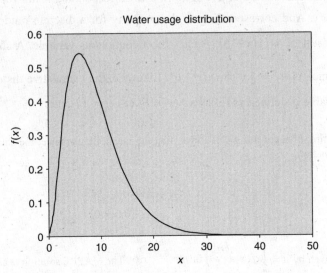

Figure 4.5   Probability density function for daily water usage.

```
> evalf(int((x/9)*exp(-x/3),x=0..infinity));      1
```

The above MAPLE integration shows that the function is a pdf.

The mean daily water usage in hundreds of thousands of gallons is

$$\mu = \int_{-\infty}^{\infty} xf(x)\,dx = \int_{0}^{\infty} \frac{x^2}{9} e^{-\frac{x}{3}}\,dx$$

The MAPLE evaluation of the integral is

```
> evalf(int((x^2/9)*exp(-x/3),x=0..infinity));     6.
```

We see that the mean is six hundred thousand gallons.

Consider the solution using EXCEL. By considering Figure 4.5, we see that even though the curve is defined for all x, if we go to 50, we have all the values of practical significance. Suppose first of all that we wish to confirm that the total area under the curve is 1. Break the interval from 0 to 50 into 50 sub-intervals each one unit long. Enter the numbers at the middle of the sub-intervals into column A of the worksheet. Enter 0.5 into A1 and 1.5 into A2 and then perform a click-and-drag. The function values =(A1/9)*EXP(−A1/3) are entered into B1 and a click-and-drag is performed. Then =B1*1 is entered into C1. This is the area of the first rectangle from 0 to 1 and height of the function at 0.5. A click-and-drag from C1 to C50 gives the areas of the first 50 rectangles. The expression =SUM(C1:C50) gives 1.004584, the approximation of the area under the curve. If 100 or 1000 rectangles are used instead, the area approximation gets closer to 1.

EXAMPLE 4.4   A variable has the following cumulative probability function. Find the pdf for X.

$$F(x) = \begin{cases} 0, & x < 0 \\ 1 - e^{-x^3}, & x \geq 0 \end{cases} \qquad f(x) = \frac{dF(x)}{dx} = \begin{cases} 0, & x < 0 \\ 3x^2 e^{-x^3}, & x \geq 0 \end{cases}$$

Or, we may have MAPLE find the derivative.

```
> diff(1-exp(-x^3), x); 3x²e^(-x³)
```

Figure 4.6 looks at both $f(x)$ and $F(x)$ in separate panels of the same graph.

Sometimes we know the shape for the probability density function from experience with the variable. All that is needed is to adjust the function by multiplying by a constant to make the total area under the curve equal to 1.

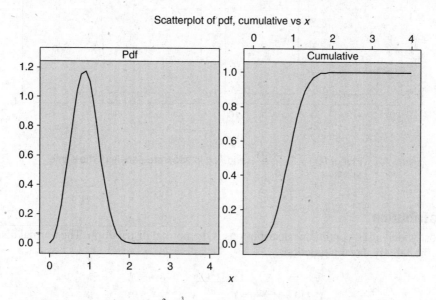

Figure 4.6   Plot of $f(x) = 3x^2 e^{-x^3}$ and $F(x)$ in separate panels of the same graph.

EXAMPLE 4.5   Find the value of K, using MAPLE, that makes the following function a pdf. Then plot the probability density function and the cumulative distribution function using MINITAB.

$$f(x) = \frac{Kx^2}{e^x}, \ 0 < x < 2$$

```
> evalf(int(K*x^2/exp(x),x=0..2));    0.646647168K
```

This must be equal to 1 to be a pdf. Setting 0.647K equal to 1, we find that K = 1.546. Therefore, $f(x) = \dfrac{1.546x^2}{e^x}$, $0 < x < 2$ is a pdf. The cumulative distribution function for $0 < x < 2$ is found as follows:

```
> evalf(int(1.546*t^2/exp(t),t=0..x));
1.546000000(-2.e^x + 2 + 2.x + x²)e^(-1.x)
```

$$F(x) = \begin{cases} 0, x < 0 \\ -1.546(-2e^x + 2 + 2x + x^2)e^{-x}, 0 < x < 2 \\ 1, x > 2 \end{cases}$$

A plot of $f(x)$ and $F(x)$ in separate panels of the same graph is shown in Figure 4.7.

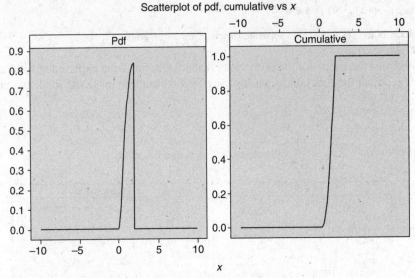

Figure 4.7   Plot of $f(x) = \dfrac{1.546x^2}{e^x}$ and $F(x)$ in separate panels of the same graph.

Normal Distribution

One of the most widely used continuous distributions is the normal distribution. The normal distribution has the following probability density function:

$$f(x) = \frac{1}{\sqrt{2\pi}\sigma} e^{-\frac{(x-\mu)^2}{2\sigma^2}}, \quad -\infty < x < \infty$$

The graph of the normal curve may be constructed using any of the many software packages. We will illustrate how to construct it using EXCEL and MINITAB.

EXAMPLE 4.6   Electrical engineers find that video game players have lifetimes that are normally distributed with a mean equal to 48 months and a standard deviation equal to 6 months. First we will describe how to plot the graph using EXCEL. First put the x values in the worksheet from 4 standard deviations below the mean to 4 above the mean. In this case, the numbers would extend from $48 - 4(6) = 24$ to $48 + 4(6) = 72$ months in the A column. The expression =NORMDIST(A1,48,6,0) is entered into cell B1 and a click-and-drag is performed from B1 to B49. The function =NORMDIST(A1,48,6,0) has four parameters. The first is a value on the x-axis, the second is the mean of the normal curve, the third is the standard deviation of the normal curve, and the fourth is 0 or 1. If 0 is entered, the height up to the normal curve is computed. If 1 is entered, the area to the left of the x-value entered is computed. This will create the points on the normal curve. All that remains is to plot the points using the chart wizard. The result is shown in Figure 4.8.

What are some of the characteristics of the normal curve?

1. The total area under the curve is 1.

2. The curve is asymptotic to the x-axis. That is, it gets closer and closer, but never touches as you go to plus infinity and minus infinity.

3. The curve is symmetrical about the mean. The mean, μ, locates the center of the curve.

4. The EXCEL function =NORMDIST(x, μ, σ, 1) may be used to find the area to the left of x.

5. The standard deviation, σ, determines the shape of the curve. If σ is small, the curve is tall and skinny. If σ is large, the curve is more spread out.

6. There is 68% of the area under the curve within 1 standard deviation of the mean, 95% within 2 standard deviations of the mean, and 99.7% within 3 standard deviations of the mean.

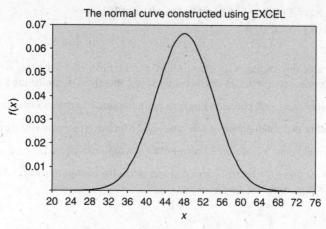

Figure 4.8   Lifetimes of video game players.

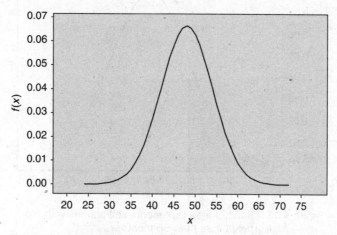

Figure 4.9   MINITAB plot of lifetimes of Xbox 360s.

If the values from columns A and B are copied from EXCEL into C1 and C2 of MINITAB, then MINITAB may be used to graph the normal curve. This is shown in Figure 4.9. Use the pulldown **Graph ⇒ Scatterplot**.

Figure 4.10 shows the effects of μ and σ on shape and location of the normal curve. The mean locates the center of the curve and the standard deviation affects the shape of the curve.

MAPLE may be used to find the total area under the curve and show that it indeed equals 1.

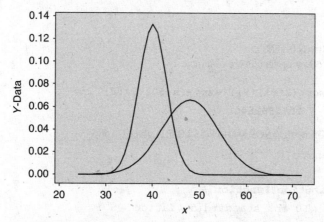

Figure 4.10   Normal curve with $\mu = 40$, $\sigma = 3$ and normal curve with $\mu = 48$, $\sigma = 6$.

```
> evalf(int((1/(sqrt(2*Pi)*6))*exp(-(x-48)^2/72),x=-
infinity..infinity));     1
```

EXAMPLE 4.7   Communications engineers have determined that lifetimes of Ace cell phones are normally distributed with a mean equal to 60 months and a standard deviation equal to 5 months. Determine the following:

(a) What is the probability that a cell phone of this type has a lifetime less than 55 months?

(b) What is the probability that a cell phone of this type has a lifetime greater than 70 months?

(c) Find a lifetime such that only 5% of the cell phones last this long or longer.

Figure 4.11 shows the area under the normal curve that represents the answer to part (a). Figure 4.12 shows the area under the normal curve that represents the answer to part (b).

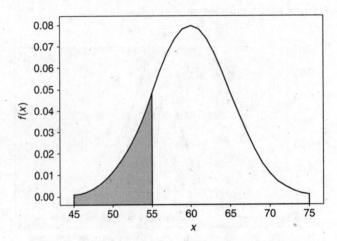

Figure 4.11   Shaded area represents cell phones with life-times less than 55 months.

(a) If MINITAB is used to find the shaded area, the pulldown **Calc ⇒ Probability Distributions ⇒ Normal** is used to produce the following output.

Cumulative Distribution Function

Normal with mean = 60 and standard deviation = 5

x P(X <= x)

55 0.158655

We see that the probability is 0.158655.
If MAPLE is used, the following results are obtained.

```
> evalf(int((1/(sqrt(2*Pi)*5))*exp(-(x-60)^2/50), x=-
infinity..55));     0.1586552540
```

If EXCEL is used, the following =NORMDIST(55,60,5,1) gives 0.158655.

(b) The MINITAB solution is

Cumulative Distribution Function

Normal with mean = 60 and standard deviation = 5

x P(X <= x)

70   0.977250

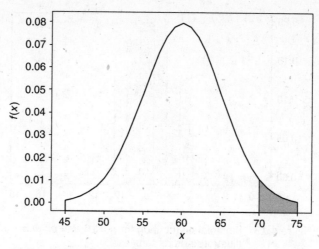

Figure 4.12   Shaded area represents cell phones with lifetimes greater than 70 months.

The probability asked for is $P(X > 70)$ which equals $1 - P(X < 70) = 0.02275$.
The MAPLE solution is

```
> evalf(int((1/(sqrt(2*Pi)*5))*exp(-(x-
60)^2/50), x = 70..infinity));    0.0227501320.
```

The EXCEL solution is $=$1-NORMDIST(70,60,5,1) $= 0.02275$.

(c) This part is worked by using the inverse function of the normal distribution.

If MINITAB is used, the solution is

```
Inverse Cumulative Distribution Function
Normal with mean = 60 and standard deviation = 5
P( X <= x )     x
   0.95  68.2243
```

The EXCEL solution is $=$NORMINV(0.95,60,5) $= 68.2243$.

Normal Approximation to the Binomial Distribution

A coin is flipped 16 times and the random variable of interest is $X =$ the number of heads to appear in the 16 flips. This distribution is a binomial distribution with $n = 16$ trials and $p = 0.5$. The binomial probabilities are shown in Figure 4.13. In this figure, the area of each rectangle is equal to the binomial probability of the number of heads shown on the x-axis. Each rectangle extends 0.5 units on either side of the x-value and the height of the rectangle is the probability that goes with that value. For example, the rectangle at $x = 10$ extends from 9.5 to 10.5 and it has height equal to 0.122192. If the area of the rectangle is found, it is $1 \times 0.122192 = 0.122192$. The sum of the 17 rectangle areas is 1. The mean of the distribution is $\mu = np = 16(0.5) = 8$. The standard deviation of the distribution is $\sigma = \sqrt{npq} = \sqrt{16(0.5)(0.5)} = \sqrt{4} = 2$. This is basically a review of what was learned in the section on binomial distribution in the previous chapter. From the histogram of the binomial distribution, it is clear that the most likely outcome is 8 heads in the 16 flips and the least likely outcome is either 0 or 16 heads in the 16 flips.

Figure 4.14 shows a normal distribution with mean $\mu = 8$ and standard deviation $\sigma = 2$. The two distributions are strikingly similar. However, one is discrete and the other one is continuous. The normal distribution may be used to approximate the binomial distribution when $np > 5$ and $nq > 5$.

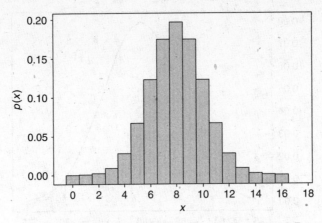

Figure 4.13   Binomial distribution for $n = 16$ and $p = 0.5$ illustrated as rectangles.

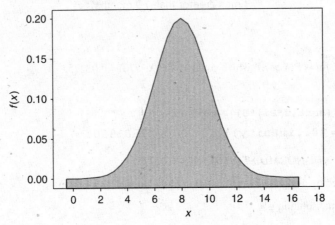

Figure 4.14   Normal distribution with mean $= 8$ and standard deviation $= 2$.

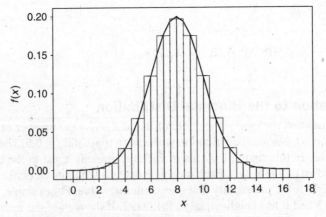

Figure 4.15   Binomial distribution and the normal distribution shown together.

Figure 4.15 shows the two distributions together. Sometimes the normal curve is lower than the rectangle and sometimes the normal curve is higher than the rectangle.

The application of this connection is shown in Figure 4.16. Suppose our interest is in the probability of flipping 6, 7, 8, or 9 heads. That is we are interested in finding $P(X = 6, 7, 8,$ or $9)$. This would be the area under the rectangles located at 6, 7, 8, or 9. By studying Figure 4.16, it is clear that this would be approximately

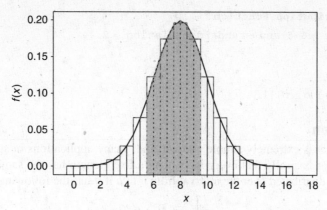

Figure 4.16 Area under the rectangles at 6, 7, 8, and 9 is approximated by the area under the normal curve from $x = 5.5$ to $x = 9.5$.

the area under the normal curve from where $x = 5.5$ to where $x = 9.5$. Let us compare the area under the rectangles and the area under the normal curve. Using EXCEL, the binomial probabilities or the areas under the rectangles is =BINOMDIST(6,16,0.5,0) + BINOMDIST(7,16,0.5,0) + BINOMDIST(8,16,0.5,0) + BINOMDIST(9,16,0.5,0) = 0.667694. The EXCEL area under the normal curve is =NORMDIST(9.5,8,2,1)- NORMDIST(5.5,8,2,1) = 0.667723. Notice how amazingly close the two areas are to each other.

Before the modern software packages such as EXCEL, MINITAB, SAS, SPSS, and STATISTIX, this approximation was often used to work applied problems. It is still an important approximation.

EXAMPLE 4.8 A manufacturer knows that 1% of their personal digital assistants will need repair within a year. What is the probability that 80 or fewer will need repair in 10,000 that are shipped to Wall Market within a year? The basic variable is a binomial where X = the number needing repair in the 10,000 PDAs sent to Wall Market. First make sure the normal approximation is appropriate: $n(p) = 10000(0.01) = 100 > 5$ and $n(q) = 10000(0.99) = 9900 > 5$. The approximation is appropriate. We are asked to find $P(X \leq 80)$. A normal curve with mean $\mu = n(p) = 10000(0.01) = 100$ and $\sigma = \sqrt{npq} = \sqrt{10000(0.01)(0.99)} = \sqrt{99} = 9.95$ is fit to the binomial distribution. The EXCEL solution is given by the following: =NORMDIST(80.5,100,9.95,1) = 0.025. The exact answer is =BINOMDIST(80,10000,0.01,1) = 0.022. The binomial answer was not readily available just 25 years ago. The only choice was to do the normal approximation. Computers and their abilities have changed the nature of statistics!

EXAMPLE 4.9 A production line makes surgical masks. Five percent are defective. Approximate the probability that a sample of 125 will contain 10 or more defectives. The normal approximation can be made since $np > 5$ and $nq > 5$. We are asked to approximate $P(X \geq 10)$. We fit a normal curve through the binomial distribution. The normal curve has center 6.25 and standard deviation 2.44. We need to find the area to the right of 9.5 under the normal curve. Figure 4.17 shows the area under the normal curve that we need to find. The MINITAB area is

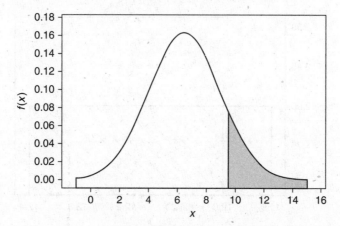

Figure 4.17 Normal curve area that approximates binomial probability $P(X \geq 10)$.

```
Cumulative Distribution Function
Normal with mean = 6.5 and standard deviation = 2.44
x P( X <= x )
9.5    0.890559
```

Now subtract this from 1 to get 0.11.

Uniform Distribution

A continuous model that is extremely simple and yet finds many applications in statistics is the uniform model. This model states basically that every interval of a fixed length has the same probability. Random variable X is uniformly distributed over the interval from a to b if it has the following pdf.

$$f(x) = \begin{cases} 0, x < a \\ \dfrac{1}{b-a}, a \le x \le b \\ 0, x > b \end{cases}$$

The mean of a uniformly distributed random variable is $\mu = \int_a^b \dfrac{x}{b-a}\,dx$. If this integral is evaluated using MAPLE, the following is obtained:

> `evalf(int(x/(b-a),x=a..b));` $\dfrac{0.5000000000\,(b^2 - 1.\,a^2)}{b - 1.\,a}$

This may be simplified to $\mu = \dfrac{a+b}{2}$. The variance is given by $\sigma^2 = \int_a^b \left(x - \dfrac{a+b}{2}\right)^2 \dfrac{1}{(b-a)}\,dx$. Again, letting MAPLE do the integration,

> `evalf(int(x^2/(b-a),x=a..b)-(a+b)^2/4);`

$$\dfrac{0.3333333333\,(b^3 - 1.\,a^3)}{b - 1.\,a} - 0.2500000000\,(a+b)^2.$$

The students are asked to see how good their algebra is by showing that this expression simplifies to

$$\sigma^2 = \dfrac{(b-a)^2}{12}.$$

EXAMPLE 4.10   A quality control engineer knows that the amount of fill that a machine puts into 12 ounce bottles is actually uniformly distributed between 11.90 and 12.40 ounces. Use MINITAB to draw $P(X > 12.0)$. This is shown in Figure 4.18.

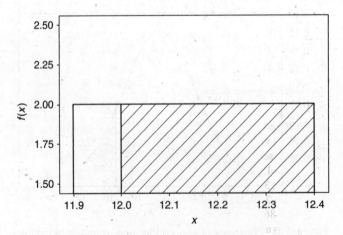

Figure 4.18   Uniform distribution between 11.9 and 12.4.

The average fill per container is $(11.9 + 12.4)/2 = 12.15$ ounces. The percentage of containers with more than 12.0 ounces is $\int\limits_{12.0}^{12.4} \frac{1}{12.4-11.9}\, dx = \int\limits_{12.0}^{12.4} 2dx = 2(.4) = 0.8$ or 80%. The standard deviation of fills is

$$\sigma = \sqrt{\frac{(12.4-11.9)^2}{12}} = 0.14 \text{ ounces.}$$

Log-Normal Distribution

The log-normal random variable X has a probability density function as follows:

$$f(x) = \begin{cases} \dfrac{1}{\sqrt{2\pi}\beta} x^{-1} e^{-\frac{(\ln x - \alpha)^2}{2\beta^2}}, & x > 0, \beta > 0 \\[2mm] 0, & \text{otherwise} \end{cases}$$

If this variable is transformed by taking its logarithm, i.e. $Y = \ln(X)$, the random variable Y has a normal distribution with mean α and standard deviation β. It can be shown that the density of Y is

$$f(y) = \frac{1}{\sqrt{2\pi}\beta} e^{-\frac{(y-\alpha)^2}{2\beta^2}}, \quad -\infty < y < \infty.$$

Going back to the log-normal variable X, the mean of the log-normal may be shown to equal

$$\mu = e^{\alpha + \beta^2/2}.$$

And the variance may be shown to equal

$$\sigma^2 = e^{2\alpha + \beta^2}(e^{\beta^2} - 1).$$

Remember, if you are in a mathematical statistics course, your interest is in deriving the equation for the mean and variance of X. If you are in an engineering statistics course, your interest is in applying and using the results and understanding how you may apply the results to engineering. We shall now look at how to use the software to apply the above results.

To obtain some idea of what this log-normal density looks like, consider it when α is 0 and β is 1 and when α is 1 and β is 1. In Figure 4.19, C2 is the log-normal graph when $\alpha = 0$ and $\beta = 1$ and C3 is the log-normal graph when $\alpha = 1$ and $\beta = 1$.

The mean of the graph, where $\alpha = 0$ and $\beta = 1$, is $\mu = e^{0+1^2/2} = e^{0.5} = 1.65$, the variance is $\sigma^2 = e^{2(0)+1^2}(e^{1^2} - 1) = e(e-1) = 4.67$, and the standard deviation is 2.16. The mean of the graph, where $\alpha = 1$ and $\beta = 1$, is $\mu = e^{1+1^2/2} = e^{1.5} = 4.48$, the variance is $\sigma^2 = e^{2(1)+1^2}(e^{1^2} - 1) = e^3(e-1) = 34.51$, and the standard deviation is 5.88.

EXAMPLE 4.11 Engineers for the EPA have found that fluoride in a large city's water supply has concentrations that follow a log-normal distribution with parameters $\alpha = 3.2$ and $\beta = 1$. What is the probability that the concentration exceeds 8 ppm? Suppose we look at several solutions to this problem.

EXCEL solution: The command =LOGNORMDIST(8,3.2,1) is entered into any cell. The parameters are 8, the value in the probability statement $P(X < 8)$; 3.2, the mean of $\ln(X)$; and 1, the standard deviation of $\ln(X)$. Recall that $\ln(X)$ has a normal distribution with mean = 3.2 and standard deviation = 1. The value 0.131238 is subtracted from 1 to get the answer. $P(X > 8) = 1 - P(X < 8) = 0.868$.

MINITAB solution: Figure 4.20 is a MINITAB representation of $P(X > 8)$. The area is found as follows: The pulldown **Calc $\Rightarrow$ Probability Distributions $\Rightarrow$ Log-normal** gives a dialog box that allows you to find the area as follows:

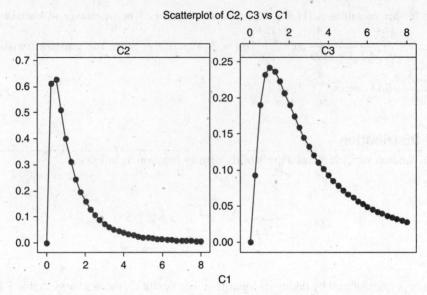

Figure 4.19   C2 is the log-normal graph when $\alpha = 0$ and $\beta = 1$ and C3 is the log-normal graph when $\alpha = 1$ and $\beta = 1$.

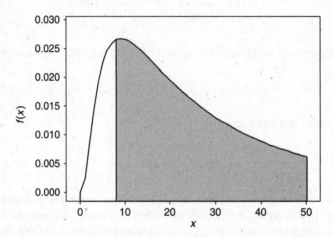

Figure 4.20   Probability that fluoride concentration exceeds 8 ppm.

Cumulative Distribution Function

```
Lognormal with location = 3.2 and scale = 1
x  P( X <= x )
8    0.131238
```

The probability is found as $1 - 0.131238 = 0.868$.

Normal Curve solution: Recall that $Y = \ln(X)$ has a normal distribution with $\mu = 3.2$ and $\sigma = 1$. We are seeking $P(X > 8)$ which is the same as $P(\ln(X) > \ln(8)) = P(\ln(X) > 2.0794)$. Therefore, if we find the area to the right of 2.0794 under a normal curve having a mean equal to 3.2 and a standard deviation equal to 1, we should find the same area as in Figure 4.20. Using EXCEL to find this area under the normal curve, we give the command =1-NORMDIST(2.0794, 3.2,1,1) which gives the answer 0.868, the same as was obtained using the log-normal distribution.

EXAMPLE 4.12   Zonyl RP grease-resistant coating leaches out chemicals according to a log-normal distribution. The parameters $\alpha = 0.62$ and $\beta = 0.25$. Find the mean and standard deviation of the log-normal distribution.

The mean amount of leaching is $\mu = e^{0.62+0.25^2/2} = e^{0.65} = 1.92$ and the variance is equal to $\sigma^2 = e^{2(0.62)+0.25^2}(e^{0.25^2}-1) = e^{1.3025}(e^{0.0625}-1) = 0.24$. The standard deviation is 0.49. The amount of probability within 3 standard deviations of the mean is $P(0.45 < X < 3.39)$. Using EXCEL, the answer is given by =LOGNORMDIST(3.39,0.62,0.25)-LOGNORMDIST(0.45,0.62,0.25) = 0.991877.

The log-normal distribution has been found useful in describing failure times of equipment, risk analysis of nuclear power plants, and current gains in transistors.

Gamma Distribution

The gamma function is an applied mathematics function that is defined as an integral with limits from 0 to infinity. Since the gamma distribution has a probability density function that contains the gamma function, it is necessary to study the gamma function before studying the gamma distribution. The gamma function is defined as follows:

$$\Gamma(\alpha) = \int_0^\infty x^{\alpha-1}e^{-x}dx, \ \alpha > 0$$

This improper integral has certain properties that we shall consider. An integration by parts where $u = x^{\alpha-1}$ and $dv = e^{-x}\,dx$ yields $du = (\alpha-1)x^{\alpha-2}\,dx$ and $v = -e^{-x}$. An application of the integration by parts technique yields the following recursive formula:

$$\Gamma(\alpha) = \int_0^\infty x^{\alpha-1}e^{-x}dx = (\alpha-1)\int_0^\infty x^{\alpha-2}e^{-x}dx = (\alpha-1)\Gamma(\alpha-1)$$

For $\alpha = 2$, $\Gamma(2) = (2-1)\Gamma(2-1) = 1\Gamma(1)$. The gamma function evaluated at $\alpha = 1$ is $\Gamma(1) = \int_0^\infty x^{1-1}e^{-x}\,dx = \int_0^\infty e^{-x}\,dx = 1$. Therefore $\Gamma(2) = 1$. For $\alpha = 3$, $\Gamma(3) = (3-1)\Gamma(3-1) = 2\,\Gamma(2) = 2(1) = 2!$. For $\alpha = 4$, $\Gamma(4) = (4-1)\Gamma(4-1) = 3\,\Gamma(3) = 3(2!) = 3!$. Continuing in this manner, we find that for any positive integer n, $\Gamma(n) = (n-1)!$ It may also be shown that $\Gamma(1/2) = \sqrt{\pi}$. The gamma function has many interesting properties. The actual numerical values of the gamma function are found in EXCEL.

EXAMPLE 4.13　Evaluate $\Gamma(0.1)$, $\Gamma(0.5)$, $\Gamma(5)$, and $\Gamma(3.46465)$ using EXCEL.

The EXCEL function =GAMMALN(0.1) gives 2.2527 in Figure 4.21. Using the property that $y = \ln(x)$ is equivalent to $x = e^y$. We have ln $(\Gamma(0.1)) = 2.2527$ is equivalent to $\Gamma(0.1) = e^{2.2527}$ or using EXCEL again, we obtain =EXP(GAMMALN(0.1)) = 9.5135. Similarly =EXP(GAMMALN(0.5)) gives 1.7725, =EXP(GAMMALN(5)) gives 24,

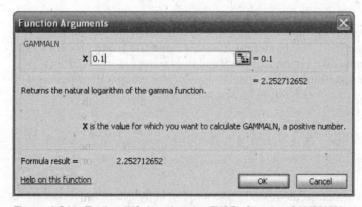

Figure 4.21　Finding $\Gamma(0.1)$ using the EXCEL function GAMMALN.

x	f(x)	x	f(x)	x	f(x)	x	f(x)
0.01	99.43259	0.6	1.489192	1.6	0.893515	3.5	3.323351
0.03	32.785	0.7	1.298055	1.7	0.908639	3.75	4.422988
0.05	19.47009	0.8	1.16423	1.8	0.931384	4	6
0.07	13.7736	0.9	1.068629	1.9	0.961766	4.25	8.285085
0.09	10.61622	1	1	2	1	4.5	11.63173
0.1	9.513508	1.1	0.951351	2.25	1.133003	4.75	16.58621
0.2	4.590844	1.2	0.918169	2.5	1.32934	5	24
0.3	2.991569	1.3	0.897471	2.75	1.608359	5.25	35.21161
0.4	2.21816	1.4	0.887264	3	2	5.75	78.78448
0.5	1.772454	1.5	0.886227	3.25	2.549257	6	120

Figure 4.22   Values of the gamma function, $\Gamma(x)$.

and =EXP(GAMMALN(3.46465)) gives 3.1969. Figure 4.22 gives the EXCEL generated values for the gamma function. Figure 4.23 gives the MINITAB plot for this function.

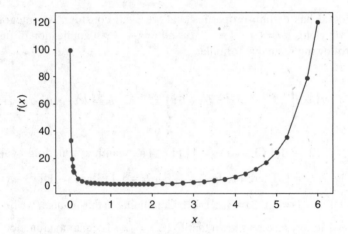

Figure 4.23   Plot of the gamma function.

Now that the gamma function has been discussed, the gamma distribution will be defined.
The **gamma distribution** with parameters α and β is defined as follows:

$$f(x) = \frac{1}{\beta^{\alpha}\Gamma(\alpha)} x^{\alpha-1} e^{-x/\beta}, \; x > 0, \; \alpha > 0, \; \beta > 0$$

Figure 4.24 shows four different gamma distributions. For the variable C2, $\alpha = 1$ and $\beta = 1$, for the variable C3, $\alpha = 2$ and $\beta = 1$, for the variable C4, $\alpha = 3$ and $\beta = 1$, and for the variable C5, $\alpha = 4$ and $\beta = 1$. The mean of the gamma distribution is $\mu = \alpha\beta$ and the variance is $\sigma^2 = \alpha\beta^2$.

The **exponential distribution** is a special case of the gamma distribution with $\alpha = 1$. The density becomes

$$f(x) = \frac{1}{\beta^{\alpha}\Gamma(\alpha)} x^{\alpha-1} e^{-x/\beta}, \; x > 0, \; \alpha = 1, \; \beta > 0 = \frac{1}{\beta^{1}\Gamma(1)} x^{1-1} e^{-\frac{x}{\beta}}, \; x > 0, \; f(x) = \frac{1}{\beta} e^{-\frac{x}{\beta}}, \; x > 0.$$

The mean of the exponential distribution is $\mu = \alpha\beta = (1)\beta = \beta$ and the variance is $\sigma^2 = \alpha\beta^2 = (1)\beta^2$ and $\sigma = \beta$.

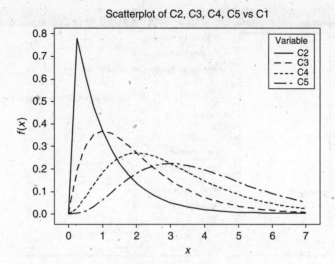

Figure 4.24   Different gamma distributions.

Summarizing, the exponential has probability density function

$$f(x) = \frac{1}{\beta} e^{-\frac{x}{\beta}}, \; x > 0.$$

and $\mu = \beta$ and $\sigma = \beta$.

EXAMPLE 4.14   The time to failure for mp3s follows an exponential distribution with mean time to failure equal to 3 years. Find the percentage of mp3s that have times to failure equal to 4 years or more. Solve using EXCEL, MINITAB, and MAPLE.

EXCEL solution: Sometimes the pdf for the exponential is expressed as $f(x) = \lambda e^{-\lambda x}$. The mean and standard deviation is $1/\lambda$. The dialog box for the exponential is shown in Figure 4.25.

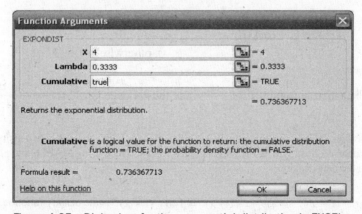

Figure 4.25   Dialog box for the exponential distribution in EXCEL.

The parameter $\lambda = 1/3 = 0.3333$. The **true** indicates cumulative distribution or $P(X < 4)$. Therefore $P(X > 4) = 1 - 0.737 = 0.263$.

MINITAB solution: The pulldown **Calc ⇒ Probability Distributions ⇒ Exponential** gives the dialog box shown in Figure 4.26.

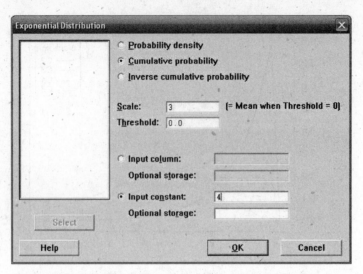

Figure 4.26   MINITAB dialog box for exponential distribution.

Cumulative Distribution Function

Exponential with mean = 3

x P(X <= x)

4 0.736403

$P(X > 4) = 1 - 0.736 = 0.264$. The MAPLE solution is as follows:
MAPLE solutions:

```
> evalf(int((1/3)*exp(-x/3),x=4..infinity));     0.2635971382
```

Beta Distribution

The beta distribution is defined on the unit interval. It has a probability density function that is a function of the gamma function. The beta distribution has the following probability density function:

$$f(x) = \begin{cases} \dfrac{\Gamma(\alpha+\beta)}{\Gamma(\alpha)\Gamma(\beta)} x^{\alpha-1}(1-x)^{\beta-1}, & 0 < x < 1, \alpha > 0, \beta > 0 \\ 0, & \text{otherwise} \end{cases}$$

The mean of the beta distribution is $\mu = \dfrac{\alpha}{\alpha + \beta}$ and the variance is $\sigma^2 = \dfrac{\alpha\beta}{(\alpha + \beta)^2(\alpha + \beta + 1)}$.

EXAMPLE 4.15   Consider the beta distribution when $\alpha = 4$ and $\beta = 4$. Verify that the beta density is a density function by showing that it integrates to 1 using MAPLE. Find the mean and variance using MAPLE and also by using the above expression. Also graph the beta density.

$$f(x) = \frac{\Gamma(4 + 4)}{\Gamma(4)\Gamma(4)} x^{4-1}(1 - x)^{4-1} = \frac{7!}{3!3!} x^3(1 - x)^3 = 140x^3(1 - x)^3 \text{ is the density function.}$$

(Note that we made use of $\Gamma(n) = (n - 1)!$)
Integrating by the use of MAPLE, we find that $f(x)$ is a density function since the integral equals 1.

```
> evalf(int(140*x^3*(1-x)^3,x=0..1));     1
```

Likewise the mean is found to be 0.5.

```
> evalf(int(140*x*x^3*(1-x)^3,x=0..1));     0.5000000000
```

Using the expression $\mu = \dfrac{4}{4+4}$, also yields 0.5 for the mean. The variance is found using MAPLE as follows:

```
> evalf(int(140*x^2*x^3*(1-x)^3,x=0..1))-0.5^2;    0.0277777778
```

Using the expression $\sigma^2 = \dfrac{\alpha\beta}{(\alpha+\beta)^2(\alpha+\beta+1)} = \dfrac{16}{64(9)}$ gives 0.02777 for the variance.

The beta density is shown in Figure 4.27.

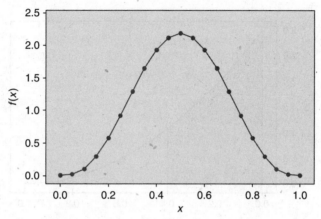

Figure 4.27   Beta density function $f(x) = 140x^3(1-x)^3$.

EXAMPLE 4.16   A civil engineer knows from experience that the proportion of highway sections requiring repairs in any given year in Douglas County is a random variable with a beta distribution having $\alpha = 1$ and $\beta = 3$. Use EXCEL, MINITAB, and MAPLE to determine the probability that at most 40% of the highway sections will require repairs in any given year.

EXCEL solution: The filled in dialog box is shown in Figure 4.28. $(P(X < 0.4) = 0.784)$

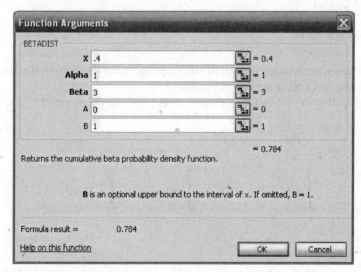

Figure 4.28   EXCEL dialog box for $P(X < 0.4)$ beta distribution with $\alpha = 1$ and $\beta = 3$.

MINITAB solution: The MINITAB output is as follows:

Cumulative Distribution Function

Beta with first shape parameter = 1 and second = 3

x P(X <= x)

0.4   0.784

A MINITAB graph of the area is given in Figure 4.29.

MAPLE solution: The density function is $f(x) = 3(1-x)^2$, $0 < x < 1$. The MAPLE solution is

```
> evalf(int(3*(1-x)^2,x=0..(0.4)));    0.7840000000
```

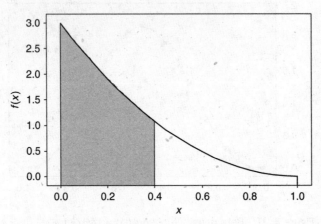

Figure 4.29 $P(X < 0.4)$ for beta distribution with $\alpha = 1$ and $\beta = 3$.

Weibull Distribution

Another important probability density distribution for engineers is the Weibull distribution. The Weibull density function has several forms.

First form $f(x) = \alpha\beta x^{(\beta-1)}e^{-\alpha x^\beta}$, $x > 0$, $\alpha > 0$, $\beta > 0$

Even though we have recommended the use of statistical software throughout the book, "by hand" techniques learned in the calculus course are still acceptable.

For example, to show that the Weibull is a density, make the change of variable $y = x^\beta$. The differential change is $dy = \beta x^{\beta-1}dx$ or $dx = \dfrac{dy}{\beta x^{\beta-1}}$.

$$\int_0^\infty \alpha\beta x^{(\beta-1)}dx\, e^{-\alpha x^\beta} = \int_0^\infty \alpha dy\, e^{-\alpha y} = \int_0^\infty e^{-\alpha y}\alpha dy = 1$$

The mean of the Weibull is

$$\mu = \int_0^\infty x\alpha\beta x^{\beta-1}e^{-\alpha x^\beta}dx.$$

Make the change of variable $y = \alpha x^{\beta}$, the limits remain 0 and ∞, $dy = \alpha \beta x^{\beta-1} dx$ or $dx = \dfrac{dy}{\alpha \beta x^{\beta-1}}$ and $x = \left(\dfrac{y}{\alpha}\right)^{\frac{1}{\beta}}$. The result is

$$\mu = \alpha^{-\frac{1}{\beta}} \Gamma\left(1 + \frac{1}{\beta}\right).$$

Similarly, it can be shown that the variance is

$$\sigma^2 = \alpha^{-\frac{2}{\beta}} \left\{ \Gamma\left(1 + \frac{2}{\beta}\right) - \left[\Gamma\left(1 + \frac{1}{\beta}\right)\right]^2 \right\}.$$

EXAMPLE 4.17 The length of life of a cell phone battery has a first form Weibull distribution with parameters $\alpha = 1$ and $\beta = 2$. The mean length of life for this phone battery is $\mu = 1^{-\frac{1}{2}} \Gamma\left(1 + \frac{1}{2}\right) = \Gamma(1.5)$. The EXCEL function =EXP(GAMMALN(1.5)) gives 0.886 years for μ. The probability that it has a life over 2 years is $\int_{2}^{\infty} 1(2)x^{2-1}e^{-1x^2} dx$. Evaluating this integral using MAPLE, we get:

```
> evalf(int(2*x*exp(-x^2),x=2..infinity));    0.01831563889
```

Very few of these batteries have lifetimes exceeding 2 years.

A second form that you will sometimes see for the Weibull distribution is the following density function.

$$\textbf{Second form} \qquad f(x) = \frac{\alpha}{\beta} x^{\alpha-1} e^{-\frac{x^{\alpha}}{\beta}}, \; x > 0, \; \alpha > 0, \; \beta > 0$$

For this functional form for the density, it can be shown that the mean and variance are as follows:

$$\mu = \beta^{\frac{1}{\alpha}} \Gamma\left(1 + \frac{1}{\alpha}\right) \text{ and } \sigma^2 = \beta^{\frac{2}{\alpha}} \left\{ \Gamma\left(1 + \frac{2}{\alpha}\right) - \left[\Gamma\left(1 + \frac{1}{\alpha}\right)\right]^2 \right\}$$

A third form that you will see for the Weibull distribution is the following density function.

$$\textbf{Third form} \qquad f(x) = \frac{\alpha}{\beta^{\alpha}} x^{\alpha-1} e^{-\left(\frac{x}{\beta}\right)^{\alpha}}, \; x > 0, \; \alpha > 0, \; \beta > 0$$

For this functional form for the density, it can be shown that the mean and variance are as follows:

$$\mu = \beta \Gamma\left(1 + \frac{1}{\alpha}\right) \text{ and } \sigma^2 = \beta^2 \left\{ \Gamma\left(1 + \frac{2}{\alpha}\right) - \left[\Gamma\left(1 + \frac{1}{\alpha}\right)\right]^2 \right\}$$

It is this third form that is utilized by MINITAB and EXCEL.

EXAMPLE 4.18 An industrial engineer finds that the life of a bearing (in hundreds of hours) follows a Weibull distribution, third form with $\alpha = 2$ and $\beta = 50$. Use the third form of the Weibull distribution and let X stand for bearing life in hundreds of hours. Find the probability that X is less than 1000 hours using EXCEL, MINITAB, and MAPLE. Also give the mean and standard deviation of the lives of the bearings.

EXCEL solution: Figure 4.30 gives the dialog box for finding $P(X < 10)$ using the third form of the Weibull model with $\alpha = 2$ and $\beta = 50$. We see that the probability is very small that the life will be less than 1000 hours. The mean

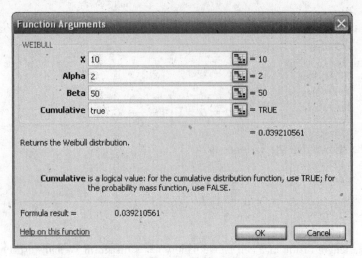

Figure 4.30   EXCEL dialog box for finding Weibull probabilities.

lifetime of such bearings is $\mu = \beta\Gamma\left(1 + \dfrac{1}{\alpha}\right) = 50\Gamma(1.5)$. The gamma function of 1.5 is 0.886. The mean lifetime of such bearings is equal to 44.3 hundred or 4430 hours. The MAPLE validation of the mean is:

```
> evalf(int(0.0008*x^2*exp(-x^2/2500), x = 0..infinity));    44.31134627
```

The variance is $\sigma^2 = \beta^2\left\{\Gamma\left(1 + \dfrac{2}{\alpha}\right) - \left[\Gamma\left(1 + \dfrac{1}{\alpha}\right)\right]^2\right\} = 2500(\Gamma(2) - \Gamma(1.5)^2) = 537.51$ and the standard deviation is 23.18. The MAPLE verification of this variance is

```
> evalf(int(0.0008*x^3*exp(-x^2/2500),x=0..infinity))-44.311^2;    536.535279
```

MINITAB solution: Using the pulldown **Calc ⇒ Probability Distribution ⇒ Weibull** gives the dialog box shown in Figure 4.31 in which the shape parameter is the alpha value and the scale parameter is the beta value. The output is shown below.

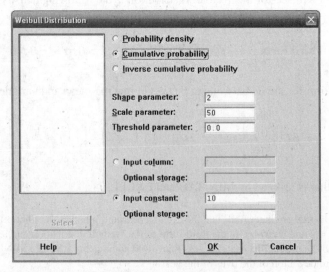

Figure 4.31   Dialog box for the Weibull distribution with $\alpha = 2$ and $\beta = 50$.

Cumulative Distribution Function

```
Weibull with shape = 2 and scale = 50
xP( X <= x )
10 0.0392106
```

MAPLE solution:

```
> evalf(int(0.0008*x*exp(-x^2/2500),x=0..10));  0.03921056085
```

Joint Distributions—Discrete and Continuous

A cell phone receives up to 2 calls per hour and makes up to 2 calls per hour. X represents the number of calls received and Y represents the number of calls made. An engineer developed the joint probability distribution shown in the table from historical data. This table gives the **bivariate probabilities** $p(x, y) = P(X = x, Y = y)$ for $X = 0$, 1, and 2 and $Y = 0$, 1, and 2.

	$X = 0$	$X = 1$	$X = 2$
$Y = 0$	0.02	0.04	0.03
$Y = 1$	0.04	0.15	0.17
$Y = 2$	0.05	0.23	0.27

A bivariate probability function satisfies two requirements: (1) $0 \leq p(x, y) \leq 1$ for all values of x and y. (2) $\sum_x \sum_y p(x, y) = 1$. Note that $\sum_x \sum_y p(x, y) = 1$ for the above bivariate distribution.

$P(X = 0) = 0.02 + 0.04 + 0.05 = 0.11$, $P(X = 1) = 0.04 + 0.15 + 0.23 = 0.42$, and $P(X = 2) = 0.03 + 0.17 + 0.27 = 0.47$. This is called the **marginal probability** of X and can be summarized as follows.

x	0	1	2
$p_1(x)$	0.11	0.42	0.47

Similarly, Y has a marginal probability distribution as follows.

y	0	1	2
$p_2(y)$	0.09	0.36	0.55

Recalling the conditional probability formula $P(A/B) = \dfrac{P(A \cap B)}{P(B)}$ and applying it to this example, $p(X = 0/Y = 0) = \dfrac{p(X = 0, Y = 0)}{p(Y = 0)} = \dfrac{0.02}{0.09} = 0.222$. Similarly, $p(1/0) = 0.444$ and $p(2/0) = 0.333$. The conditional distribution of x given $Y = 0$ is

x	0	1	2
$p(x/0)$	0.222	0.444	0.333

There are two other conditional distributions for X given $Y = 1$ and for $Y = 2$. Similarly, there are three conditional distributions for Y given $X = 0$, $X = 1$, and $X = 2$.

Two continuous random variables have a **bivariate joint probability density function** if they have the following properties: (1) $f(x, y) \geq 0$ for all x and y. (2) $\int\limits_{-\infty}^{\infty}\int\limits_{-\infty}^{\infty} f(x, y)dxdy = 1$.

EXAMPLE 4.19 Use MAPLE to determine the value of c that makes the following a joint pdf.

$$f(x, y) = \begin{cases} cxy^2, & 0 < x < 2, 0 < y < 2 \\ 0, & \text{elsewhere} \end{cases}$$

```
> evalf(int(int(c*x*y^2,x=0..2),y=0..2));    5.333333333c
```

Since the integral must equal 1 to be a joint pdf, set $5.333c = 1$ or $c = 0.1875$.
The marginal density functions are found by integrating the other variable out of the joint density function.

EXAMPLE 4.20 Find the marginal of X and the marginal of Y for the joint pdf.

$$f(x, y) = \begin{cases} 0.1875xy^2, & 0 < x < 2, 0 < y < 2 \\ 0, & \text{elsewhere} \end{cases}$$

The marginal of X is $f_1(x) = \int\limits_0^2 0.1875xy^2 dy$. The MAPLE solution is

```
> evalf(int(0.1875*x*y^2,y=0..2));    0.5000000000x
```

$$f_1(x) = 0.5x, \ 0 < x < 2$$

The marginal of Y is $f_2(y) = \int\limits_0^2 0.1875xy^2 dx$. The MAPLE solution is

```
> evalf(int(0.1875*x*y^2,x=0..2));    0.3750000000y^2
```

$$f_2(y) = 0.375y^2, \ 0 < y < 2$$

Note that for this example $f(x, y) = f_1(x) f_2(y)$. When $f(x, y) = f_1(x) f_2(y)$, we say that X and Y are **independent random variables**.
The **conditional density function for X given Y** is $f_1(x/y) = \dfrac{f(x, y)}{f_2(y)}$ and the **conditional density function for Y given X** is $f_2(y/x) = \dfrac{f(x, y)}{f_1(x)}$.

We have considered bivariate distributions thus far in this section. Any number of variables may be considered. The following example considers a **trivariate distribution**.

EXAMPLE 4.21 Random variables X, Y, and Z have the following joint probability density function. Determine the value that c would need to be for it to be a density.

$$f(x, y, z) = \begin{cases} cxy^2z^3, & 0 < x < 3, 0 < y < 2, 0 < z < 1 \\ 0, & \text{elsewhere} \end{cases}$$

We need to solve the following for c:

$$\int\limits_0^1\int\limits_0^2\int\limits_0^3 cxy^2z^3 dxdydz = 1$$

Once again, we turn to MAPLE for help. The integration is not hard, but it takes a lot of time.

```
> evalf(int(int(int(c*x*y^2*z^3,x=0..3),y=0..2),z=0..1));    3c
```

Set $3c = 1$ and get $c = 1/3$.

$$f(x, y, z) = \begin{cases} 0.3333xy^2z^3, 0 < x < 3, 0 < y < 2, 0 < z < 1 \\ 0, \text{elsewhere} \end{cases}$$

Next, find the three marginal densities and see if the variables are independent. Three double integrations will be needed.

```
> evalf(int(int(0.333*x*y^2*z^3,y=0..2),z=0..1));    0.2220000000x
```

$$f_1(x) = 0.222x, 0 < x < 3$$

```
> evalf(int(int(0.333*x*y^2*z^3,x=0..3),z=0..1));    0.3746250000y^2
```

$$f_2(y) = 0.375y^2, 0 < y < 2$$

```
> evalf(int(int(0.333*x*y^2*z^3,x=0..3),y=0..2));    3.996000000z^3
```

$$f_3(y) = 3.996z^3, 0 < z < 1$$

Since $f(x, y, z) = f_1(x)f_2(y)f_3(y)$, the three variables are independent.

Checking Data for Normality

When presented with a set of sample data, it is of interest to determine the form of the population distribution. In particular, it will be of interest in later chapters to test whether a set of data was taken from a population having a normal distribution. Several of the software packages have **normality tests**. These are tests that help answer the question "Is it reasonable to assume that this data was taken from a population that is normally distributed?" The author prefers to use one of the following tests to answer this question: the Shapiro–Wilk normality test, the Anderson–Darling test of normality, the Ryan–Joiner test of normality, or the Kolmogorov–Smirnov test of normality. All the tests are basically conducted the same way. A test statistic is calculated and it is assumed that the data is taken from a normally distributed population. The probability that the sample came from a normally distributed population is computed. It is accepted as reasonable that the data came from a normally distributed population unless the probability is unusually small (usually less than 0.05) that it came from a normal population.

EXAMPLE 4.22　An engineer wishes to test that the following pressure readings shown in Figure 4.32 were taken from a normal distribution.

15.31	15.15	14.64
14.66	15.66	15.18
14.59	15.62	15.18
14.71	14.92	14.44
15.80	15.45	14.89
14.95	15.64	14.44
14.47	14.57	15.71
14.26	14.30	14.76
15.32	15.11	14.22
14.87	14.54	15.04

Figure 4.32　Pressure readings (from normal population).

Using MINITAB and the pulldown **Stat ⇒ Basic Statistics ⇒ Normality Tests** gives a dialog box with a choice of the Anderson–Darling test of normality, the Ryan–Joiner test of normality, and the Kolmogorov–Smirnov test of normality. The Kolmogorov–Smirnov test is shown in Figure 4.33.

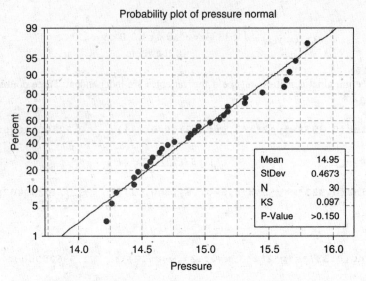

Figure 4.33   Kolmogorov–Smirnov test of normality for data in Figure 4.32.

The important number in Figure 4.33 is the *p*-value > 0.150. If this value is less than 0.05 then reject normality. Otherwise accept that the data came from a normally distributed population. The *p*-value given by the Anderson–Darling test is 0.369 and the *p*-value given by the Ryan–Joiner test is *p*-value > 0.10.

If the package STATISTIX is used the following results are obtained.

The pulldown **Statistics ⇒ Randomness Normality tests ⇒ Shapiro–Wilk test** with the data from Figure 4.32 gives the following output:

```
Statistix 8.0

Shapiro-Wilk Normality Test
Variable        N       W          P
pressure        30      0.9539     0.2144
```

The *p*-value here is 0.2144. It is reasonable that the data came from a normally distributed population.

EXAMPLE 4.23   Figure 4.35 gives the Kolmogorov–Smirnov results from the set of data given in Figure 4.34. The *p*-value is 0.014. This is less than 0.05. The chances of getting such a sample from a normal distribution is very small (0.014) and so we reject that the data came from a normal distribution.

4.56	46.26	48.16
12.55	14.51	24.29
5.89	22.33	37.22
21.04	6.75	4.21
1.04	4.11	20.43
4.22	4.19	17.84
20.79	6.34	5.96
20.20	8.22	1.17
13.79	59.01	0.25
4.01	14.94	5.04

Figure 4.34   Pressure readings (from non-normal population).

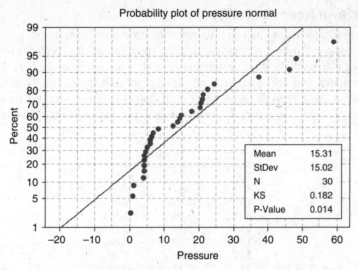

Figure 4.35 Kolmogorov–Smirnov test of normality for data in Figure 4.34.

The *p*-value given by the Anderson–Darling test is less than 0,005 and the *p*-value given by the Ryan–Joiner test is *p*-value < 0.01. The STATISTIX analyses gave the following output.

```
Statistix 8.0
Shapiro-Wilk Normality Test
Variable      N       W         P
pressure      30      0.8183    0.0001
```

Notice the *p*-value is 0.0001. The chances of getting this sample data from a normal distribution are very, very small.

Figures 4.33 and 4.35 reveal the following: If the data are from a normal population, the points tend to be close to the straight line. If the data are not from a normal population, the points do not fall close to the straight line. The *p*-values are reflective of how near the points are to the line.

Figure 4.36 gives dot plots for the data in Figures 4.32 and 4.34. The four tests of normality of sample data given in this section take the guess work out of testing for normality. They represent reasonable techniques that will lead to a conclusion. Some practitioners recommend looking at a histogram, a box plot, or a dot plot. If the histogram, box plot, or dot plot reveals a normal shape then conclude that you have normality. If the shape is not normal, then conclude you do not have normality. This, of course, leads to different conclusions from different practitioners for the same set of data.

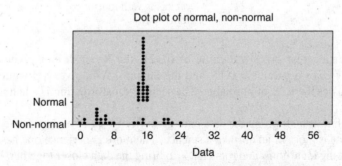

Figure 4.36 Dot plot of normal data and non-normal data.

Transforming Data to Near Normality

What do you do if a statistical test, to be valid, requires the normality assumption, and you have concluded that your sample did not come from a normally distributed population? It is possible to transform your data by replacing it with logarithms of the data, or replacing it with square roots of the data. The transformed data may well pass the normality test, while the original data did not pass the normality test.

The data in Figure 4.34 is reproduced here. Recall that this data failed all four normality tests as given in the previous section. First, transform this data using natural logs. This new data is shown in Figure 4.37.

1.52	3.83	3.87
2.53	2.67	3.19
1.77	3.11	3.62
3.05	1.91	1.44
0.04	1.41	3.02
1.44	1.43	2.88
3.03	1.85	1.79
3.01	2.11	0.16
2.62	4.08	−1.39
1.39	2.70	1.62

Figure 4.37   Data in Figure 4.34 transformed by replacing with natural logs.

The Anderson–Darling test gives a p-value of 0.102, the Ryan–Joiner gives a p-value of 0.049, the Kolmogorov–Smirnov gives a p-value of 0.065, and the Shapiro–Wilk gives a p-value of 0.0599. The transformed data basically passes the tests of normality.

Consider one more transformation. Suppose the fourth root transformation is used. That is, the data is replaced by taking the fourth root of each number and forming a new set of data consisting of fourth roots. This transformed data is shown in Figure 4.38.

1.46	2.61	2.63
1.88	1.95	2.22
1.56	2.17	2.47
2.14	1.61	1.43
1.01	1.42	2.13
1.43	1.43	2.06
2.14	1.59	1.56
2.12	1.69	1.04
1.93	2.77	0.71
1.42	1.97	1.50

Figure 4.38   Data in Figure 4.34 transformed by replacing with fourth roots.

The Anderson–Darling test gives a p-value of 0.321, the Ryan–Joiner gives a p-value > 0.1, the Kolmogorov–Smirnov gives a p-value > 0.15, and the Shapiro–Wilk gives a p-value of 0.538. The transformed data clearly passes the tests of normality. The second transformation is a better one than the first for this set of data.

Figure 4.39 shows how transforming the data changes it from a set of highly skewed data to more normal data. Transforming data is more of an art than a science. The more experience one has at transforming data, the better one gets. Taking logarithms and roots tends to bring the data closer together. It takes the skew away and causes the data to move closer to the center as a general rule.

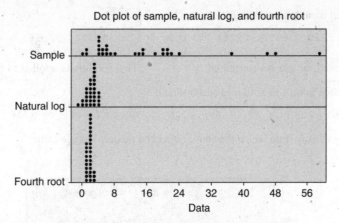

Figure 4.39 Dot plot of original data, natural log, and fourth root of original data.

Simulation

The need to form a simulated sample often arises in engineering applications. For one thing, it may be expensive to obtain a sample and for another, it may be very difficult to obtain a real sample.

EXAMPLE 4.24 Consider a random variable X, having the following pdf:

$$f(x) = \left\{ \begin{array}{l} 0.08x, 0 < x < 5 \\ 0, \text{elsewhere} \end{array} \right\}$$

The variable has the following cumulative distribution function:

$$F(x) = \left\{ \begin{array}{l} 0, x < 0 \\ 0.04x^2, 0 < x < 5 \\ 1, x > 5 \end{array} \right\}$$

Now, consider a uniform variable over the interval (0, 1). Most software packages have a uniform variable built into the package capable of giving uniform values over the unit interval. Let u be a uniformly generated value between 0 and 1. Set $F(x) = u$ and solve the resulting equation for x in terms of u. This gives the following $0.04x^2 = u$ or $x = \sqrt{\dfrac{u}{0.04}}$.

EXAMPLE 4.25 Give a simulated random sample from the distribution in the above example of size 10. Using the following pulldown **Calc $\Rightarrow$ Random Data $\Rightarrow$ Uniform** in MINITAB gives the dialog box shown in Figure 4.40.

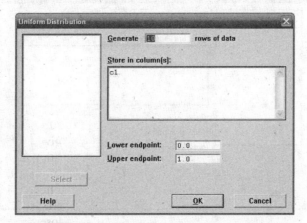

Figure 4.40 Generating a random sample of size 10 from the uniform distribution.

The following uniform numbers are created:

0.822   0.229   0.478   0.284   0.489   0.128   0.296   0.958   0.425   0.042

The following simulated sample is determined: $x = \sqrt{\dfrac{u}{0.04}}$. This formula is applied to the uniform values. The following simulated random sample of x values is determined:

4.53   2.39   3.46   2.67   3.49   1.79   2.72   4.90   3.26   1.03

Figure 4.41 shows the density from which the above simulated random sample came.

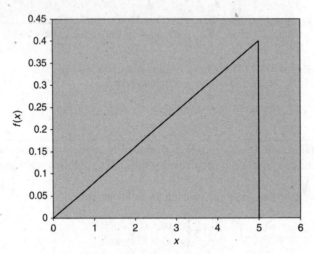

Figure 4.41   Graph of density from which the simulated
sample was obtained.

If this simulation were repeated, a different but typical sample would result.

In addition to the technique described above, MINITAB has the ability to simulate samples from some of the distributions that we have studied previously in the past two chapters.

EXAMPLE 4.26   Simulate the outcome of tossing a coin 10 times and observing the number of heads to occur in the 10 tosses. Obtain a sample of size 15 of this random variable.

The pulldown **Calc $\Rightarrow$ Random Data $\Rightarrow$ Binomial** gives the dialog box shown in Figure 4.42. The request is for a sample of size 15 to be placed into column C1. The variable is the binomial based on $n = 10$ and $p = 0.5$.

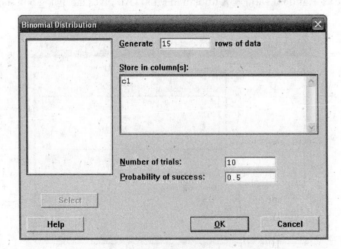

Figure 4.42   Simulated binomial random variable with $n = 10$
and $p = 0.5$ $(n = 15)$.

The following random sample is simulated:

5　4　6　7　6　6　6　7　3　4　7　6　4　7　7

If the command is given again the following new simulated sample is obtained:

3　6　4　3　5　4　4　6　5　4　8　5　2　5　4

EXAMPLE 4.27　A software engineer is interested in how long it takes to learn to use a new set of software. The engineer believes the time required to be proficient with the software is normally distributed with mean equal to 20 hours and standard deviation equal to 3.5 hours. Create a simulated sample for the times required for 25 employees of Ace Manufacturing to become proficient with the software.

The pulldown **Calc ⇒ Random Data ⇒ Normal** gives the dialog box which is filled out as shown in Figure 4.43. The following simulated sample was given by MINITAB.

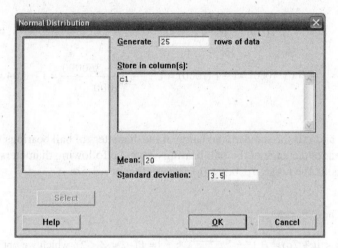

Figure 4.43　Simulated normal distribution with $\mu = 20$ and $\sigma = 3.5$.

```
18.6   15.2   19.8   19.7   19.6   24.6   23.8   16.2   23.1   16.0   19.2
26.0   14.4   16.7   27.7   21.5   24.5   15.1   20.5   19.6   18.0   24.7
14.4   26.6   21.9
```

The data represents a typical sample of size 25.

SOLVED PROBLEMS

Continuous random variables

4.1. A component in an engineering system has a life (in hours) with a probability density $p(x) = c/x^3$ when $x \geq 600$ hours and $p(x) = 0$ when $x < 600$. (a) Find the value of c. (b) Calculate the probability that a component will last at least 1000 hours. (c) Calculate the mean component life in hours.

SOLUTION

(a)　$> evalf\left(int\left(\dfrac{c}{x^3}, x = 600..infinity \right) \right);$

　　$0.00000138888c$

This integral is set equal to 1 and solved for c. We find $c = 720000$.

(b)　We integrate the probability density function from 1000 hours to infinity. The probability is 0.36.

　　$> evalf\left(int\left(\dfrac{720000}{x^3}, x = 1000..infinity \right) \right);$

　　0.36

(c) $\mu = \int_{-\infty}^{\infty} xp(x)dx$. The mean is 1200 hours.

$$> evalf\left(int\left(\frac{x.720000}{x^3}, x = 600..infinity \right) \right);$$

1200

4.2. (a) Find an expression for the cumulative distribution function in Problem 4.1. (b) Calculate the probability that a component will last at least 1000 hours using the cumulative distribution function.

SOLUTION

(a) $F(x) = 0$, if $x < 600$

$$F(x) = \int_{600}^{x} f(t)\,dt = \int_{600}^{x} \frac{720000}{t^3}\,dt = \frac{-360000}{x^2} + 1 = \frac{x^2 - 360000}{x^2} \text{, if } x > 600$$

(b) $P(X \geq 1000) = 1 - P(X < 1000) = 1 - F(1000) = 1 - \dfrac{1000^2 - 360000}{1000^2} = 1 - 0.64 = 0.36$

Normal distribution

4.3. If μ is the mean and σ is the standard deviation of the diameters of ball bearings which follow a normal distribution, what percentage of the ball bearings have the following diameters? (a) Within the range $\mu \pm 2\sigma$. (b) Outside the range $\mu \pm 0.9\sigma$. (c) Greater than $\mu - 1.5\sigma$.

SOLUTION

X is the ball bearings' diameters.

(a) $P(\mu - 2\sigma < X < \mu + 2\sigma) = P\left(-2 < \dfrac{X - \mu}{\sigma} < 2 \right) = P(-2 < Z < 2)$, which we obtain by subtracting μ and dividing by σ from each part of the inequality. The Z is a standard normal having mean 0 and standard deviation 1. We now go to EXCEL to evaluate the probability. = NORMSDIST(2)-NORMSDIST(-2) = 0.9545. The function NORMSDIST gives the cumulative area under a standard normal curve. The answer is 95.4% of the ball bearings have diameters within 2 standard deviations of the mean.

(b) $P(X < \mu - 0.9\sigma) + P(X > \mu + 0.9\sigma) = P(Z < -0.9) + P(Z > 0.9) = 2*P(Z < -0.9) = 2*$NORMSDIST(-0.9) = 0.3681 or 36.8%.

(c) $P(X > \mu - 1.5\sigma) = P(Z > -1.5)$. =1-NORMSDIST(-1.5) = 0.9332 or 93.3%.

4.4. The finished inside diameter of a piston ring is normally distributed with a mean of 4.50 cm and a standard deviation of 0.005 cm. What is the probability of a diameter exceeding 4.51 cm? Find the probability using MAPLE.

SOLUTION

We use MAPLE to integrate the normal density function with $\mu = 4.50$ and $\sigma = 0.005$.

```
>evalp(int((1/(sqrt(2*Pi)*0.005))*exp(-(x-4.5)^2/0.00005),x=4.51..infinity));
```

0.0227501319

The answer that MAPLE finds is 0.0228.

Normal approximation to the binomial distribution

4.5. Determine whether the normal approximation to the binomial distribution is appropriate in the following cases. (a) A binomial variable has $n = 20$ and $p = 0.05$. (b) A binomial variable has $n = 100$ and $p = 0.30$. (c) A binomial variable has $n = 150$ and $p = 0.07$.

SOLUTION

(a) Is not appropriate since $np = 1$ is less than 5.

(b) Yes, since $np = 30$ and $nq = 70$.

(c) Yes, since $np > 5$ and $nq > 5$.

4.6. At Midwestern University, 15% of the engineering students believe that online degrees in engineering are valid. In a poll, 500 Midwestern engineering students are asked, "Do you think online degrees in engineering are valid?" What is the probability that 100 or more answer yes?

Use the normal approximation to the binomial to answer. Compare this approximation to the answer using the binomial distribution.

SOLUTION

The normal approximation is, using EXCEL, = 1-NORMDIST(99.5,75,7.984,1) = 0.001075. The answer using the binomial distribution is =1-BINOMDIST(99,500,0.15, 1) = 0.00153.

Uniform distribution

4.7. A microcomputer firm finds that intrastate contracts for shipping contracts have low bids that are uniformly distributed between $20,000 and $25,000. Find the probability that the low bid on the next intrastate shipping contract is (a) below $22,000, (b) in excess of $24,000. (c) Find the average cost of low bids on contracts of this type.

SOLUTION

(a) $P(X < \$22,000) = \int_{20000}^{22000} \frac{1}{5000} dx = \frac{2000}{5000} = 0.4$

(b) $P(X > \$24,000) = \int_{24000}^{25000} \frac{1}{5000} dx = \frac{1000}{5000} = 0.2$

(c) $\mu = (20000 + 25000)/2 = \$22,500$

4.8. Suppose that X has a uniform distribution over the interval (a, b). Find the following:

(a) The cumulative distribution function.
(b) Find $P(X > c)$ for some point c, between a and b.

SOLUTION

(a) $F(x) = 0$ if $x < a$ $F(x) = 1$ if $X > b$ $F(x) = \int_{a}^{x} \frac{1}{(b-a)} dx = \frac{x-a}{b-a}$ if $a < x < b$

(b) $P(X > c) = 1 - P(X < c) = 1 - \int_{a}^{c} \frac{1}{(b-a)} dx = 1 - \frac{c-a}{b-a} = \frac{b-c}{b-a}$

Log-normal distribution

4.9. Engineers for the EPA have found that fluoride in a large city's water supply have concentrations that follow a log-normal distribution with parameters $\alpha = 3.2$ and $\beta = 1$. What is the probability that the concentration exceeds 8 ppm? Set up the integral that gives the answer and evaluate the integral using MAPLE.

SOLUTION

The log-normal distribution is $f(x) = \begin{cases} \dfrac{1}{\sqrt{2\pi}\beta} x^{-1} e^{-\frac{(\ln x - \alpha)^2}{2\beta^2}}, & x > 0, \beta > 0 \\ 0, & \text{otherwise} \end{cases}$. We can get the answer by integrating

the density from 8 to infinity. The MAPLE command is as follows.

$$> \text{evalf}\left(\text{int}\left(\frac{0.398}{x} \exp\left(-\frac{(\ln(x) - 3.2)^2}{2} \right), x = 8..\text{infinity} \right) \right);$$

0.8667.

This answer is the same as obtained in the example in the log-normal section using EXCEL and MINITAB.

4.10. In Problem 4.9, find the mean and standard deviation of the fluoride in the city's water supply. Use MAPLE and MINITAB to find percentage of the fluoride readings within 1 standard deviation of the mean.

SOLUTION

The mean is $\mu = e^{\alpha+\beta^2/2} = e^{3.2+0.5} = e^{3.7} = 40.447$ and the variance is $\sigma^2 = e^{2\alpha+\beta^2}(e^{\beta^2}-1) = e^{7.4}(1.718) = 2810.621$ and the standard deviation equals 53.015. One standard deviation below is -12.568 and one standard deviation above is 93.462. The integral from 0 to 93.462 will give the percentage of readings in a one standard deviation interval above and below the mean. The MAPLE answer is given as follows:

$$> \text{evalf}\left(\text{int}\left(\frac{0.398}{x}.\exp\left(-\frac{(\ln(x)-3.2)^2}{2}\right), x = 0..93.462\right)\right);$$

0.9073.

The MINITAB answer is found as follows:

```
Cumulative Distribution Function

Lognormal with location = 3.2 and scale = 1

xP( X <= x )
93.462    0.909479
```

$P(X < 93.462) = 0.909$

Gamma distribution

4.11. The lifetime of an electronic component is known to be exponentially distributed with a mean life of 1000 hours. What proportion of such components will fail before 500 hours? Find the answer using MINITAB and MAPLE.

SOLUTION

The density $f(x) = 0.001e^{-0.001x}$, $x > 0$. The proportion failing before 500 hours is given by

$$> \text{evalf}(\text{int}(0.001.\exp(-0.001x), x = 0..500));$$

0.393.

Using MINITAB,

```
Cumulative Distribution Function

Exponential with mean = 1000

x  P(X <= x)
500    0.393469
```

4.12. In a certain city, the daily consumption of electric power (in millions of kilowatt-hours) can be treated as a random variable having a gamma distribution with $\alpha = 3$ and $\beta = 2$. If the power plant of this city has a daily capacity of 12 million kilowatt-hours, what is the probability that this power supply will be inadequate on any given day?

SOLUTION

The density is $f(x) = \dfrac{1}{\beta^\alpha \Gamma(\alpha)} x^{\alpha-1} e^{-x/\beta}$, $x > 0$, $\alpha > 0$, $\beta > 0$ or

$f(x) = \dfrac{1}{16} x^2 e - \dfrac{x}{2}$, $x > 0$. The probability we seek is $P(X > 12)$. Using MAPLE, we have

$$> \text{evalf}\left(\text{int}\left(0.0625. \quad x^2.\exp\left(\frac{-x}{2}\right), x = 12..\text{infinity}\right)\right);$$

0.0619.

Beta distribution

4.13. A factory that produces cell phones produces products for which the proportion defective has a beta distribution with $\alpha = 1$ and $\beta = 19$. Use EXCEL to sketch the density function for X = proportion defective. Find the mean and standard deviation of the distribution.

SOLUTION

The density is $f(x) = 19(1 - x)^{18}$, $0 < x < 1$. The EXCEL graph for this function is:

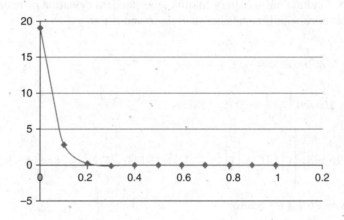

The mean is $\mu = \dfrac{\alpha}{\alpha + \beta} = 1/20 = 0.05$. The variance is $\sigma^2 = \dfrac{\alpha\beta}{(\alpha + \beta)^2(\alpha + \beta + 1)}$ which equals 0.00226 and the standard deviation is 0.0476.

4.14. Refer to Problem 4.13. Find the probability that the number defective is within 1 standard deviation of the mean.

SOLUTION

One standard deviation below the mean is 0.0024 and 1 standard deviation above the mean is 0.0976. The MAPLE solution is

> evalf(int(19. $(1-x)^{18}$, $x = 0.0024.0.0976$));

0.813.

Weibull distribution

4.15. The lifetime of an electronic component has a Weibull distribution of the first form with parameters $\alpha = 1$ and $\beta = 2$. Find the mean lifetime, the standard deviation of lifetimes, and the percentage of components with lifetimes between $\mu - 2\sigma$ and $\mu + 2\sigma$.

SOLUTION

$$\mu = \alpha^{-\frac{1}{\beta}}\Gamma\left(1 + \frac{1}{\beta}\right) \text{ and } \sigma^2 = \alpha^{-\frac{2}{\beta}}\left\{\Gamma\left(1 + \frac{2}{\beta}\right) - \left[\Gamma\left(1 + \frac{1}{\beta}\right)\right]^2\right\}$$

$$\mu = \alpha^{-\frac{1}{\beta}}\Gamma\left(1 + \frac{1}{\beta}\right) = 1^{-\frac{1}{2}}\Gamma\left(\frac{3}{2}\right) = 0.886 \text{ years} \quad \text{and}$$

$$\sigma^2 = 1^{-\frac{2}{2}}\left\{\Gamma\left(1 + \frac{2}{2}\right) - \left[\Gamma\left(1 + \frac{1}{2}\right)\right]^2\right\} = \Gamma(2) - \Gamma\left(\frac{3}{2}\right)^2 = 0.215 \text{ and } \sigma = 0.464.$$

Now, $\mu - 2\sigma = -0.042$ and $\mu + 2\sigma = 1.814$. Since lifetimes cannot be negative, we integrate from 0 to 1.814. The result, using MAPLE is

> evalf(int(2.x.exp((x²), x = 0..1814));

0.96.

Therefore, 96% of the components have lifetimes within 2 standard deviations of the mean.

4.16. Batteries have lifetimes that have a Weibull distribution of the second form having parameter values $\alpha = 1$ and $\beta = 5$. Find the mean battery lifetime, the standard deviation of battery lifetimes, and the percentage of batteries with lifetimes between $\mu - \sigma$ and $\mu + \sigma$.

SOLUTION

The mean and variance are as follows:

$$\mu = \beta^{\frac{1}{\alpha}}\Gamma\left(1+\frac{1}{\alpha}\right) = 5\Gamma(2) = 5 \text{ years}$$

$$\sigma^2 = \beta^{\frac{2}{\alpha}}\left\{\Gamma\left(1+\frac{2}{\alpha}\right)-\left[\Gamma\left(1+\frac{1}{\alpha}\right)\right]^2\right\} = 25(\Gamma(3)-(\Gamma(2)^2) = 25 \text{ and } \sigma = 5$$

$$\mu - \sigma = 0 \text{ and } \mu + \sigma = 10$$

The MAPLE solution is given by

> evalf$\left(\text{int}\left(0.2.\exp\left(-\frac{x}{5}\right), x = 0..10\right)\right)$; 0.865.

That is, 86.5% have lifetimes between $\mu - \sigma$ and $\mu + \sigma$.

Joint distributions–discrete and continuous

4.17. X and Y have the following joint distribution.

	$X = 2$	$X = 4$
$Y = 1$	0.10	0.15
$Y = 2$	0.20	0.30
$Y = 3$	0.10	0.15

Find the marginal probability distributions and determine whether X and Y are independent.

SOLUTION

The marginal distribution for X is

x	2	4
$P_1(x)$	0.4	0.6

and the marginal distribution for Y is

y	1	2	3
$P_2(y)$	0.25	0.5	0.25

Note that $P(x, y) = P_1(x)P_2(y)$ for all x and y. Therefore, X and Y are independent random variables.

4.18. X and Y have the following joint density function. Find the marginal density of X and the marginal density of Y. Determine whether X and Y are independent.

$$f(x,y) = \begin{cases} \dfrac{x(1+3y^2)}{4}, & 0 < x < 2, 0 < y < 1, \\ 0, & \text{otherwise} \end{cases}$$

SOLUTION

The marginal of x is

$$> \text{evalf}\left(\text{int}\left(\frac{x(1+3y^2)}{4}, y = 0..1 \right) \right);$$

$0.5000000000x$

The marginal of y is

$$> \text{evalf}\left(\text{int}\left(\frac{x(1+3y^2)}{4}, x = 0..2 \right) \right);$$

$0.5000000000C \quad 1.5000000000\, y^2$

Note that the product of the marginals is $f_1(x)f_2(y) = 0.25x + 0.75xy^2 = f(x,y)$. Therefore, X and Y are independent.

Checking data for normality

4.19. Sample pressures from a system were measured and gave the results shown in the following table. Test to see if the pressures may be assumed to be from a normal random variable. Use the Shapiro–Wilk test of normality.

13.46	13.26	13.90
13.04	13.48	13.79
13.54	13.83	13.60
13.82	12.57	12.74
13.09	13.47	12.73

SOLUTION

The STATISTIX analysis for the Shapiro–Wilk test is:

```
Shapiro-Wilk Normality Test
Variable      N      W          P
pressure      15     0.9189     0.1856
```

Because the p-value is 0.1856, do not reject the normality hypothesis. The pressure readings may be assumed to be from a normal random variable.

4.20. A sample of the lifetimes of system components were recorded for 15 systems. Test the data for normality using the Shapiro–Wilk test from the package STATISTIX. The data are:

18.0	15.7	15.2
17.4	15.2	22.5
15.1	15.8	15.5
16.4	15.7	16.9
17.5	15.1	15.3

SOLUTION

```
Shapiro-Wilk Normality Test
Variable      N      W          P
lifetimes     15     0.7091     0.0003
```

The hypothesis of normality for the data is rejected at $\alpha = 0.05$ since the p-value is less than $\alpha = 0.05$.

Transforming data to near normality

4.21. The following data is the total carbon in micrograms per cubic meter of air in an 8-hour workday at a salt mine in which diesel exhaust is present for 15 selected days.

14	410	3	25	129
180	17	58	7	9
26	147	37	187	55

Determine which, if any, of the four normality tests these data pass. The four tests are (1) the Anderson–Darling test, (2) the Ryan–Joiner test, (3) the Kolmogorov–Smirnov test, and (4) the Shapiro–Wilk test.

SOLUTION

The data fail all four normality tests.

4.22. The data in Problem 4.21 are replaced by the natural log of the values and is given below.

2.6	6.0	1.1	3.2	4.9
5.2	2.8	4.1	1.9	2.2
3.3	5.0	3.6	5.2	4.0

Determine which, if any, of the four normality tests these data pass. The four tests are (1) the Anderson–Darling test, (2) the Ryan–Joiner test, (3) the Kolmogorov–Smirnov test, and (4) the Shapiro–Wilk test.

SOLUTION

The data pass all four of the normality tests.

Simulation

4.23. It is of interest to simulate the lifetimes of a component that it is believed that the lifetimes have a normal distribution with mean equal to 5 hours and a standard deviation of 0.25 hours. Use MINITAB to generate the simulated lifetimes of a sample of 10 such components.

SOLUTION

The MINITAB pulldown **Calc** $\Rightarrow$ **Random Data** $\Rightarrow$ **Normal** gives the following dialog box which is filled out as shown.

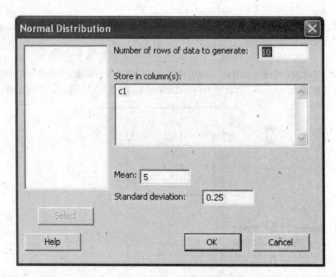

The following simulated lifetimes are created:

Data Display

```
lifetimes
4.9  5.1  4.9  4.8  4.9  4.8  5.4  5.1  5.1  4.9
```

4.24. Random variable X has the density $f(x) = \begin{cases} 0, x < 0 \\ 2xe^{-x^2}, x \geq 0 \end{cases}$.

(a) Plot the graph of the density.

(b) Select a simulated sample from this density of size 20.

SOLUTION

(a) An EXCEL plot of $f(x)$ is as follows.

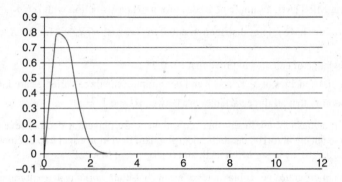

(b) The cumulative distribution function is

$F(x) = 0$, if $x < 0 = 1 - e^{-x^2}$, $x > 0$. Now generate 20 uniform values between 0 and 1. Let u represent one of these uniform values. Set $u = F(x)$ and solve for x.

$x =$ square root $(-\ln(1 - u))$. The 20 uniform numbers are:

0.665	0.526	0.528	0.178	0.284
0.985	0.520	0.134	0.668	0.131
0.094	0.020	0.072	0.876	0.298
0.629	0.581	0.891	0.211	0.287

The random values for x are:

1.045	0.864	0.866	0.443	0.578
2.047	0.857	0.380	1.050	0.375
0.315	0.141	0.274	1.446	0.594
0.996	0.933	1.490	0.487	0.581

SUPPLEMENTARY PROBLEMS

4.25. The voltage in an engineering experiment X varies according to the following probability density function.

$$f(x) = 4x^3 e^{-x^4}, \; x > 0$$

(a) Graph the function $f(x)$ over the interval from 0 to 2 using EXCEL.

(b) Verify that $f(x)$ is a pdf by exact integration.

(c) Use MAPLE to verify that $f(x)$ is a pdf.

(d) Use EXCEL and the definition of the integral as a limiting sum to verify that $f(x)$ is a pdf.

(e) Find the mean and standard deviation.

(f) Find $P(X < 1)$.

(g) Find $P(X > 2)$.

(h) Find $P(0.5 < X < 1.5)$.

4.26. The proportion of defective ammunition produced by Bullets Inc. has a defective proportion per day that follows a beta distribution with $\alpha = 1$ and $\beta = 19$.

(a) Give the density and graph the pdf of X, the proportion of defectives produced per day.

(b) Use MAPLE to verify that it is a pdf.

(c) Find the mean and standard deviation of X.

(d) Find the probability that the percentage defective on a given day is 10% or less.

(e) Simulate, using MINITAB, a sample of 10 days for a beta distribution with $\alpha = 1$ and $\beta = 19$.

4.27. Three random variables have a joint distribution equal to $f(x, y, z) = cxy^2e^{-z}$ for $0 < X < 1, 0 < Y < 2, Z > 0$ and 0 elsewhere.

(a) Use your knowledge of calculus to prove that $c = 0.75$.

(b) Find the marginal densities of X, Y, and Z and determine if the three variables are independent.

(c) Find the probability that all three variables are between 0 and 1.

4.28. It is determined by acoustical engineers that mp3 players are played at decibel levels (X) that are normally distributed with a mean equal to 100 decibels and a standard deviation equal to 5 decibels. Find $P(90 < X < 105)$ (a) by using EXCEL, (b) by using MINITAB, and (c) by using MAPLE.

4.29. The total carbon in micrograms per cubic meter of air in an 8-hour workday at a salt mine in which diesel exhaust is present, X, is the variable of interest. It is found that $\ln(X)$ has a normal distribution with mean equal to 3.5 and standard deviation equal to 1.5. This may also be stated as follows: X has a log-normal distribution with parameters $\alpha = 3.5$ and $\beta = 1.5$.

(a) Plot the distribution of X.

(b) Find the mean and standard deviation of the total carbon in an 8-hour shift.

(c) Find the probability $X < 150$.

4.30. The chi-square distribution is a gamma distribution with $\alpha = v/2$ and $\beta = 2$. The parameter v is called the number of **degrees of freedom** for the chi-square distribution.

(a) Give the chi-square pdf by substituting these values for α and β into the gamma pdf.

(b) Take the expression for the mean of a gamma and substitute the above values for α and β into the mean of the gamma distribution to find the mean of the chi-square.

(c) Take the expression for the variance of a gamma and substitute the above values for α and β into the variance of the gamma distribution to find the variance of the chi-square.

(d) Draw MINITAB graphs of the chi-square in separate panels of the same graph for 2, 5, 10, and 15 degrees of freedom.

CHAPTER 5

Sampling Distributions

Populations and Samples

A **population** is a set of data that is of interest to the engineer or scientist. The population may be described by a probability distribution function or a probability density function. There are **parameters** of the populations in which we are interested. The **parameters** are population measurements such as the mean, the standard deviation, the population proportion, and so forth. The parameters are usually unknown. For example, the normal population has two parameters that are usually unknown. Until the mean μ and standard deviation σ are known, the normal population's form is known but the exact density is unknown $\left(f(x) = \dfrac{1}{\sqrt{2\pi}\sigma} e^{-\frac{(x-\mu)^2}{2\sigma^2}}, -\infty < x < \infty \right)$. We must estimate the mean and standard deviation before any probabilities concerning the population can be made.

The population is sampled in order to find out something about it. The most common sample is a random sample. A set of n observations constitutes a **simple random sample** if it is chosen in such a manner so that each subset of n elements has the same probability of being chosen. One way of selecting a simple random sample is to write the names of the population items on a piece of paper and pull your sample from a box containing the names. A random number generator in a software package is the modern-day technique of obtaining a random sample from a finite population. In a large population the sample is chosen in as haphazard a way as possible since it may not be possible to number the elements.

The parameters such as the mean μ and standard deviation σ are estimated by the sample mean $\bar{x}$ and the sample standard deviation $s = \sqrt{\dfrac{\sum(x - \bar{x})^2}{(n-1)}}$ which are calculated from the sample observations. The sample mean and sample standard deviation are referred to as statistics. A **statistic** is a measurement made on the sample. A statistic is actually a random variable. Its value changes from sample to sample. The parameter is a constant and the statistic is a variable.

The **sampling distribution** of a sample statistic, calculated from a sample of n measurements, is the probability distribution of the statistic. We will find in the next section that under certain conditions $\bar{X}$ has a normal distribution. We also find that $\dfrac{(n-1)S^2}{\sigma^2}$ has a chi-square distribution with $(n-1)$ degrees of freedom.

Sampling Distribution of the Mean (σ Known)

The results in this section are best illustrated by an example. We illustrate the central limit theorem by an example and then we state the central limit theorem more generally.

EXAMPLE 5.1   We are interested in taking various sample sizes from the students at Midwestern University. The random variable of interest X is the credit hours the student is taking. The probability distribution of X is shown in Figure 5.1.

The distribution is uniform with the possible values 6, 9, 12, and 15. The mean of the distribution is $\mu = 6(0.25) + 9(0.25) + 12(0.25) + 15(0.25) = 10.5$ and the variance is $\sigma^2 = \sum x_i^2 p(x_i) - \mu^2 = 36(0.25) + 81(0.25) + 144(0.25) + 225(0.25) - 10.5^2 = 11.25$. The population mean and variance are therefore known to be $\mu = 10.5$ and $\sigma^2 = 11.25$.

x	6	9	12	15
p(x)	0.25	0.25	0.25	0.25

Figure 5.1 Distribution of credit hours per student at Midwestern University.

Now consider the distribution of the sample mean constructed by taking all samples of size $n = 2$, then $n = 3$, and so forth. A pattern will soon emerge. Start by considering all samples of size 2. In Figure 5.2, the sample mean will be represented by Xbar. The symbols Xbar and $\bar{x}$ will be used interchangeably throughout the discussion.

A	B	C	D	E	F	G	H
6	6	6	Distribution of sample mean for $n = 2$				
6	9	7.5					
6	12	9	Xbar	P(Xbar)	Xbar*p(Xbar)	Xbar^2*p(Xbar)	
6	15	10.5	6	0.0625	0.375	2.25	
9	6	7.5	7.5	0.125	0.9375	7.03125	
9	9	9	9	0.1875	1.6875	15.1875	
9	12	10.5	10.5	0.25	2.625	27.5625	
9	15	12	12	0.1875	2.25	27	
12	6	9	13.5	0.125	1.6875	22.78125	
12	9	10.5	15	0.0625	0.9375	14.0625	
12	12	12		sum = 1	sum = 10.5	sum = 115.88	
12	15	13.5					
15	6	10.5					
15	9	12					
15	12	13.5					
15	15	15					

Figure 5.2 Constructing the distribution of Xbar ($\bar{X}$), for $n = 2$.

Figure 5.2 shows all possible samples from Midwestern University of size $n = 2$ in A1:B16. For example, 6 and 6 represents a sample in which the first selected student was taking 6 credit hours and so was the second. The mean of each sample is shown in C1:C16. Each of the sample means in Column C is equally likely (since the samples are taken from a uniform distribution) and the distribution of Xbar is given in Columns E and F. The sample mean, 10.5 for example, occurs 4 times in 16 and has probability 4/16 or 0.25. A plot of the distribution of Xbar is given in Figure 5.3. Column G of Figure 5.2 illustrates that $\mu_{\bar{x}} = \sum \bar{x} p(\bar{x}) = 10.5 = \mu$. That is, the mean of the sample means equals the

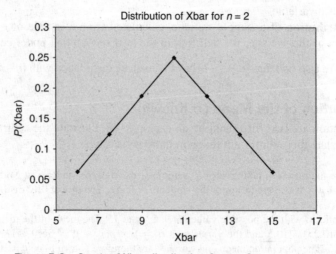

Figure 5.3 Graph of Xbar distribution for $n = 2$.

population mean for $n = 2$. A sample statistic whose expected value equals the parameter it is estimating is said to be an **unbiased estimator** of the parameter. $(E(\bar{x}) = \mu_{\bar{x}} = \mu)$ $\bar{x}$ is said to be an unbiased estimator of μ. The variance of the sample mean is $\sigma_{\bar{x}}^2 = \sum \bar{x}^2 P(\bar{x}) - \mu_{\bar{x}}^2 = 115.875 - 10.5^2 = 5.625$. The computations are shown in Figure 5.2.

For $n = 2$, $\mu_{\bar{x}} = \mu$ and $\sigma_{\bar{x}}^2 = \dfrac{\sigma^2}{2}$.

Now suppose we list all samples of size 3 and see what we may discover. Figure 5.4 lists all possible samples of size 3. The 64 samples of size 3 are all listed as well as their means. From Figure 5.4, the distribution of Xbar, for samples of size $n = 3$, may be built. The distribution of Xbar is shown in Figure 5.5. We also see from Figure 5.5 that Xbar is an unbiased estimator of μ for samples of size 3.

$$\mu_{\bar{x}} = \sum \bar{x}p(\bar{x}) = 10.5 = \mu$$

A First	B Second	C Third	D Mean	E	F First	G Second	H Third	I Mean
6	6	6	6		12	6	6	8
6	6	9	7		12	6	9	9
6	6	12	8		12	6	12	10
6	6	15	9		12	6	15	11
6	9	6	7		12	9	6	9
6	9	9	8		12	9	9	10
6	9	12	9		12	9	12	11
6	9	15	10		12	9	15	12
6	12	6	8		12	12	6	10
6	12	9	9		12	12	9	11
6	12	12	10		12	12	12	12
6	12	15	11		12	12	15	13
6	15	6	9		12	15	6	11
6	15	9	10		12	15	9	12
6	15	12	11		12	15	12	13
6	15	15	12		12	15	15	14
9	6	6	7		15	6	6	9
9	6	9	8		15	6	9	10
9	6	12	9		15	6	12	11
9	6	15	10		15	6	15	12
9	9	6	8		15	9	6	10
9	9	9	9		15	9	9	11
9	9	12	10		15	9	12	12
9	9	15	11		15	9	15	13
9	12	6	9		15	12	6	11
9	12	9	10		15	12	9	12
9	12	12	11		15	12	12	13
9	12	15	12		15	12	15	14
9	15	6	10		15	15	6	12
9	15	9	11		15	15	9	13
9	15	12	12		15	15	12	14
9	15	15	13		15	15	15	15

Figure 5.4   Constructing the distribution of Xbar for $n = 3$.

Xbar	p(Xbar)	Xbar*p(Xbar)	Xbar^2*p(Xbar)
6	0.015625	0.09375	0.5625
7	0.046875	0.328125	2.296875
8	0.09375	0.75	6
9	0.15625	1.40625	12.65625
10	0.1875	1.875	18.75
11	0.1875	2.0625	22.6875
12	0.15625	1.875	22.5
13	0.09375	1.21875	15.84375
14	0.046875	0.65625	9.1875
15	0.015625	0.234375	3.515625
	sum = 1	sum = 10.5	sum = 114

Figure 5.5   Distribution of Xbar for $n = 3$.

Referring again to Figure 5.5, we find that

$$\sigma_{\bar{x}}^2 = \sum \bar{x}^2 P(\bar{x}) - \mu_{\bar{x}}^2 = 114 - 10.5^2 = 3.75.$$

A plot of the distribution of Xbar for samples of size $n = 3$ is given in Figure 5.6.
Or the relationship between the population variance and the variance of the sample means is $\sigma_{\bar{x}}^2 = \dfrac{\sigma^2}{3}$.

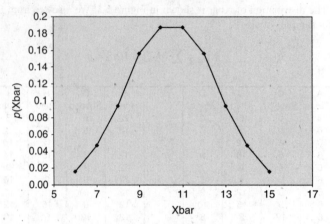

Figure 5.6   Graph of Xbar distribution for $n = 3$.

Figure 5.7 summarizes the discussion so far. The original population distribution for the credit hours taken by students at Midwestern University is given in the first two columns and is labeled X and $p(X)$. The distribution for the sample mean based on samples of size $n = 2$ is labeled X1bar and $p(\text{X1bar})$. The distribution for the sample mean based on samples of size $n = 3$ is labeled X2bar and $p(\text{X2bar})$. The distribution for the sample mean based on samples of size $n = 4$ is labeled X3bar and $p(\text{X3bar})$. To make certain that you comprehend the discussion, it is suggested that you use EXCEL and build the distribution given in C7 and C8.

C1	C2	C3	C4	C5	C6	C7	C8
X	p(X)	X1bar	p(X1bar)	X2bar	p(X2bar)	X3bar	p(X3bar)
6	0.25	6.0	0.0625	6	0.015625	6.00	0.003906
9	0.25	7.5	0.1250	7	0.046875	6.75	0.015625
12	0.25	9.0	0.1875	8	0.093750	7.50	0.039063
15	0.25	10.5	0.2500	9	0.156250	8.25	0.078125
		12.0	0.1875	10	0.187500	9.00	0.121094
		13.5	0.1250	11	0.187500	9.75	0.156250
		15.0	0.0625	12	0.156250	10.50	0.171875
				13	0.093750	11.25	0.156250
				14	0.046875	12.00	0.121094
				15	0.015625	12.75	0.078125
						13.50	0.039063
						14.25	0.015625
						15.00	0.003906

Figure 5.7   Original population distribution and sample mean distribution for $n = 2$, 3, and 4.

The sampling distributions given in Figure 5.7 are graphed in Figure 5.8. MINITAB plots of C2 vs C1, C4 vs C3, C6 vs C5, and C8 vs C7 are shown in separate panels of the same graph. Figure 5.8 shows the original population and the distribution of Xbar for $n = 2$, 3, and 4.

When this process is continued, the listing of all possible samples becomes very difficult. For $n = 2$, the number of possible samples is $4^2 = 16$, for $n = 3$, the number of possible samples is $4^3 = 64$. For samples of size 10, the worksheet to

Scatterplot of C2 vs C1, C4 vs C3, C6 vs C5, C8 vs C7

Figure 5.8　Original population distribution and sample mean distribution for $n = 2$, 3, and 4.

list all samples would require $4^{10} = 1,048,576$ rows. But the same pattern would occur for samples of size 10. However, if we did continue, we would find that the mean of the sample means is the mean of the population from which the samples are selected, that is, $\mu_{\bar{x}} = \mu$. We would find that $\sigma_{\bar{x}}^2 = \dfrac{\sigma^2}{10} = \dfrac{11.25}{10} = 1.125$. We would also find that the graph of the probability distribution would move closer to a normal curve for $n = 10$ than for $n = 2$ or $n = 3$, etc.

These results are referred to as the **central limit theorem**. The central limit theorem is one of the fundamental theorems of statistics. **It is extremely important that the student understand this theorem**.

Central limit theorem: Let $X_1, X_2, \ldots, X_n$ be a simple random sample from a population with mean μ and variance σ^2. Let $\bar{X} = \dfrac{\sum X_i}{n}$ be the sample mean. Then if n is sufficiently large, the sample mean has an approximately normal distribution with mean μ and standard deviation $\sigma_{\bar{x}} = \dfrac{\sigma}{\sqrt{n}}$. The standard deviation $\sigma_{\bar{x}} = \dfrac{\sigma}{\sqrt{n}}$ is called the **standard error of the mean** to keep it distinct from the population standard deviation σ. The standard error of the mean measures the variability of the sample mean. The standard error of the mean depends on the variability of the original population and the sample size. How quickly the distribution of the sample mean approaches normality depends on the original population. In our opening example in this section, the original population was uniform and the normality appeared for fairly small n. **For most populations, if the sample size is 30 or greater, the central limit approximation is good.** Let us consider some additional examples of the implications of the central limit theorem.

EXAMPLE 5.2   The number of flaws on DVDs produced by a company has the following distribution. (Note that this distribution is not uniform.)

x	0	1	2	3
$p(x)$	0.60	0.25	0.10	0.05

Thirty-six of the DVDs are sampled from this population. What is the probability that the average number of flaws per DVD in this sample is less than 0.5? We are interested in $P(\bar{X} < 0.5)$. We may treat $\bar{X}$ as a normal random variable having mean μ and standard error $= \sigma/6$, where μ and σ are the population mean and standard deviation, respectively. $\mu = 0(0.60) + 1(0.25) + 2(0.10) + 3(0.05) = 0.6$ and $\sigma^2 = 0 + 0.25 + 0.40 + 0.45 - 0.36 = 0.74$. The standard error of the mean is $\sigma_{\bar{x}} = \dfrac{\sigma}{\sqrt{n}} = \dfrac{\sqrt{0.74}}{\sqrt{36}} = 0.143$.

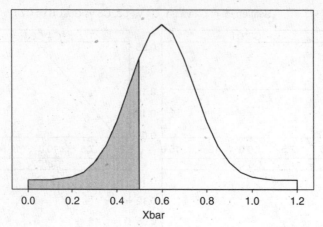

Figure 5.9 P(Xbar < 0.5) is represented by the gray area under the curve.

The gray area under the normal curve shown in Figure 5.9 is given by the EXCEL command =NORMDIST (0.5,0.6,0.143,1) which is 0.242182. The probability is 0.242 that the sample mean will be 0.5 or less. Recall the parameters in the NORMDIST function: the first is the value of X, the second is the mean, the third is the standard error of the distribution, and the 1 in the fourth position tells EXCEL to accumulate the area from 0.5 to the left of 0.5.

EXAMPLE 5.3   An engineer claimed that wood stoves emit an average of 40 million micrograms of fine particle pollution per hour. The standard deviation is 15 million micrograms per hour. To test the engineer's claim, 30 wood stoves were tested for fine particle pollution and it was found that the sample average for the 30 stoves was 50 million micrograms of fine particle pollution per hour. If the engineer's statement is correct, what is the probability of obtaining a sample of 30 with a mean of 50 or larger?

Xbar has a normal distribution with mean equal to 40 and standard error equal to $\sigma_{\bar{x}} = \dfrac{\sigma}{\sqrt{n}} = \dfrac{15}{\sqrt{30}} = 2.74$, assuming the engineer's statement is correct. The probability that the sample mean exceeds 50, assuming the engineer is correct, is given by the EXCEL command =1-NORMDIST(50,40,2.74,1). This represents the area in the right hand tail past 50 of a normal curve centered at 40 and having a standard error equal to 2.74. This gives P(Xbar > 50) = 0.000131. Since this probability involving Xbar is extremely small, Xbar > 50 would cast doubt on the engineer's claim.

Sampling Distribution of the Mean (σ Unknown)

In the previous section, it was assumed that σ was known. When σ is unknown, S is calculated from the sample and is used in place of σ. The proof of the following theorem is beyond the scope of this book. Theorem: When the sample is taken from a normal distribution, having population mean μ, the variable $t_v = \dfrac{\bar{x} - \mu}{s/\sqrt{n}}$ has a t distribution with v **degrees of freedom** and $v = n - 1$. The Greek letter v is used to represent degrees of freedom. S is the standard deviation of the sample. If the sample does not come from a normal distribution, it is not known what distribution the variable t_v has. One of the four normality tests may be used to determine if it may be assumed that the sample was taken from a normal distribution. Note that different sample sizes determine different t distributions.

Some Properties of the *t* Distribution

1. The t distribution is a mound-shaped symmetrical distribution much like the normal distribution. However, its shape is dependent on the sample size. As the sample size is increased, it approaches the standard normal distribution. A **standard normal distribution** is a normal distribution with $\mu = 0$ and $\sigma = 1$. When the degrees of freedom is 30 or more, there is very little difference between a t distribution

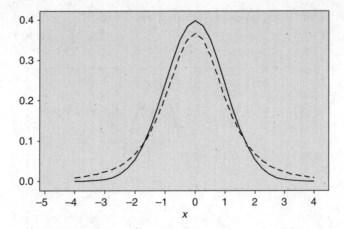

Figure 5.10   Standard normal as a solid curve and *t* with $v = 5$ as a dashed curve.

and a standard normal distribution. Figure 5.10 shows a standard normal curve as a solid line and a *t* distribution with $v = 5$ degrees of freedom as a dashed line. The *t* distribution is thicker in the tails than the normal curve. Both curves have a mean of 0. The standard deviation of the *t* distribution is greater than 1.

2. The total area under any *t* distribution is 1, with half the area to the right of 0 and half to the left of 0.

3. The *t* distribution was developed by William Gosset in 1908. Gosset worked for the Guinness Brewing Company in Dublin, Ireland. He was a chemist who was good at mathematics. The company forbade their employees to publish results of research, so Gosset published the results concerning the *t* distribution using the pen name Student. Hence the distribution is often referred to as the Student's *t* distribution.

4. Most statistics texts contain tables of the Student's *t* distribution with various degrees of freedom. However, most statistical software packages have the distribution as part of the software and the tables in books are not really necessary. Consider the following examples.

EXAMPLE 5.4   A sample of size 6 is selected from a normal distribution. Find the following probabilities using EXCEL: (a) $P(t_5 > 2)$ (b) $P(t_5 < -2.5)$ (c) $P(-1.5 < t_5 < 1.5)$

(a) The paste function =TDIST of EXCEL gives the dialog box shown in Figure 5.11. When the dialog box is filled in as shown, it gives the area to the right of 2 under the Student's *t* distribution with 5 degrees of freedom.

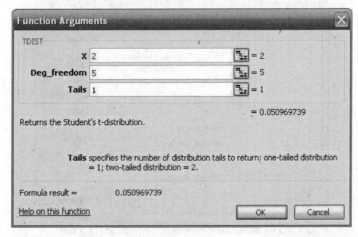

Figure 5.11   Dialog box for finding $P(t_5 > 2)$.

Figure 5.12 shows the area described above. The actual area is 0.05097.

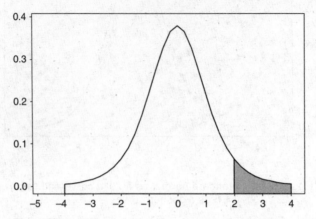

Figure 5.12   Picture of the area found in Figure 5.11.

(b) The paste function =TDIST(2.5,5,1) gives the area to the right of 2.5 which equals 0.027245. But because of symmetry, this is the same as the area to the left of −2.5. Therefore, $P(t_5 < -2.5) = 0.027245$.

(c) =TDIST(1.5,5,2) gives the area beyond −1.5 and 1.5 which equals 0.193904. This must be subtracted from 1 to give the correct answer. $P(-1.5 < t_5 < 1.5) = 1 - 0.193904$ or 0.806096.

EXAMPLE 5.5   Find the following probabilities using MINITAB. (a) $P(t_5 > 2)$ (b) $P(t_5 < -2.5)$ (c) $P(-1.5 < t_5 < 1.5)$

(a) The pulldown **Calc ⇒ Probability distributions ⇒ *t* Distribution** gives the dialog box in Figure 5.13. It is filled in as shown.

The output produced is as follows:

```
Student's t distribution with 5 DF
x P(X <= x)
2 0.949030
```

$$P(t_5 > 2) = 1 - 0.949030 = 0.05097$$

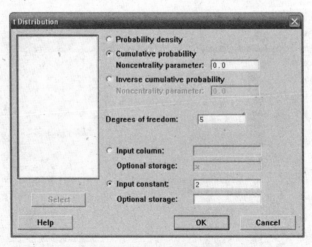

Figure 5.13   MINITAB dialog box for finding $P(t_5 < 2)$.

(b) The solution here is given directly.

```
Student's t distribution with 5 DF
x P(X <= x)
-2.5 0.0272450
```

$$P(t_5 < -2.5) = 0.027245$$

(c) The solution here is obtained by subtraction.

Cumulative Distribution Function

```
Student's t distribution with 5 DF
x P( X <= x )
1.5 0.903048
```

Cumulative Distribution Function

```
Student's t distribution with 5 DF
x P(X <= x)
-1.5 0.0969518
```

Subtracting, we have

$$P(-1.5 < t_5 < 1.5) = 0.903048 - 0.0969518 = 0.806096.$$

EXAMPLE 5.6 Find the following probabilities using STATISTIX. (a) $P(t_5 > 2)$ (b) $P(t_5 < -2.5)$ (c) $P(-1.5 < t_5 < 1.5)$

The pulldown **Statistics ⇒ Probability Functions** produces the dialog box in Figure 5.14. The results of three different commands are shown in the lower portion of the dialog box.

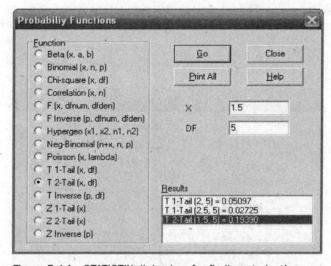

Figure 5.14   STATISTIX dialog box for finding student's
t probabilities.

Figure 5.14 shows that (a) $P(t_5 > 2) = 0.05079$, (b) $P(t_5 < -2.5) = 0.02725$ (using the symmetry of the distribution), and (c) $P(-1.5 < t_5 < 1.5) = 1 - 0.19390 = 0.8061$.

Properties of the *t* Distribution (continued)

5. The probability density function for a Student's *t* distribution, with v degrees of freedom, is

$$f(x) = \frac{\Gamma\left(\dfrac{v+1}{2}\right)}{\sqrt{\pi v}\,\Gamma\left(\dfrac{v}{2}\right)}\left(1+\frac{t^2}{v}\right)^{-\frac{v+1}{2}}, \quad -\infty < x < \infty.$$

The mean is 0 and for $v > 2$, the variance of the Student's *t* distribution is $\sigma^2 = \dfrac{v}{v-2}$.

EXAMPLE 5.7 An engineer wishes to test the claim that a new high-intensity light on aircraft has a mean lifetime of 1250 hours. She tests 10 of the lights and finds the following lifetimes in hours:

$$962 \quad 1127 \quad 1089 \quad 1132 \quad 1282 \quad 955 \quad 1071 \quad 1319 \quad 1121 \quad 1141$$

First, she uses STATISTIX to perform the Shapiro–Wilk test of normality to see if the normality assumption is reasonable. She obtains the following results.

```
Shapiro-Wilk Normality Test

Variable     N      W          P
lifetime     10     0.9232     0.3848
```

Because of the large *p*-value, she feels the normality assumption is reasonable.
Performing a descriptive statistics using STATISTIX, gives the following results.

```
Descriptive Statistics

Variable     Mean      SD        SE Mean
lifetime     1109.9    104.67    33.101
```

Assuming the claim is true, (i.e. $\mu = 1250$ hours) she calculates $t_v = \dfrac{\bar{x} - \mu}{\dfrac{s}{\sqrt{n}}} = \dfrac{1109.9 - 1250}{33.101} = -4.23$. If the

mean lifetime claim is true (i.e. $\mu = 1250$ hours), she has obtained a highly unlikely value for t_9. In fact the chances of obtaining such a low value for t_9 or one that is smaller is only 0.0011. This calculation is shown in the dialog box of Figure 5.15.

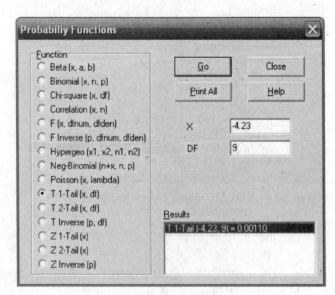

Figure 5.15   Calculation of $P(t_9 < -4.23)$.

If the claim that $\mu = 1250$ hours is true, then there are only 11 chances out of 10,000 of obtaining such a low sample mean. She would likely reject the claim as being too high.

Sampling Distribution of the Proportion

Consider the population shown in Figure 5.16. It contains 15 defectives (*D*) and 35 non-defectives (*N*). The items are contained in 50 numbered locations. For example, location 1 contains a defective item and location 31 contains a non-defective item, and so forth. Samples are selected at random with replacement.

1D	11N	21N	31N	41D
2N	12D	22N	32N	42D
3N	13N	23D	33N	43N
4N	14D	24N	34N	44N
5N	15N	25N	35N	45D
6N	16N	26N	36D	46N
7N	17N	27N	37N	47D
8N	18N	28D	38N	48D
9N	19D	29N	39N	49D
10D	20D	30N	40N	50N

Figure 5.16   Population with each cell numbered and status of item.

The population has 15 out of 50 or proportion $p = 0.30$ that are defective. Ten samples of size 5 are taken from the population using a random number generator. The 5 random numbers are shown in the columns of the following table.

9	14	16	14	31	2	49	4	48	47
32	1	26	43	14	1	28	5	21	50
13	28	22	46	41	10	25	16	33	25
18	1	13	39	14	42	3	38	5	7
45	41	11	33	27	39	15	19	42	46

These random numbers are used to select defectives and non-defectives from the population in Figure 5.16 as follows:

N	D	N	D	N	N	D	N	D	D
N	D	N	N	D	D	D	N	N	N
N	D	N	N	D	D	N	N	N	N
N	D	N	N	D	D	N	N	N	N
D	D	N	N	N	N	N	D	D	N

The 10 sample proportion defectives are: 0.2, 1, 0, 0.2, 0.6, 0.6, 0.4, 0.2, 0.4, and 0.2. The sample proportion defective is represented by $\hat{p}$ and the population proportion by p. Generally p is constant but $\hat{p}$ is variable. The average value of $\hat{p}$ for the 10 samples is 0.38 and the value of p is 0.30. Now, suppose this situation is magnified and the population is much larger and all possible samples are taken. What is the connection between $\hat{p}$ and p? In other words, what is the sampling distribution of $\hat{p}$?

If n is large enough to insure that $\hat{p} \pm 3\,\sigma_{\hat{p}}$ does not include 0 or 1, then the distribution of $\hat{p}$ may be assumed to be normal. The sample proportion defective $\hat{p}$ is an unbiased estimator of p. The standard error of $\hat{p}$ is $\sigma_{\hat{p}} = \sqrt{\dfrac{pq}{n}}$ where $q = 1 - p$. It is customary to replace p and q by $\hat{p}$ and $\hat{q}$ when calculating $\sigma_{\hat{p}} = \sqrt{\dfrac{pq}{n}}$. Thus, when n is large enough, $\hat{p}$ has a sampling distribution that is normal.

EXAMPLE 5.8   A sample of 100 DVDs is taken and it is found that 10 have places where the movie stops and will not go forward until some action is taken. It is of interest to know something about p, the population proportion that has this problem. The sample proportion $\hat{p}$ is $10/100 = 0.1$ for this sample. Calculate the probability that $\hat{p}$ is as small as 0.10 or smaller if the population proportion is 0.15.

First lets ask if $\hat{p}$ may be treated as if it is normally distributed. Suppose we calculate $\hat{p} \pm 3\,\sigma_{\hat{p}}$ and see if it does not include 0 or 1. To calculate $\sigma_{\hat{p}} = \sqrt{\dfrac{pq}{n}}$, assume $p = \hat{p} = 0.1$ and $q = 0.9$ since these are our only estimates of p and q. The standard deviation of $\hat{p}$ is $\sigma_{\hat{p}} = \sqrt{\dfrac{.1(.9)}{100}} = 0.03$; $\hat{p} \pm 3\,\sigma_{\hat{p}}$ is and $0.1 \pm 3(0.03)$ is $(0.01, 0.19)$. This interval does not include 0 or 1 and the distribution of $\hat{p}$ may be assumed to be normal.

Assume $p = 0.15$ and calculate the chances that $\hat{p} \leq 0.1$. That is, find $P(\hat{p} \leq 0.1)$ assuming $p = 0.15$. The sample proportion defective $\hat{p}$ may be treated as normally distributed with mean 0.15 and standard deviation $\sigma_{\hat{p}} = \sqrt{\dfrac{pq}{n}} = \sqrt{\dfrac{0.15(0.85)}{100}} = 0.0357$. Using EXCEL, $P(\hat{p} \leq 0.1)$ is given by =NORMDIST(0.1,0.15,0.0357,1), which equals 0.0807.

Summarizing, if the sample size is large enough, then $\hat{p}$ may be assumed to have a normal distribution. If the population has proportion p_o that has the characteristic of interest, then under repeated sampling, $\hat{p}$ has mean equal to p_o and standard deviation $\sigma_{\hat{p}} = \sqrt{\dfrac{p_o q_o}{n}}$.

EXAMPLE 5.9   Ten percent of the cell phones manufactured by Cell Phones Inc. have minor flaws. If you selected a sample of 50 of their phones, you would expect to find 5 of them with minor flaws. To check if $\hat{p}$ has an approximate normal distribution, see if $p_o \pm 3\sqrt{\dfrac{p_o q_o}{n}}$ contains 0 or 1 where $p_o = 0.1$. The interval is $0.1 \pm 3\sqrt{\dfrac{.1(.9)}{50}}$ or 0.1 ± 0.13 which is the interval $(-0.3, 0.23)$. Since this interval contains 0, either increase the sample size and then assume normality or use small sample techniques, which rely on the binomial distribution.

Sampling Distribution of the Variance

Consider a tire manufacturer that produces a tire that has lifetimes that are normally distributed with a mean lifetime equal to 60,000 miles and a standard deviation equal to 5,000 miles. Samples of size 4 are taken each day for 30 days. The MINITAB simulated samples and their variances are shown in Figure 5.17. The first sample is 63, 68, 59, and 65. The sample variance for that sample is 14.3. This is repeated 30 times.

1	2	3	4	5	6	7	8	9	10	11	12	13	14	15
63	67	62	53	64	65	69	66	62	62	61	62	52	59	62
68	63	60	62	53	58	61	55	66	65	57	45	59	60	69
59	65	55	64	56	63	57	59	58	55	65	58	65	59	57
65	56	59	58	63	53	60	64	54	48	65	55	56	68	57
14.3	22.9	8.67	23.6	28.7	28.9	26.3	24.7	26.7	57.7	14.7	52.7	30	19	32.3

16	17	18	19	20	21	22	23	24	25	26	27	28	29	30
59	59	67	60	59	48	60	64	63	57	53	62	57	54	60
56	60	57	58	56	55	61	66	48	57	59	58	60	62	64
54	63	64	60	59	64	54	59	55	58	59	64	65	57	52
64	63	60	55	64	58	63	59	49	66	61	69	62	62	63
18.9	4.25	19.3	5.58	11	44.3	15	12.7	47.6	19	12	20.9	11.3	16	29.6

Figure 5.17   Thirty samples of size 4 and variances of tire mileages in thousands.

It can be proved that $(n-1)S^2/\sigma^2$ has a chi-square sampling distribution with $(4-1)$ degrees of freedom. The sample size n is 4, the population variance is $5^2 = 25$, and the values of S^2 are given in Figure 5.17. The variable $(n-1)S^2/\sigma^2$ is equal $3S^2/25$. If each value of S^2 is multiplied by 3 and divided by 25 and then plotted, a histogram resembling a chi-square distribution with 3 degrees of freedom will result. Figure 5.18 shows a plot of the 30 values of $3S^2/25$.

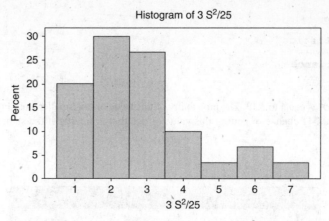

Figure 5.18   Histogram of the 30 values of $3S^2/25$.

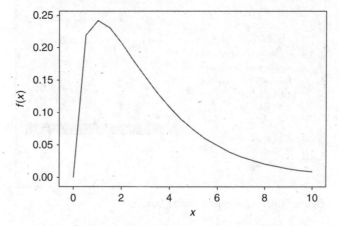

Figure 5.19   Chi-square distribution with 3 degrees of freedom.

If the number of samples is increased from 30 to a very large number, the shape of the histogram in Figure 5.18 approaches the shape of the probability density in Figure 5.19.

The above discussion illustrates the sampling distribution of S^2. In summary, this is what we know. Suppose a sample of size n is taken from a normal distribution having variance σ^2. The variable $(n-1)S^2/\sigma^2$ has a chi-square distribution with $(n-1)$ degrees of freedom. The normality assumption is necessary. That is, the sample has to come from a normal distribution for the sampling distribution to be chi-square. We have previously discussed how to check the normality assumption.

EXAMPLE 5.10   The standard deviation in years experience at a large engineering consulting firm is stated to be 8 years. A sample of size $n = 10$ is taken and the following data (in years) is obtained: 20, 15, 27, 16, 10, 13, 17, 20, 15, and 20. If the 8 years claim is correct, what is the probability of getting a set of data as contradictory to the 8 year claim as this one or more so?

First of all check out normality using the Shapiro–Wilk test of STATISTIX.

Shapiro-Wilk Normality Test

Variable	N	W	P
years	10	0.9492	0.6591

The normality of the data is accepted because the p-value is greater than 0.05.
The variance using STATISTIX is found to be

```
Statistix 8.0

Descriptive Statistics

Variable      Variance
years         22.233
```

The quantity $(n-1)S^2/\sigma^2$ is equal to 3.13. The probability of this value or one smaller from Figure 5.20 is $1 - 0.95892 = 0.04108$. There is only a 0.041 chance of getting this value or one that is smaller if the standard deviation of all years of experience is 8 years.

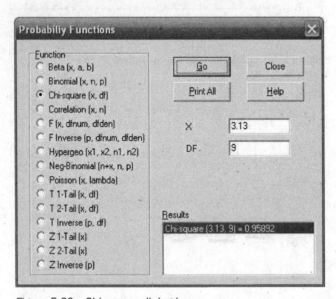

Figure 5.20   Chi-square dialog box.

Note: Even though we take only one sample, if it is random, we know what values are likely for $(n-1)S^2/\sigma^2$, and what values are unlikely.

EXAMPLE 5.11   An industrial engineer is concerned about the variability of fill in 340-gram containers of coffee. The process is deemed to be in control if σ is less than or equal to 1 gram. Corrective action (the process is stopped and the variability reduced) is taken only if the standard deviation is judged to be greater than 1 gram. A sample of 20 of the 340-gram containers is selected at random from production and the following weights are found. Use MINITAB to assist in your decision making.

338   339   337   342   342   343   342   339   338   341   340   342   340   344   341   342   337   340   341   343

First, a Kolmogorov–Smirnov test is conducted for normality. When the Kolmogorov–Smirnov is performed, the result is a p-value > 0.15. This result tells us that that we may assume normality. That is, we may assume that our sample came from a normally distributed population.

```
Descriptive Statistics: weight

Variable      Mean       Variance
weight        340.56     3.91
```

A descriptive statistics analysis where only the mean and variance is requested is performed using MINITAB. The mean is a little over 340 grams, $S^2 = 3.91$, and the chi-square variable is $(n-1)S^2/\sigma^2$ or $19(3.91)/1 = 74.3$. Now the question is: If $\sigma = 1$, how likely are we to get a chi-square equal to 74.3 or larger? We wish to compute the probability of obtaining a chi-square with 19 degrees of freedom that is 74.3 or larger. Using the pulldown **Calc $\Rightarrow$ Probability Distributions $\Rightarrow$ Chi-square** we find the following results:

```
Cumulative Distribution Function
Chi-Square with 19 DF
x P( X <= x )
74.3 1.00000
```

And $P(X \geq 74.3) = 0.00000$. If $\sigma \leq 1$, the probability of obtaining a chi-square as large as 74.3 is practically 0. Therefore we conclude $\sigma > 1$. The process would be stopped and corrective action taken.

SOLVED PROBLEMS

Populations and samples

5.1. Consider the population of lifetimes of motors manufactured by ACME Engines Inc. A quality control engineer has determined that the lifetimes are normally distributed with mean equal to 5.3 years and a standard deviation equal to 0.4 years. (a) Give the density function that describes the population of lifetimes. (b) Give the integral that gives the probability that a motor has a lifetime exceeding 6.0 years. (c) Use EXCEL to evaluate the integral.

SOLUTION

(a) The normal probability model is $f(x) = \dfrac{1}{\sqrt{2\pi}(0.4)} e^{-\frac{(x-5.3)^2}{2(0.16)}}$, $-\infty < x < \infty$.

(b) $\displaystyle\int_{6}^{\infty} \dfrac{1}{\sqrt{2\pi}(0.4)} e^{-\frac{(x-5.3)^2}{2(0.16)}} \, dx$, $-\infty < x < \infty$.

(c) 1-NORMDIST(6,5.3,0.4,1) = 0.04. Only 4% last more than 6 years.

5.2. In Problem 5.1 generate a simulated random sample of lifetimes of the motors.

SOLUTION

Use the pull down menu **Calc** $\Rightarrow$ **Random data** $\Rightarrow$ **Normal**. Fill out the dialog box as shown below.

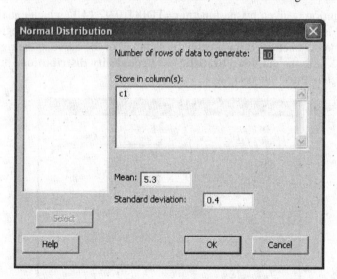

The random sample generated is: 5.4   4.0   4.9   5.1   5.1   5.5   5.4   5.7   5.3   5.5.

Sampling distribution of the mean (σ known)

5.3. Bags of concrete mix labeled as containing 100 pounds have a population mean of 100 pounds and a population standard deviation of 0.5 pounds. (a) What is the probability that the mean weight of a random sample of 50 bags is less than 99.9 pounds? (b) If the population mean weight is increased to 100.15 pounds, what is the probability the mean weight of a sample of size 50 will be less than 100 pounds?

SOLUTION

(a) Xbar will have a normal distribution because the sample size exceeds 30. The mean of Xbar will be the mean of the population or 100 pounds. The standard error of Xbar will be $0.5/\sqrt{50} = 0.5/7.071 = 0.0707$.

$P(\bar{X} < 99.9) = P(Z < -1.414) = 0.0787$.

(b) $P(\bar{X} < 100) = P(Z < -2.12) = .0170$

5.4. The lifetimes of a particular integrated circuit have a mean of 500.5 hours with a standard deviation equal to 5.7 hours. A sample of size 36 is taken of these integrated circuits. What is the probability the sample mean exceeds 501.0 hours?

SOLUTION

$$P(\bar{X} > 501.0) = P\left(\frac{\bar{X} - 500.5}{0.95} > \frac{501.0 - 500.5}{0.95}\right) = P(Z > 0.526) = 0.2994$$

Sampling distribution of the mean (σ unknown)

5.5. Bags of concrete mix labeled as containing 100 pounds have a population mean of 100 pounds and an unknown population standard deviation. (a) What is the probability that the mean weight of a random sample of 15 bags is less than 99.9 pounds? The sample standard deviation of the sample is found to be 0.4 pounds. Use EXCEL, MINITAB, and STATISTIX to evaluate the probability.

SOLUTION

We wish to find $P(\bar{X} < 99.9) = P\left(\frac{\bar{X} - 100}{0.103} < \frac{99.9 - 100}{0.103}\right) = P(T_{14} < -0.97)$. The MINITAB solution is found using the pulldown **Calc $\Rightarrow$ Probability distribution $\Rightarrow$ t**. The output is as follows:

```
Cumulative Distribution Function
Student's t distribution with 14 DF
x P( X <= x )
-0.97 0.174254
```

The EXCEL solution is given by the function =TDIST(0.97,14,1) which equals 0.1742. The area to the right of 0.97 is the same as to the left of −0.97. The parameter 14 is the degrees of freedom for the *t* distribution. The parameter 1 is the area in the right tail of the curve to the right of 0.97.

In STATISTIX, use the pulldown **Statistics $\Rightarrow$ Probability distributions** that gives the following dialog box, which is filled out as shown:

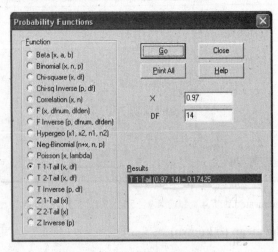

Note that all three software packages give the same answer.

5.6. A manufacturer of fuses claims that with a 15% overload, his fuses will blow in 12.00 minutes on the average. To test this claim, a sample of 10 of the fuses was subjected to a 15% overload and the times it took them to blow had a mean of 10.5 minutes and a standard deviation of 2.10 minutes. On the assumption that these times may be looked upon as a random sample from an approximate normal population with population mean equal to 12.00 minutes, what is the probability of getting a sample mean equal to or smaller than 10.5 minutes?

SOLUTION

The statistic $t = \dfrac{10.5 - 12.00}{0.66} = -2.27$. The probability of getting a t value this small or smaller is given by all three of the packages MINITAB, EXCEL, and STATISTIX as 0.024 if $\mu = 12.00$. Since the probability of getting a sample mean this small is so small, the claim would be rejected in favor of the claim that μ equals a value smaller than 12.00.

Sampling distribution of the proportion

5.7. Concentrations of atmospheric pollutants such as carbon monoxide (CO) can be measured with a spectrophotometer. In a calibration test, 30 measurements were taken of a laboratory gas sample that is known to have CO concentrations of 70 parts per million (ppm). A measurement is considered to be satisfactory if it is within 5 ppm of the true concentration. Of the 30 measurements, 21 were satisfactory. (a) What proportion of the sample measurements was satisfactory? (b) Find a 95% confidence interval for the proportion of measurements made by this instrument that will be satisfactory.

SOLUTION

(a) The proportion of sample measurements that were satisfactory was $\hat{p} = 21/30 = 0.7$.

(b) The 95% confidence interval on p, the proportion in the population, is $\hat{p} \pm 1.96\sqrt{\dfrac{\hat{p}(1-\hat{p})}{n}}$. This interval is 0.7 ± 0.16 or between 0.54 and 0.86. We are 95% confident that the population percentage is between 54% and 86%.

5.8. On a certain day, a large number of fuses were manufactured, each rated at 15A. A sample of 100 fuses is drawn from the day's production, and 15 of them were found to have burnout amperages greater than 15A. (a) What proportion of the sample were found to have burnout amperage greater than 15A? (b) Find a 95% confidence interval for the proportion of these fuses that will have a burnout percentage greater than 15A.

SOLUTION

(a) The proportion of sample measurements having burnout percentage greater than 15A was $\hat{p} = 15/100 = 0.15$.

(b) The 95% confidence interval on p, the proportion in the population, is $\hat{p} \pm 1.96\sqrt{\dfrac{\hat{p}(1-\hat{p})}{n}}$. This interval is 0.15 ± 0.07 or between 0.08 and 0.22. We are 95% confident that the population percentage is between 8% and 22%.

Sampling distribution of the variance

5.9. Tests on 10 concrete tension specimens yielded an estimate of the population variance of 40,000 (standard deviation of 200 kN/m^2). Use MINITAB to set 90% confidence limits on σ.

SOLUTION

Give the MINITAB command **Stat** $\Rightarrow$ **Basic Statistics** $\Rightarrow$ **1 variance** which gives the following dialog box filled in as shown.

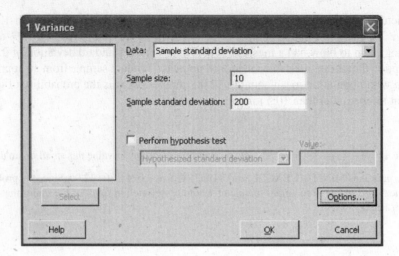

The pulldown menu gives the following output.

```
Statistics
N       StDev     Variance
10      200       40000

90% Confidence Intervals
            CI for          CI for
Method      StDev           Variance
Standard    (146, 329)      (21278, 108267)
```

5.10. Suppose the refractive indices of 20 pieces of glass (randomly selected from a large shipment purchased by an optical firm) had a sample standard deviation equal to 0.011. Use MINITAB to construct a 95% confidence interval for σ.

SOLUTION

See Problem 5.9 for the pulldown menu that is used.

```
Statistics
N       StDev     Variance
20      0.0110    0.000121

95% Confidence Intervals

Method      CI for StDev        CI for Variance
Standard    (0.0084, 0.0161)    (0.000070, 0.000258)
```

We are 95% confident that σ is between 0.0084 and 0.0161.

SUPPLEMENTARY PROBLEMS

5.11. The number of TVs per household has the distribution shown in the following table:

x	1	2	3	4	5
$p(x)$	0.2	0.2	0.2	0.2	0.2

(a) For this population, find the mean number of TVs per household and find the standard deviation of the number of TVs per household.

(b) What shape describes this population distribution?

(c) Give the distribution of the sample mean for $n = 2$, find $\mu_{\bar{x}}$ and $\sigma_{\bar{x}}$.

(d) Give the distribution of the sample mean for $n = 3$, find $\mu_{\bar{x}}$ and $\sigma_{\bar{x}}$.

(e) Give the distribution of the sample mean for $n = 4$, find $\mu_{\bar{x}}$ and $\sigma_{\bar{x}}$.

5.12. Plot the distributions in Problem 5.11 (c), (d), and (e). Plot the three distributions on the same graph.

5.13. A city claims that the mean fluoride is 2.5 milligrams per liter in its drinking water. The standard deviation is known to be 1.0 milligrams per liter. In order to check the claim, the EPA collects a sample of size 35 and finds the mean amount of fluoride to be 3.1 milligrams per liter. The EPA does not expect the sample mean to be the same as the claimed population mean. The claim that $\mu = 2.5$ is in doubt, since the sample mean was 3.1 based on a sampling of 35 locations randomly chosen around the city. The EPA assumes that $\mu = 2.5$ and will calculate the probability of obtaining a sample whose mean is 3.1 or larger based on 35 observations. The probability $P(\bar{x} > 3.1$ assuming that $\mu = 2.5)$ will be calculated and if that probability is less than 0.05, the EPA will reject the city engineers claim and claim that $\mu > 2.5$. Calculate this probability and state your conclusion.

5.14. Suppose in Problem 5.13 that a small sample is used instead of a large sample. A sample of size $n = 10$ is obtained from water supplies around the city. The sample values are as follows:

$$3.7 \quad 2.0 \quad 1.8 \quad 2.6 \quad 3.8 \quad 1.6 \quad 2.8 \quad 4.5 \quad 3.3 \quad 1.3$$

Test to see if 2.5 is a reasonable value for the population mean. If there is less than a 5% chance that a sample with a mean equal to the one obtained from this sample, then reject the claim that $\mu = 2.5$. Calculate the value of t_9 and base your decision on this calculated value. Check out the one assumption that is critical when working with the t distribution.

5.15. In addition to the rule given in the section on sampling distribution of the proportion to test if $\hat{p}$ may be assumed to have an approximate normal distribution, there is another rule that is often used. The first rule given in the section states that if the interval $\hat{p} \pm 3\sqrt{\dfrac{\hat{p}\hat{q}}{n}}$ does not contain 0 or 1, then $\hat{p}$ may be approximated by a normal distribution. The second rule states that if $n\hat{p} > 5$ and $n\hat{q} > 5$, where $\hat{q} = 1 - \hat{p}$, then the distribution of $\hat{p}$ may be approximated by a normal distribution. Using both rules, determine if the normal approximation is appropriate in the following cases:

(a) $n = 150$, $\hat{p} = 0.10$

(b) $n = 500$, $\hat{p} = 0.05$

(c) $n = 50$, $\hat{p} = 0.15$

(d) $n = 1000$, $\hat{p} = 0.001$

5.16. A quality engineer is interested in estimating the proportion of personal digital assistants (PDAs) that have a minor surface flaw. She selects 100 at random from the production process and finds that 15 have a minor surface flaw. The estimate of the proportion from the process with a minor flaw is 0.15. Since $n\hat{p} > 5$ and $n\hat{q} > 5$, the distribution may be approximated by the normal distribution. Because of the property of the normal distribution that 95% of the distribution is within 2 standard deviations of the mean of the distribution, there is close to a 0.95 probability that $\hat{p}$ will be within 2 standard errors of the true unknown value of p. (**Note: To attain 0.95 probability, 1.96 standard errors would be needed. We will address this in the next chapter.**)

The standard error of $\hat{p}$ is $\sigma_{\hat{p}} = \sqrt{\dfrac{pq}{n}}$. This standard error is estimated by $\sigma_{\hat{p}} = \sqrt{\dfrac{\hat{p}\hat{q}}{n}} = \sqrt{\dfrac{0.15(0.85)}{100}} = 0.036$ and 2 standard errors is 0.072. At this point, we estimate that 15% of the PDAs have minor flaws. We are approximately 95% sure that our **error of estimate** is off by no more than 7.2%. This means the proportion defective may be as low as 7.8% or as high as 22.2%. Suppose management directs the engineer to reduce the 95% error of estimate from 7.2% to 5%. What sample size would need to be taken to have a 5% error of estimate?

5.17. A population consists of 6 items: A, B, C, and D are defective and E and F are non-defective. The parameter p is the proportion defective in the population and equals $4/6 = 0.67$. Sample size is $n = 4$ and the statistic PHAT is the sample proportion defective. Give the sampling distribution of PHAT and find the mean of PHAT.

5.18. The number of TVs per household has the distribution shown in the following table:

x	1	2	3	4	5
$p(x)$	0.2	0.2	0.2	0.2	0.2

(a) For this population, find the variance and standard deviation of the number of TVs per household.

(b) Give the sampling distribution of the sample variance and the sampling distribution of the sample standard deviation for $n = 2$, and find the expected value of both.

(c) Give the sampling distribution of the sample variance and the sampling distribution of the sample standard deviation for $n = 3$, and find the expected value of both.

(d) Give the sampling distribution of the sample variance and the sampling distribution of the sample standard deviation for $n = 4$, and find the expected value of both.

5.19. In Problem 5.18, plot the sampling distributions of S^2 for $n = 2, 3,$ and 4 overlaid on the same graph. Plot the sampling distributions of S for $n = 2, 3,$ and 4 overlaid on the same graph.

5.20. Too much variability in shaft lengths results in some being too short and some being too long. It is found that if σ is no more than 3.5 millimeters, the shaft lengths are acceptable. The shaft lengths in millimeters are as follows:

103.704   98.311   98.895   97.373   98.506   96.565   97.506   104.004   96.411   96.923

Perform the test of normality using STATISTIX.

CHAPTER 6

Inferences Concerning Means

Point Estimation

Some definitions and properties of estimators were discussed in the previous chapter. It is time to formalize these discussions. First let us develop some properties of expectation and variance. Assume random variable X is continuous. Let $Y = a + bX$ and consider the expectation (or mean) of Y and the variance of Y.

$$E(Y) = E(a + bX) = \int_{-\infty}^{\infty} (a + bx)f(x)dx = a\int_{-\infty}^{\infty} f(x)dx + b\int_{-\infty}^{\infty} xf(x)dx = a + bE(x)$$

$$Var(a + bX) = \int_{-\infty}^{\infty} (a + bx - (a + bE(x)))^2 f(x)dx = \int_{-\infty}^{\infty} (bx - bE(x))^2 f(x)dx = b^2 Var(X)$$

Note: If X is discrete rather than continuous, sums replace integrals and the proofs are similar.

Notation: μ or $E(X)$ is used to represent the mean of X. σ^2 or $Var(X)$ is used to represent the variance of X.

A **point estimator** is a single number used to estimate a parameter. $\bar{X}$ is a point estimator of the parameter μ. Before the sample is taken, $\bar{X}$ is regarded a linear combination of random variables. After the sample is taken and the numbers averaged, $\bar{x}$ is a single number having no distribution. If the relationship $E(a + bX) = a + bE(X)$ is extended, it becomes $E(a_1 X_1 + \ldots + a_n X_n) = a_1 E(X_1) + \ldots + a_n E(X_n)$. This is true because the integral of the sum is the sum of the integrals. Applying this to $\bar{X}$, we have

$E(\bar{X}) = E\left(\dfrac{X_1 + \ldots + X_n}{n}\right) = \dfrac{1}{n}(E(X_1) + \ldots + E(X_n)) = \mu$, since each expected value of X_i is μ and there are n

μ's. When the expected value of a statistic is equal to the parameter, the statistic is intended to estimate, the statistic is said to be an **unbiased estimator** of the parameter. Therefore $\bar{X}$ is an unbiased estimator of the population mean, μ.

It can be shown that S^2 is an unbiased estimator of σ^2. $\dfrac{(n-1)S^2}{\sigma^2}$ has a chi-square distribution with

$(n-1)$ degrees of freedom as stated in the last chapter. Now, $E\left(\dfrac{(n-1)S^2}{\sigma^2}\right) = n - 1$ since the mean of a

chi-square is its degrees of freedom. $E\left(\dfrac{(n-1)S^2}{\sigma^2}\right) = \dfrac{(n-1)}{\sigma^2}E(S^2) = n - 1$. Solving for $E(S^2)$, we obtain

$E(S^2) = \dfrac{n-1}{n-1}\sigma^2 = \sigma^2$ and hence S^2 is an unbiased estimator of σ^2.

Similarly, it can be shown that $E(\hat{P}) = P$, that is the sample proportion is an unbiased estimator of the population proportion. It can be shown that S is a biased estimator of σ. Reiterating a point made earlier, whether using $\bar{X}$ as an estimator of μ, $\hat{P}$ as an estimator of P, or S^2 as an estimator of σ^2, one can view the estimator as a random variable or the estimate as a single number. Before taking the sample, we know certain properties of the estimator. After the sample has been taken, we have confidence in the point estimate because of the properties the estimator has.

Recalling the central limit theorem, we know that for large $n(n \geq 30)$, $\bar{X}$ has a normal distribution with mean μ and standard error, $\sigma_{\bar{x}} = \dfrac{\sigma}{\sqrt{n}}$. The random variable $\dfrac{\bar{X} - \mu}{\sigma/\sqrt{n}}$ has a standard normal distribution that is represented by the letter Z. The notation $z_{\alpha/2}$ represents the standard normal value, with area $\alpha/2$ under the standard normal curve to its right. There will be another value $-z_{\alpha/2}$ with $\alpha/2$ area to its left. There will be $(1 - \alpha)$ area under the standard normal curve between $-z_{\alpha/2}$ and $z_{\alpha/2}$. An example will help clarify this notation.

EXAMPLE 6.1   Suppose $1 - \alpha = 0.90$, then $\alpha = 0.10$ and $\alpha/2 = 0.05$. Using EXCEL, =NORMSINV(0.05) gives -1.645 and =NORMSINV(0.95) gives 1.645. There are two values (-1.645 and 1.645) with $1 - \alpha = 0.90$ between them.

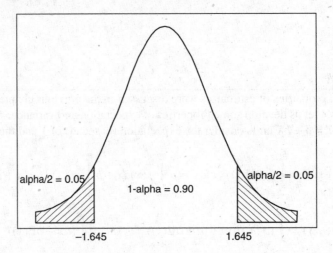

Note that $-Z_{0.05} = -1.645$ and $Z_{0.05} = 1.645$.

We have $-z_{\alpha/2} < \dfrac{\bar{X} - \mu}{\sigma/\sqrt{n}} < z_{\alpha/2}$ with probability $(1 - \alpha)$. The inequality is equivalent to $|\bar{X} - \mu| < z_{\alpha/2}\dfrac{\sigma}{\sqrt{n}}$. The quantity $|\bar{X} - \mu|$ is called the **maximum error of estimate**. It is the absolute difference between the population mean and the estimate of it. The maximum error of estimate E is $z_{\alpha/2}\dfrac{\sigma}{\sqrt{n}}$. If the value of $z_{\alpha/2}$ is determined for $1 - \alpha = 0.95$, it is found that $z_{0.025} = 1.96$, and if the value of $z_{\alpha/2}$ is determined for $1 - \alpha = 0.99$, it is found that $z_{0.005} = 2.575$. (The student should confirm these values using EXCEL.)

EXAMPLE 6.2   It is of interest to find the average time spent on the Internet per week by engineers. Fifty engineers are randomly selected and on the basis of experience and other studies it is felt that $\sigma = 5$ hours. What can be asserted with probability 0.95 about the maximum error of estimate of the survey?

$$E = 1.96\frac{5}{\sqrt{50}} = 1.39$$

The survey will have, with probability 0.95, an error of at most 1.39 hours.

What is the Difference Between Confidence and Probability?

Consider the past example further. To speak of taking samples and to say that the probability is 0.95 that the maximum error will be 1.39 is to talk about the probabilistic properties of this technique. Now if the sample is actually taken and $\bar{X} = 20.5$ hours is found, then $-z_{\alpha/2} < \dfrac{\bar{X} - \mu}{\sigma/\sqrt{n}} < z_{\alpha/2}$ is equivalent to $19.11 < \mu < 21.89$ and this statement does not have 95% probability of being true. The statement is either false or true and has probability either 0 or 1. Rather than say the probability is 95% that $19.11 < \mu < 21.89$, we say that we are 95% confident about the statement $19.11 < \mu < 21.89$. In general, probability statements are made about future values of the random variable. Confidence statements are made once the data have been collected.

Suppose we wish to specify the maximum error and ask the question "What sample size will give this error?" The answer is found in the solution of the equation $E = z_{\alpha/2} \dfrac{\sigma}{\sqrt{n}}$ for n. If both sides of the equation are squared and solved for n, the solution is

$$n = \left[\frac{z_{\alpha/2}\sigma}{E} \right]^2$$

EXAMPLE 6.3 It is of interest to find the average time spent on the Internet per week by engineers. Use 5 as the estimated value of σ. Find the sample size needed to be 95% confident that the sample mean will be within 1 hour of the population mean.

$$n = \left[\frac{z_{\alpha/2}\sigma}{E} \right]^2 = \left[\frac{1.96(5)}{1} \right]^2 = 96.04$$

Use a sample of size 97. Always round up to the next integer.

Now consider that only small samples are available and that sampling from a normal population is a reasonable assumption. Also assume the sample standard deviation is used in place of the population standard deviation. The random variable $t_\nu = \dfrac{\bar{x} - \mu}{s/\sqrt{n}}$ may be shown to have a Student's t distribution with $\nu = n - 1$ degrees of freedom. The probability will be $(1 - \alpha)$ that $|\bar{x} - \mu| < t_{\alpha/2} \dfrac{S}{\sqrt{n}}$ or that the maximum error of estimate is $E = t_{\alpha/2} \dfrac{S}{\sqrt{n}}$. Now, when the sample is actually taken, and the numbers computed, it can be asserted with $(1 - \alpha) \times 100$ % confidence that the maximum error of estimate is $E = t_{\alpha/2} \dfrac{s}{\sqrt{n}}$.

EXAMPLE 6.4 A computer engineer wished to estimate the mean time between failures of a disk drive. The engineer recorded the time between failures for a sample of 10 disk drive failures. Give the estimated maximum error of estimate with 99% confidence. The following times between failures were recorded in hours:

$$1870 \quad 2140 \quad 1711 \quad 2039 \quad 1894 \quad 1677 \quad 1640 \quad 1700 \quad 1969 \quad 1750$$

First, the data is tested to see if it is reasonable to assume that the data came from a normally distributed population. The Kolmogorov–Smirnov test gives a p-value > 0.15. Thus, it is reasonable to assume normality.

A descriptive statistics routine gives the sample mean equal to 1839.1 hours, a standard deviation equal to 170.3 hours, and a standard error equal to 53.9. The 0.01 Student's t-value with 9 degrees of freedom is given by =TINV(0.01,9) = 3.25. The TINV function takes two parameters. The first is the area in the two tails. The second is the degrees of freedom.

The 99% confidence maximum error of estimate is $3.25(53.9) = 175.2$ hours.

Interval Estimation

Rather than give a point estimate, it is often preferable to give an interval. The interval not only gives a feeling for the central value but also gives an idea for how far away from the central value the parameter values might lie. First, consider the case when n is large and σ is known. If the inequality $-z_{\alpha/2} < \dfrac{\bar{X} - \mu}{\sigma/\sqrt{n}} < z_{\alpha/2}$ is solved for μ using simple algebra, we get

$$\bar{x} - z_{\alpha/2} \frac{\sigma}{\sqrt{n}} < \mu < \bar{x} + z_{\alpha/2} \frac{\sigma}{\sqrt{n}}$$

This is generally referred to as the **large sample confidence interval** or the **Z-interval**. If σ is unknown, it is replaced by the sample standard deviation and the interval takes on the following form:

$$\bar{x} - z_{\alpha/2} \frac{s}{\sqrt{n}} < \mu < \bar{x} + z_{\alpha/2} \frac{s}{\sqrt{n}}$$

The interval is referred to as a **confidence interval**. The $(1 - \alpha)$ is called the **degree of confidence**. The endpoints of the interval are called **confidence limits**.

EXAMPLE 6.5 The lifetime of drill bits is of interest. The lifetime of a drill bit is defined as the number of holes drilled before failure. The data in Figure 6.1 gives the lifetimes of 40 drill bits.

174	149	169	150
101	145	134	104
137	131	113	161
124	123	135	124
147	139	137	144
106	128	131	137
143	118	135	109
153	122	111	137
132	140	149	138
141	150	140	118

Figure 6.1   Lifetime of 40 drill bits.

Find a 90% confidence interval for the mean lifetime using EXCEL, MINITAB, SAS, SPSS, and STATISTIX.

EXCEL   Figure 6.2 is a dialog box to be filled out for finding a $(1 - \alpha)$ confidence interval for the population mean. If you are interested in a 90% confidence interval, the alpha is equal to 0.10. The standard deviation is found to be 16.77. The sample size is 40. The EXCEL function CONFIDENCE finds $E = z_{\alpha/2} \dfrac{\sigma}{\sqrt{n}} = 1.645(16.77)/\sqrt{40} = 4.361$. To find the 90% interval, add and subtract E from the sample mean which equals 134.42. The 90% confidence interval thus extends from $134.42 - 4.36 = 130.06$ to $134.42 + 4.36 = 138.78$.

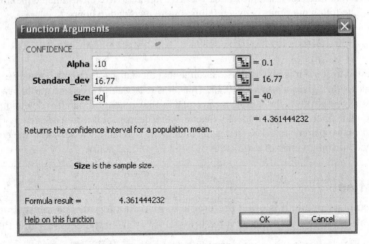

Figure 6.2   EXCEL dialog box for computing E.

MINITAB   Give the pulldown **Stat** $\Rightarrow$ **Basic Statistics** $\Rightarrow$ **1-sample z**. This gives Figure 6.3 which is filled out as shown.

Under options choose confidence level equal to 90%. This gives the following output:

```
One-Sample Z: lifetime

The assumed standard deviation = 16.77

Variable    N     Mean      StDev     SE Mean     90% CI

lifetime    40    134.420   16.766    2.652       130.058, 138.781)
```

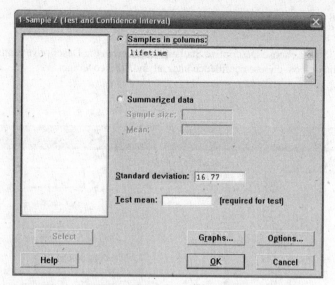

Figure 6.3   MINITAB dialog box for computing confidence
interval for μ.

SAS

```
One Sample Z Test for a Mean
Sample Statistics for lifetime

  N      Mean      Std. Dev.    Std. Error
-------------------------------------------

  40     134.48    16.75        2.65
```

With a specified known standard deviation of 16.77

90% Confidence Interval for the Mean

```
   Lower Limit      Upper Limit
   -----------      -----------

   130.11           138.84
```

The 90% confidence interval is shown in bold.

SPSS   The pulldown **Analyze $\Rightarrow$ Compare means $\Rightarrow$ 1 sample t test** gives the 1-sample *t*-test dialog box. In that box choose options and as an option choose a 90% confidence interval. This gives the following output:

One-Sample Statistics

	N	Mean	Std. Deviation	Std. Error Mean
Lifetime	40	134.48	16.750	2.648

One-Sample Test

			Test Value = 0			
					90% Confidence Interval of the Difference	
	t	df	Sig. (2-tailed)	Mean Difference	Lower	Upper
Lifetime	50.776	39	.000	134.475	130.01	138.94

The reason the interval differs some from EXCEL and MINITAB is that *t* rather than *z* is used in the formula for the confidence interval.

STATISTIX   In STATISTIX, choose Descriptive Statistics. That gives the Descriptive Statistics dialog box shown in Figure 6.4. In the dialog box, choose confidence interval and 90% coverage.

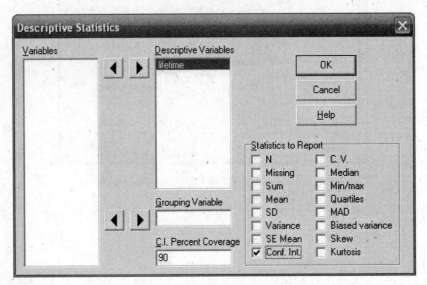

Figure 6.4   STATISTIX dialog box for 90% confidence interval.

The following output is given.

```
Statistix 8

Descriptive Statistics

Variable       Lo 90% CI       Up 90% CI
lifetime       130.01          138.94
```

The package STATISTIX is, like SPSS, using the *t* distribution, and so the answer is a little different from MINITAB and EXCEL.

When working with a small sample from a normally distributed population and using *S* as an estimate of σ, we have the following as a confidence interval for μ:

$$\bar{x} - t_{\alpha/2}\frac{S}{\sqrt{n}} < \mu < \bar{x} + t_{\alpha/2}\frac{S}{\sqrt{n}}$$

This is generally referred to as the **small sample confidence interval** or the ***t*-interval**. When the sample size is small, the *t*-interval must be used. If a *z*-value rather than a *t*-value is used in the *t*-interval, the interval will be too narrow.

EXAMPLE 6.6   The lifetime of drill bits is of interest. The lifetime of a drill bit is defined as the number of holes drilled before failure. The data in Figure 6.5 is the lifetimes of 12 drill bits. Find a 95% confidence interval for μ using EXCEL, MINITAB, SAS, SPSS, and STATISTIX.

174	149	169	150
101	145	134	104
137	131	113	161

Figure 6.5   Lifetime of 12 drill bits.

First, test for normality. Remember, the sample is small and the theory underlying the confidence interval requires the normality assumption.

```
Statistix 8.0

Shapiro-Wilk Normality Test

Variable      N     W          P
lifetime      12    0.9524     0.6725
```

Using the Shapiro–Wilk test of STATISTIX, we do not reject normality because of the large p-value.

Figure 6.6 shows the EXCEL computations for the 95% confidence interval for the population mean. The drill bit lifetimes are shown in A2:A13, the average is computed in C2, the standard deviation is computed in C3, the standard error in C4, the 95% t-value in C5, the lower 95% confidence limit in C7, and the upper 95% confidence limit in C8.

	A	B	C	D
1	Lifetime			
2	174		139	=AVERAGE(A2:A13)
3	101		23.93932	=STDEV(A2:A13)
4	137		6.910686	=C3/SQRT(12)
5	149		2.200985	=TINV(0.05,11)
6	145			
7	131		123.7897	=C2-C4*C5
8	169		154.2103	=C2+C4*C5
9	134			
10	113			
11	150			
12	104			
13	161			

Figure 6.6   EXCEL computations for 95% confidence interval for μ.

Figure 6.7 illustrates an **incorrect use** of EXCEL. The dialog box shown in this figure is intended only for large samples, that is samples > 30. The quantity E is computed and given as 13.545. Here, $E = z_{\alpha/2} \dfrac{\sigma}{\sqrt{n}} = 1.96(6.911) = 13.545$. The small sample confidence interval requires $E = t_{\alpha/2} \dfrac{S}{\sqrt{n}} = 2.200985(6.911) = 15.211$. The interval using the value of E in Figure 6.7 would be too narrow. It would give the internal (125.455, 152.545) rather than the correct interval (123.7897, 154.2301). **Remember, the software user is responsible for using the software correctly.**

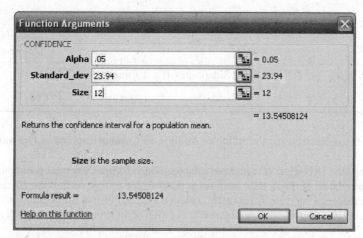

Figure 6.7   An incorrect use of an EXCEL function.

If MINITAB is used to calculate the 95% confidence interval, enter the 12 data values in column C1 and use the pulldown **Stat ⇒ Basic Statistics ⇒ 1-sample t**. The output is

```
One-Sample T: lifetime

Variable    N     Mean     StDev    SE Mean    95% CI
lifetime    12   139.000   23.939    6.911    (123.790, 154.210)
```

You have to remember that small samples is the reason you perform the 1-sample *t*. The confidence interval is the same as the one you computed using EXCEL. The STATISTIX output is

```
Statistix 8.
Descriptive Statistics

Variable    Lo 95% CI   Up 95% CI
lifetime    123.79      154.21
```

The SAS solution is

```
One Sample t-test for a Mean
Sample Statistics for lifetime

    N     Mean    Std. Dev.   Std. Error
    ------------------------------------------
    12   139.00    23.94       6.91

95 % Confidence Interval for the Mean

    Lower Limit:   123.79
    Upper Limit:   154.21
```

The SAS solution is the same as EXCEL, MINITAB, and STATISTIX. The SPSS solution is

One-Sample Statistics

	N	Mean	Std. Deviation	Std. Error Mean
Lifetime	12	139.00	23.939	6.911

One-Sample Test

			Test Value = 0		90% Confidence Interval of the Difference	
	t	*df*	Sig. (2-tailed)	Mean Difference	Lower	Upper
Lifetime	20.114	11	.000	139.000	123.79	154.21

The SPSS solution is reached by using the pulldown **Analyze ⇒ Compare means ⇒ One sample t test**.

EXAMPLE 6.7 Have MINITAB select 20 simulated samples of size 10 from a normal population with mean $\mu = 139$ and standard deviation equal to 24. Form 80% confidence intervals for μ from each of the 20 samples. Determine the percentage of the 20 confidence intervals that actually contain the mean (139) of the population.

The 20 samples of size 10 each are shown in Figure 6.8. For each of the samples, the Student's *t*-interval is calculated $\left(\bar{x} - t_{\alpha/2} \frac{S}{\sqrt{n}}, \bar{x} + t_{\alpha/2} \frac{S}{\sqrt{n}} \right)$. The 20 intervals are shown in Figure 6.9. The intervals that do not contain the value of the mean (139) are shown with an asterisk beside the sample number in Figure 6.9.

We find that 15% of the 80% confidence intervals do not contain the population mean. This means that 85% contain the mean. We say that the confidence interval captures the population mean if the population mean is found inside the interval. In practice, you will not know whether a given interval captures the mean or not. Knowing that theory predicts

Samples									
1	2	3	4	5	6	7	8	9	10
107	139	150	150	119	126	60	131	130	126
173	149	112	171	104	105	151	163	154	140
106	123	102	148	143	154	137	155	162	129
116	181	123	133	149	177	165	124	121	97
97	126	152	105	171	117	149	182	170	201
134	118	136	111	188	153	134	105	111	160
156	161	153	159	153	114	149	107	134	155
134	136	162	131	164	191	135	92	158	131
152	86	158	139	150	140	130	131	135	83
137	129	166	106	111	104	139	121	135	182

Samples									
11	12	13	14	15	16	17	18	19	20
138	141	135	128	157	107	114	173	161	108
118	175	141	176	127	108	110	128	135	153
187	134	197	142	142	146	169	172	121	151
142	200	150	168	94	111	115	111	157	158
99	133	130	130	110	127	158	151	105	137
161	147	125	141	152	162	134	143	156	126
153	132	138	161	135	104	140	171	147	129
152	121	122	156	106	109	99	138	114	97
152	198	100	147	130	124	155	136	122	137
109	109	166	167	124	117	147	144	153	128

Figure 6.8   The 20 samples ($n = 10$) from a normal population with $\mu = 139$ and $\sigma = 24$ are simulated by MINITAB.

Sample	Confidence interval	Sample	Confidence interval
1	(120.422,141.978)	11	(129.594,152.606)
2	(123.549,146.051)	12	(135.247,162.753)
3	(131.718,151.082)	13	(128.827,151.973)
4	(125.393,145.207)	14*	(144.372,158.828)
5	(133.440,156.960)	15*	(118.897,136.503)
6	(124.882,151.318)	16*	(113.173,129.827)
7	(122.514,147.286)	17	(123.782,144.418)
8	(118.796,143.404)	18	(137.742,155.658)
9	(132.648,149.352)	19	(128.195,146.005)
10	(124.696,156.104)	20	(123.914,140.886)

Figure 6.9   Of the 20 intervals, 17 contain the mean (139).

that 80% of the intervals will capture the mean is what gives the statistician the 80% confidence that he or she has. If this simulation study were carried out over thousands of times, we would see the percentage approach the 80% figure. You are invited to verify the confidence intervals given in Figure 6.9.

Basic Concepts of Testing Hypotheses

We shall introduce the concept of testing hypotheses with a very simple example involving a coin. Then, in later sections, we will move into engineering examples to test means, proportions, and variances.

EXAMPLE 6.8   Suppose we have a nickel and we think that it is loaded or biased so that there is not a 50/50 chance of heads/tails. Furthermore, we suspect that it is biased so that a head is more likely to occur than a tail. We will use the following decision-making test procedure. Let p be the probability of a head turning up when the coin is tossed once. We set up two hypotheses. One is called the **null hypothesis** and the other is called the **research** or **alternative hypothesis**. Our research hypothesis denoted H_a: is that $p > 0.5$ or that the coin is biased in favor of a head occurring. This research hypothesis is called a **one-sided** or **one-tail test**. The null hypothesis is that the coin is not biased or H_o: $p = 0.5$. The hypothesis system is H_o: $p = 0.5$ versus H_a: $p > 0.5$. In order to test this hypothesis we agree to flip the coin 16 times and if 12 or more heads occur, then we will reject the null hypothesis that the coin is fair ($p = 0.5$). Obtaining 12 or more heads will support our belief that the coin is biased. If the coin is unbiased, we would expect somewhere near 8 heads to occur. If X represents the number of heads to occur when the coin is tossed 16 times, then X has a binomial distribution. The **rejection region** is defined to be the outcomes for which we shall reject the null hypothesis that the coin is fair. The **non-rejection region** is the outcomes for which we shall not reject the null hypothesis. There are two other research hypothesis that we shall be concerned with. The hypothesis H_a: $p < 0.5$ means that we are interested in the research hypothesis that the coin is biased so that a head is less likely to occur than a tail. This is also called a **one-tail test**. Finally, there is a **two-tail test**. It is expressed as H_a: $p \neq 0.5$. This says we suspect that the coin is biased, but we are not sure which way.

We now have rejection region $X \geq 12$ and non-rejection region $X < 12$. The variable X is called our **test statistic**.

There are two types of errors that are possible using this decision-making procedure. They are illustrated in Figure 6.10.

Decision	Null true, $p = 0.5$	Null false, $p > 0.5$
Reject null	Type I error	No error made
Do not reject null	No error made	Type II error

Figure 6.10   Type I and Type II error defined.

A **Type I error** occurs when the null is true, but it is rejected. A **Type II** error occurs when the null is false, but it is not rejected. The probability of making a Type I error is represented by the Greek letter α and the probability of making a Type II error is represented by the Greek letter β. For this example, $\alpha = P(X \geq 12$ when $p = 0.5)$. Using EXCEL to evaluate this, we find $\alpha = 1 - \text{BINOMDIST}(11,16,0.5,1) = 0.038$. To calculate β, we choose $p > 0.5$. The value of β depends on which value is chosen for p. The Type II error is $P(X \leq 11$ for $p > 0.5)$. Figure 6.11 gives different values of β calculated for different values of p calculated using EXCEL. To build the results shown in Figure 6.11, place the values in the column under p from A1 to A9. Enter $=\text{BINOMDIST}(11,16,A1,1)$ in B1 and click-and-drag to B9.

p	β	$1 - \beta$
0.55	0.914691	0.085309
0.6	0.833433	0.166567
0.65	0.710793	0.289207
0.7	0.550096	0.449904
0.75	0.369814	0.630186
0.8	0.201755	0.798245
0.85	0.079051	0.920949
0.9	0.017004	0.982996
0.95	0.000857	0.999143

Figure 6.11   Calculations of Type II error and power values for H_a: $p > 0.5$.

The quantities in the column labeled $1 - \beta$ are called power values. The **power of a test** is the probability of rejecting the null for various values of p in the alternative. For example, the row where $p = 0.75$, $\beta = 0.369814$, and $1 - \beta = 0.630186$ tells us if the coin is biased so that the probability of a head is 0.75, the probability of not rejecting the null hypothesis is 0.369, and the probability of rejecting the null hypothesis is 0.630. The power curve is plotted in Figure 6.12.

It is not possible to make perfect decisions about the coin by flipping it 16 times. The coin could be balanced, but you might obtain an unusually large number of heads leading you to make a Type I error. Or the coin might be biased, but you obtain somewhere near 8 heads, leading you to make a Type II error. What the discipline of statistics helps you do is make the right decision often and to know what the probabilities of correct decisions are.

When conducting a test of hypothesis, these are the recommended steps to follow:

1. State the null and research hypotheses: H_o: $p = 0.5$ versus H_a: $p > 0.5$

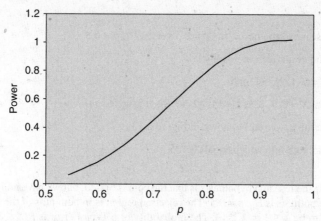

Figure 6.12 Power curve for the hypothesis test H_o: $p = 0.5$ versus H_a: $P > 0.5$.

2. Give the rejection and non-rejection regions:

Rejection region: $X = 12, 13, 14, 15$, and 16

Non-rejection region: $X = 0, 1, 2, 3, 4, 5, 6, 7, 8, 9, 10$, and 11

3. Calculate α and give the power curve shown in Figure 6.11.

$\alpha = 1\text{-BINOMDIST}(11,16,0.5,1) = 0.038$

Step 4 is performed only after the other steps have been completed.

4. Perform the experiment, give the value of the test statistic, and your conclusion.

After step 4 is performed, the **observed significance level** or **p-value** is computed. The observed significance level or p-value is the probability of observing a value of the test statistic that is at least as contradictory to the null hypothesis and supportive of the research hypothesis as the one computed from the sample data. Suppose the experiment is conducted and 11 heads are observed out of 16 flips. The p-value is the probability of 11 or more heads. Using EXCEL, the p-value is given by $=1\text{-BINOMDIST}(10,16,0.5,1)$ which equals 0.105.

When researchers follow the p-value technique for testing hypotheses, they follow this procedure: Compare the p-value with the α value. If the p-value $< \alpha$, reject the null. Otherwise, do not reject the null. In the present case, $\alpha = 0.038$ and the p-value $> \alpha$. You are unable to reject the null hypothesis using the p-value approach.

EXAMPLE 6.9 Suppose the research hypothesis is that the coin is biased so that a head is less likely to occur than a tail. In this case the research hypothesis is H_a: $p < 0.5$. In this case, the left-hand tail becomes the rejection region. This time the rejection region is chosen to be $X \leq 5$. The probability of a Type I error is $P(X \leq 5$ when $p = 0.5)$. Using EXCEL, this probability is given by $=\text{BINOMDIST}(5,16,0.5,1)$ which equals 0.105.

The value of β depends on which value is chosen for p. The Type II error is $P(X \geq 6$ for $p < 0.5)$. Figure 6.13 gives different values of β calculated for different values of p using EXCEL. To build the results shown in Figure 6.13, place the values in the column under p from A1 to A9. Enter $=\text{BINOMDIST}(11,16,A1,1)$ in B1 and click-and-drag to B9.

p	β	$1 - \beta$
0.05	8.08995E−05	0.9999191
0.1	0.003296751	0.996703249
0.15	0.023544381	0.976455619
0.2	0.081687888	0.918312112
0.25	0.189654573	0.810345427
0.3	0.340217674	0.659782326
0.35	0.510036428	0.489963572
0.4	0.671159587	0.328840413
0.45	0.80240244	0.19759756

Figure 6.13 Calculations of Type II error and power values for H_a: $p < 0.5$.

Summarizing:

1. State the null and research hypotheses: $H_o: p = 0.5$ versus $H_a: p < 0.5$

2. Give the rejection and non-rejection regions:

   Rejection region: $X = 0, 1, 2, 3, 4,$ or 5

   Non-rejection region: $X = 6, 7, 8, 9, 10, 11, 12, 13, 14, 15,$ or 16

3. Calculate α and give the power curve shown in Figure 6.13.

   $\alpha = $ =BINOMDIST(5,16,0.5,1) which equals 0.105.

EXAMPLE 6.10 Suppose the research hypothesis is that the coin is biased, but we have no idea which way it is biased. In this case the research hypothesis is $H_a: p \neq 0.5$. The rejection region is in both tails of the distribution. This time the rejection region is chosen to be $X \leq 3$ or $X \geq 13$. The probability of a Type I error is $P(X \leq 3$ or $X \geq 13$ when $p = 0.5)$. Using EXCEL, this probability is given by =BINOMDIST(3,16,0.5,1) + (1-BINOMDIST(12,16,0.5,1)) or 0.021.

The value of β depends on which value is chosen for p. The Type II error is $P(4 \leq X \leq 12$ for $p \neq 0.5)$. Figure 6.14 gives different values of β calculated for different values of p using EXCEL. To build the results shown in Figure 6.14, place the values in the column under p from A1 to A19. Enter =BINOMDIST(12,16,A1,1) – BINOMDIST(3,16,A1,1) in B1 and click-and-drag to B19.

p	β	$1 - \beta$
0.05	0.007003908	0.992996092
0.1	0.068406174	0.931593826
0.15	0.21010929	0.78989071
0.2	0.401865427	0.598134573
0.25	0.595009107	0.404990893
0.3	0.754110535	0.245889465
0.35	0.865935913	0.134064087
0.4	0.933914808	0.066085192
0.45	0.968418793	0.031581207
0.5	0.978729248	0.021270752
0.55	0.968418793	0.031581207
0.6	0.933914808	0.066085192
0.65	0.865935913	0.134064087
0.7	0.754110535	0.245889465
0.75	0.595009107	0.404990893
0.8	0.401865427	0.598134573
0.85	0.21010929	0.78989071
0.9	0.068406174	0.931593826
0.95	0.007003908	0.992996092

Figure 6.14　Calculations of Type II error and power values for $H_a: p \neq 0.5$.

The power curve for this test is shown in Figure 6.15.
Summarizing:

1. State the null and research hypotheses: $H_o: p = 0.5$ versus $H_a: p \neq 0.5$

2. Give the rejection and non-rejection regions:

   Rejection region: $X = 0, 1, 2, 3, 13, 14, 15,$ or 16

   Non-rejection region: $X = 4, 5, 6, 7, 8, 9, 10, 11, 12$

3. Calculate α and give the power curve shown in Figure 6.15.

   $\alpha = $ BINOMDIST(3,16,0.5,1) + (1-BINOMDIST(12,16,0.5,1)) or 0.021.

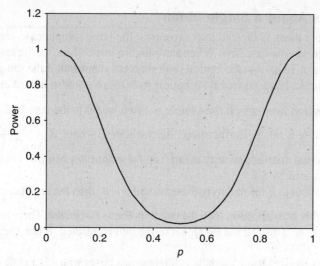

Figure 6.15   Power curve for the hypothesis test H_o: $p = 0.5$
versus Ha: $P \neq 0.5$.

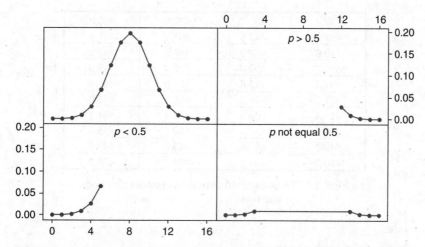

Figure 6.16   The complete binomial distribution and the one-sided and two-sided rejection regions.

Figure 6.16 gives the three cases we have considered. The upper left graph gives the complete binomial distribution for $n = 16$ and $p = 0.5$. The upper right shows the rejection region that was chosen for the research hypothesis that $p > 0.5$. The lower left shows the rejection region that was chosen for the research hypothesis that $p < 0.5$. The lower right shows the rejection that was chosen for the research hypothesis that $p \neq 0.5$. This graph shows the reason for the terms one-tail test and two-tail test. In the one-tail test either the upper tail or the lower tail of the binomial distribution was used for the rejection region. For this reason, one sometimes sees the terms **upper-tail test** and **lower-tail test**. When both tails of the distribution are used, this is called the **two-tail test** because both tails of the binomial distribution are being utilized.

The fact that the experiment is performed and outcomes in the tails occur is evidence that the coins might be biased. However, these outcomes could occur even with a fair coin. This is what the curves in Figure 6.16 show. They show the probabilities of getting outcomes in the tails with a fair coin.

You are encouraged to read and understand the logic involved in this section. The logic in this section is the same as that involved with tests involving means, proportions, and variances in later sections. There are quite a number of terms and definitions in this section that are used throughout the remainder of statistics. Remember, it is not the mathematical underpinnings that are hard in statistics but the concepts and ideas that are difficult.

Testing Hypotheses About a Single Mean

Testing hypotheses about a mean is divided into two cases: the large sample case and the small sample case. We shall discuss the large sample case first. When samples are larger than 30, the test of hypothesis is based on the central limit theorem. To review, the central limit theorem states that if the sample size, n, is larger than 30, $\overline{X}$, regarded as a variable, has a normal distribution with mean μ and standard error $\sigma_{\overline{x}} = \dfrac{\sigma}{\sqrt{n}}$, where μ is the mean of the population from which the sample is taken and σ is the standard deviation of the population from which the sample is taken. Furthermore, the variable Z where $Z = \dfrac{\overline{X} - \mu}{\sigma_{\overline{x}}}$ has a standard normal distribution, that is, a normal distribution with mean 0 and standard deviation 1. (In the large sample case, if σ is not known, it is estimated by S.)

In testing hypotheses about μ, if the null hypotheses is that $\mu = \mu_o$ then the test statistic is $Z = \dfrac{\overline{X} - \mu_o}{\sigma_{\overline{x}}}$. If the test statistic takes on a highly unusual value, then the null hypothesis is rejected. There are three possible research hypotheses: $\mu < \mu_o$, $\mu > \mu_o$, or $\mu \neq \mu_o$. If the p-value $< \alpha$, the null is rejected. If not, the null is not rejected.

EXAMPLE 6.11　It is hypothesized that a machine is filling cereal boxes with 453 grams of cereal (on the average). The research hypothesis is $H_a: \mu \neq 453$. The test is to be performed with $\alpha = 0.05$. A sample of 40 boxes gave the results shown in Figure 6.17.

452.2	450.7	453.8	450.3
451.0	447.5	451.1	447.9
449.6	452.9	448.5	449.6
450.7	450.0	449.4	449.0
451.4	453.8	451.8	451.9
451.3	454.2	452.1	448.3
453.4	448.5	448.4	451.1
448.2	449.3	450.7	455.0
442.3	451.8	449.9	447.9
450.3	447.3	451.3	453.2

Figure 6.17　Weights of 40 cereal boxes selected from production.

```
Test of mu = 453 vs not = 453

The assumed standard deviation = 2.4

Variable   N    Mean     StDev   SE Mean   95% CI                Z      P
C1         40   450.440  2.379   0.379     (449.696, 451.184)   -6.75   0.000
```

The MINITAB analysis is shown below Figure 6.17. The standard deviation of the weights is calculated and it is found that $s = 2.4$. The routine, called the Z-test, is requested by the pulldown **Stat ⇒ Basic Statistics ⇒ 1-sample Z**. The calculated test statistic is $z = -6.75$. The p-value is 0.000. It would be safe to conclude that the fill is below 453 on the average.

The EXCEL solution is shown in Figure 6.18. The data are shown in A1:D10. The computation of the test statistic is shown in F1:F4. The computation of the p-value is shown in F6.

A	B	C	D	E	F	G
452.2	450.7	453.8	450.3		450.44	AVERAGE (A1:D10)
451	447.5	451.1	447.9		2.379161671	STDEV(A1:D10)
449.6	452.9	448.5	449.6		0.37617849	F2/SQRT(40)
450.7	450	449.4	449		−6.8052801	(F1−453)/F3
451.4	453.8	451.8	451.9			
451.3	454.2	452.1	448.3		1.01049E−11	2*NORMSDIST(−6.805)
453.4	448.5	448.4	451.1			

Figure 6.18　EXCEL computation of test statistic and p-value.

EXAMPLE 6.12 It is hypothesized that the mean time per week that engineers spend on the Internet looking up job-related information is 15 hours. The research hypothesis is that the mean is greater than 15 hours. Survey results are shown in Figure 6.19.

13	12	19	17	14	14	14
14	15	15	16	17	20	17
14	12	19	14	13	15	12
17	11	14	11	19	18	15
16	12	15	17	16	11	12

Figure 6.19   Hours per week spent on the Internet.

One-Sample T Test

Null Hypothesis: mu = 15

Alternative Hyp: mu > 15

			95% Conf Interval				
Variable	Mean	SE	Lower	Upper	T	DF	P
hours	14.857	0.4224	13.999	15.715	−0.34	34	0.6314

The STATISTIX output shown above gives a *t*-value rather than a *z*. When *n* is larger than 30, the *t* distribution and the *z* distribution may be used interchangeably. Because of the large *p*-value (0.6314), there is no evidence to reject the null. Note that we have not proved that $\mu = 15$. However, the data do not tend to make us doubt that $\mu = 15$.

When samples are small, the distribution of the test statistic changes and an assumption is made that must be checked. For samples of size 30 or smaller it is assumed that the sample is taken from a normal distribution. The test statistic $t_v = \dfrac{\bar{x} - \mu}{s/\sqrt{n}}$ has a Student's *t* distribution with $v = (n-1)$ degrees of freedom. There are three possible research hypotheses: $\mu < \mu_o$, $\mu > \mu_o$, or $\mu \neq \mu_o$. If the *p*-value $< \alpha$, the null is rejected. If not, the null is not rejected.

EXAMPLE 6.13 Highway engineers were studying the causes of rising highway death rates. One study looked at the time spent on cell phones per week while operating an automobile. The time spent on cell phones per week in hours for 10 randomly chosen individuals who drive and use cell phones were recorded in a study. The data is

$$5.2 \quad 5.5 \quad 4.9 \quad 2.9 \quad 4.1 \quad 4.4 \quad 2.7 \quad 6.8 \quad 4.3 \quad 3.5$$

The research hypothesis of interest was that $\mu < 7$. Alpha was selected to be 0.01. Conduct the test of hypothesis using STATISTIX. First the Shapiro–Wilk test of normality was conducted. The *p*-value was 0.9011 indicating quite a high probability that the sample could have come from a normal distribution. Normality of the sample was not rejected.

Shapiro-Wilk Normality Test

Variable	N	W	P
time	10	0.9711	0.9011

The STATISTIX pulldown menu **Statistics ⇒ One, Two, Multi sample tests ⇒ One sample t test** gave the One sample *t*-test dialog box. Time was moved into variable slot, 7 was put into the hypothesis slot, and Less than into the alternative hypothesis slot. The results were:

One-Sample T Test

Null Hypothesis: mu = 7

Alternative Hyp: mu < 7

95% Conf Interval

Variable	Mean	SE	Lower	Upper	T	DF	P
time	4.4300	0.3930	3.5410	5.3190	6.54	9	0.0001

The computed test statistic was $t = -6.54$ with 9 degrees of freedom. There is 1 chance out of 10,000 that a sample with a t-value of -6.54 or smaller could be obtained if the mean of the population equals 7 hours. The decision is to reject the null hypothesis in favor of the research hypothesis.

Suppose the engineer is faced with this problem but has only EXCEL available. Figure 6.20 shows the variable name **time** in cell A1 of an EXCEL worksheet. The data is in A2:A11. The computation of the test statistic is shown in B2:B5 and the computation of the p-value is shown in B7.

Time	Computation	Function
5.2	4.43	AVERAGE(A2:A11)
5.5	1.242801	STDEV(A2:A11)
4.9	0.393008	C3/SQRT(10)
2.9	−6.5393	(C2–7)/C4
4.1		
4.4	5.32E–05	TDIST(6.5393,9,1)
2.7		
6.8		
4.3		
3.5		

Figure 6.20   EXCEL t-test.

Note that the EXCEL analysis results are the same as the STATISTIX analysis.
The SAS t-test for this example is as follows:

```
Sample Statistics for time

N    Mean   Std. Dev.   Std. Error
-----------------------------------

10   4.43    1.24        0.39

Hypothesis Test
  Null hypothesis:    Mean of time => 7
  Alternative:        Mean of time < 7

  t Statistic      Df   Prob > t
  -------------------------------

  -6.539            9   <.0001

99 % Confidence Interval for the Mean
(Upper Bound Only)
    Lower Limit: -infinity
    Upper Limit: 5.54
```

The SPSS t-test gives the following output:

One-Sample Test

					95% Confidence Interval of the Difference	
				Test Value = 7		
	t	df	Sig. (2-tailed)	Mean Difference	Lower	Upper
Time	−6.539	9	.000	−2.5700	−3.459	−1.681

Note that SPSS always performs a two-tail test. The p-value is designated Sig. (2-tailed). If the test is one-tailed, take half of the p-value in order to get a one-tailed p-value.

The MINITAB output is

```
One-Sample T: time
Test of mu = 7 vs < 7


                                               99%
                                               Upper
Variable   N     Mean     StDev    SE Mean    Bound     T       P
time       10    4.43000  1.24280  0.39301    5.53885  -6.54   0.000
```

Relationship Between Tests of Hypotheses and Confidence Intervals

Many statisticians/engineers use a technique that involves confidence intervals to do tests of hypotheses. This technique will be explained using examples from the previous section. We will introduce one-sided confidence intervals so that one-tail tests may be illustrated also. Some of the software packages give one-sided as well as two-sided confidence intervals if requested.

EXAMPLE 6.14   It is hypothesized that a machine is filling cereal boxes with 453 grams of cereal (on the average). That is $H_o: \mu = \mu_o = 453$. The research hypothesis is $H_a: \mu \neq 453$. The test is to be performed with $\alpha = 0.05$. Consider the MINITAB output for this example.

```
Test of mu = 453 vs not = 453
The assumed standard deviation = 2.4

Variable   N     Mean     StDev   SE Mean   95% CI                Z       P
C1         40    450.440  2.379   0.379     (449.696, 451.184)   -6.75   0.000
```

The hypothesis is rejected because the *p*-value $< \alpha = 0.05$. Note also that $\mu_o = 453$ is not contained in the 95% confidence interval for μ (449.696, 451.184). This will always be true. If a null hypothesis is being tested at α, and a $(1 - \alpha)$ confidence interval for μ is formed, then, the *p*-value $< \alpha$ **if and only if** the null hypothesis value μ_o is not contained in the $(1 - \alpha)$ confidence interval for μ.

EXAMPLE 6.15   It is hypothesized that the mean time per week that engineers spend on the Internet looking up job-related information is 15 hours. The research hypothesis is that the mean is greater than 15 hours. Alpha is chosen to be 0.10.

```
One-Sample Z: hours
Test of mu = 15 vs > 15
The assumed standard deviation = 2.5

                                               90%
                                               Lower
Variable   N     Mean     StDev    SE Mean    Bound     Z       P
hours      35    14.8571  2.4987   0.4226     14.3156  -0.34   0.632
```

In this example the *p*-value $> \alpha = 0.10$ and we do not reject. All 10% of the area under the standard normal curve is put into the lower tail for this lower bound confidence interval rather than being split half below and half above. The 90% lower bound confidence interval is $(14.3156, \infty)$. Since the interval contains $\mu_o = 15$, the null would not be rejected if the confidence interval method were used.

EXAMPLE 6.16   Highway engineers were studying the causes of rising highway death rates. One study looked at the time spent on cell phones per week while operating an automobile. The time spent on cell phones per week in hours for 10 randomly chosen individuals who drive and use cell phones were recorded in a study. The data is

$$5.2 \quad 5.5 \quad 4.9 \quad 2.9 \quad 4.1 \quad 4.4 \quad 2.7 \quad 6.8 \quad 4.3 \quad 3.5$$

The research hypothesis is $H_a: \mu < 7$ and the null hypothesis is $H_o: \mu = 7$. Test at $\alpha = 0.01$.

One-Sample T: time
Test of mu = 7 vs < 7

					99% Upper		
Variable	N	Mean	StDev	SE Mean	Bound	T	P
time	10	4.43000	1.24280	0.39301	5.53885	−6.54	0.000

Since the p-value $< \alpha = 0.01$, the null is rejected. The upper bound confidence interval is $(-\infty, 5.53885)$. Since 7 is not contained in this interval, the null is rejected. All 1% of the confidence is placed into the upper part of the t-curve.

Inferences Concerning Two Means—Independent and Large Samples

Suppose we wish to compare two populations either by setting a confidence interval on the difference in their means or by testing hypothesis about the difference in their means. Population 1 has mean μ_1 and variance σ_1^2 and population 2 has mean μ_2 and variance σ_2^2. Samples of sizes n_1 and n_2, both larger than 30 and independent of each other, are taken and the difference in the sample means are considered. Theoretical studies have shown that the difference in sample means, $\bar{X}_1 - \bar{X}_2$, has a normal distribution with a

mean equal to $\mu_1 - \mu_2$ and a variance equal to $\dfrac{\sigma_1^2}{n_1} + \dfrac{\sigma_2^2}{n_2}$. Furthermore, $\dfrac{\bar{X}_1 - \bar{X}_2 - (\mu_1 - \mu_2)}{\sqrt{\sigma_1^2/n_1 + \sigma_2^2/n_2}}$ has a standard nor-

mal distribution. To test H_o: $(\mu_1 - \mu_2) = D_o$ versus any one of the three alternative hypotheses H_a: $(\mu_1 - \mu_2) <$

D_o, H_a: $(\mu_1 - \mu_2) > D_o$, or H_a: $(\mu_1 - \mu_2) \neq D_o$, use the test statistic $Z = \dfrac{\bar{X}_1 - \bar{X}_2 - D_o}{\sqrt{\sigma_1^2/n_1 + \sigma_2^2/n_2}}$. The interval

$(\bar{X}_1 - \bar{X}_2) \pm z_{\alpha/2}\sqrt{\dfrac{\sigma_1^2}{n_1} + \dfrac{\sigma_2^2}{n_2}}$ is a $(1 - \alpha)$ confidence interval for $\mu_1 - \mu_2$. In most real world applications the

population variances are unknown and the sample variances replace them. The only assumptions needed are that the samples are independent and larger than 30.

EXAMPLE 6.17　A study was conducted to compare male and female engineers entering the work force. Their GPAs were recorded at graduation and the hypotheses that $\mu_1 - \mu_2 = 0$ versus $\mu_1 - \mu_2 \neq 0$ was tested at $\alpha = 0.05$. The GPAs for 40 males and 40 females are shown in Figure 6.21.

Male					Female			
3.18	3.18	3.02	3.15		3.29	3.73	2.98	3.32
3.40	2.71	2.76	3.38		3.12	3.72	3.64	3.16
3.19	2.88	2.58	3.40		3.01	3.10	3.56	3.54
3.13	2.86	3.18	3.22		3.64	3.82	3.26	2.99
3.24	2.57	2.92	3.37		3.24	3.39	2.94	3.18
3.30	3.06	3.29	3.42		3.11	2.74	3.33	3.40
2.67	2.69	2.74	3.30		3.44	3.41	3.22	3.80
2.75	3.09	3.05	3.21		3.57	3.30	3.62	3.05
2.74	2.95	2.89	3.30		3.64	3.14	2.88	3.08
2.94	2.77	3.18	3.42		2.80	3.43	3.39	3.49

Figure 6.21　GPAs for 40 males and 40 females.

Compare the two groups using the independent samples test and using some of the standard software.

EXCEL　The pulldown **Tools** $\Rightarrow$ **data analysis** $\Rightarrow$ **Z-test: Two sample for means** gives the dialog box shown in Figure 6.22. The ranges tell where the data is identified. The variances are computed separately and entered in the dialog box. The hypothesized difference is 0 or H_o: $(\mu_1 - \mu_2) = 0$ and the research hypothesis is H_a: $(\mu_1 - \mu_2) \neq 0$. The alpha value is 0.05.

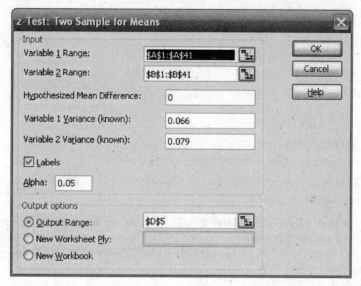

Figure 6.22   EXCEL two large independent samples *z*-test.

This dialog box gives the output shown in Figure 6.23.

	Male	Female
Mean	3.052	3.312
Known Variance	0.066	0.079
Observations	40	40
Hypothesized Mean Difference	0	
z	−4.31421	
P(Z<=z) one-tail	8.01E–06	
z Critical one-tail	1.644854	
P (Z<=z) two-tail	1.6E–05	
z Critical two-tail	1.959964	

Figure 6.23   EXCEL output for the two-sample *z*-test.

Recall the null hypothesis H_o: $(\mu_1 - \mu_2) = 0$ versus the research hypothesis H_a: $(\mu_1 - \mu_2) \neq 0$ may be tested in three different ways.

1. The two-tailed rejection region is $Z < -1.96$ or $Z > 1.96$. The computed test statistic is $z = -4.31$. Since this computed test statistic is less than −1.96 and is, therefore, in the rejection region, the null is rejected for $\alpha = 0.05$.

2. Using the *p*-value approach, compare the *p*-value = $1.6E - 05 = 0.000016$ with the pre-set $\alpha = 0.05$. Since the *p*-value is less than α, reject the null.

3. The confidence interval approach calculates the 95% two-sided confidence interval and if it does not contain $D_o = 0$, reject the null. The 95% confidence interval is $(\bar{X}_1 - \bar{X}_2) \pm z_{\alpha/2}\sqrt{\dfrac{\sigma_1^2}{n_1} + \dfrac{\sigma_2^2}{n_2}}$ or $-0.26 \pm 1.96(0.0602)$ or -0.26 ± 0.118 or $(-0.378, -0.142)$. Since 0 is not in this interval, reject the null.

The same conclusion will always be reached no matter which of the three methods of testing hypotheses is used.

When testing two means, the two-sample *t*-test to be discussed in the next section gives basically the same solution when both samples exceed 30 as the *z*-test of this section. The MINITAB two-sample *t*-test using the data in Figure 6.21 and the pulldown **Stat ⇒ Basic Statistics ⇒ 2-sample t** is as follows.

```
Two-Sample t Test and CI: male, female
Two-sample T for male vs female

          N     Mean    StDev   SE Mean
male      40    3.052   0.258   0.041
female    40    3.312   0.281   0.044

Difference = mu (male) - mu (female)
Estimate for difference: -0.259750
95% CI for difference: (-0.379727, -0.139773)
T-Test of difference = 0 (vs not =): T-Value = -4.31  P-Value = 0.000  DF = 78
Both use Pooled StDev = 0.2695
```

Comparing the output from Figure 6.23 and the MINITAB output, we see that EXCEL gives the test statistic as $z = -4.31421$ and MINITAB gives the test statistic as t-value $= -4.31$. The two are practically the same. The 95% confidence interval computed in part 3 above is $(-0.378, -0.142)$. The MINITAB 95% confidence interval is $(-0.379727, -0.139773)$. The p-value is 0 to three places in both cases.

The STATISTIX solution is as follows:

```
Statistix 8.0

Two-Sample T Tests for GPA by sex

sex           Mean      N      SD       SE
f             3.3118    40     0.2810   0.0444
m             3.0520    40     0.2576   0.0407
Difference    0.2597

Null Hypothesis: difference = 0
Alternative Hyp: difference <> 0

                                                 95% CI for Difference
Assumption              T        DF      P       Lower     Upper
Equal Variances         4.31     78      0.0000  0.1398    0.3797
Unequal Variances       4.31     77.4    0.0000  0.1398    0.3797

Test for Equality       F        DF      P
of Variances            1.19     39,39   0.2948
```

The STATISTIX data file gives the GPA in one column and the sex of the individual in the other column. The pulldown **Statistics ⇒ One, Two, Multi sample tests ⇒ Two Sample t test** is used to get to the analysis. Under the T column the value 4.31 is seen as the same that EXCEL and MINITAB produced above. The 95% confidence interval is the same as given by EXCEL and MINITAB.

The SPSS and SAS analysis give basically the same output.

Inferences Concerning Two Means—Independent and Small Samples

Equal Variances Model

When the sample sizes are small, it is necessary to make certain assumptions to compare the two population means, μ_1 and μ_2. The assumptions are: (1) The two samples are selected in an independent and random manner. (2) The two populations are normally distributed. (3) The population variances are equal, i.e. $\sigma_1^2 = \sigma_2^2 = \sigma^2$. The common population variance is estimated by pooling the two sample variances together. That estimate is $S_p^2 = \dfrac{(n_1 - 1)S_1^2 + (n_2 - 1)S_2^2}{n_1 + n_2 - 2}$ and it is called the **pooled estimate of variance**. It

can be shown that the quantity $\dfrac{\bar{X}_1 - \bar{X}_2 - (\mu_1 - \mu_2)}{\sqrt{S_p^2(1/n_1 + 1/n_2)}}$ has a Student's t distribution with $(n_1 + n_2 - 2)$ degrees of freedom. To test $H_o: (\mu_1 - \mu_2) = D_o$ versus any one of the three alternative hypotheses $H_a: (\mu_1 - \mu_2) < D_o$, $H_a: (\mu_1 - \mu_2) > D_o$, or $H_a: (\mu_1 - \mu_2) \neq D_o$, use the test statistic $T = \dfrac{\bar{X}_1 - \bar{X}_2 - D_o}{\sqrt{S_p^2(1/n_1 + 1/n_2)}}$. The interval $(\bar{X}_1 - \bar{X}_2) \pm t_{\alpha/2}\sqrt{S_p^2\left(\dfrac{1}{n_1} + \dfrac{1}{n_2}\right)}$ is a $(1 - \alpha)$ confidence interval for $\mu_1 - \mu_2$. When using this distribution theory to test hypotheses and set confidence intervals, the assumption of normality for both populations must be checked out using one of the techniques discussed earlier in this book. The assumption of equal variances must also be checked out. The next chapter discusses techniques for testing equal variances in more detail.

EXAMPLE 6.18 An engineer wishes to compare the mean lifetimes of two bulb types. She selects samples of size 10 from each and determines the lifetimes of the 20 bulbs. The lifetimes (in hours) are shown in Figure 6.24.

Bulb type 1	Bulb type 2
2164	1641
2040	1827
1926	1847
1911	1800
1973	1932
2064	1937
2018	2023
2097	1836
2105	1891
1972	1715

Figure 6.24 Bulb lifetimes, equal variability.

The Kolmogorov–Smirnov test for normality for bulb type 1 gives a p-value > 0.150 and for bulb type 2 gives a p-value > 0.15. The normality assumption is reasonable for both populations. The equal variances dialog box is shown in Figure 6.25. The columns, where the data for bulb type 1 and bulb type 2 may be found, are given in the dialog box. The output for the test of equal variances is shown in Figure 6.26. The F-test is valid if both sets of data come from normal populations. We have already tested and accepted that this is reasonable. The null hypothesis is that the variances are equal. This hypothesis is not rejected since the p-value is much greater than 0.05.

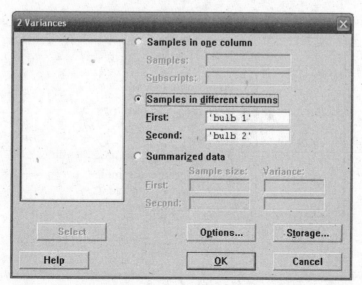

Figure 6.25 MINITAB dialog box for testing equal variances.

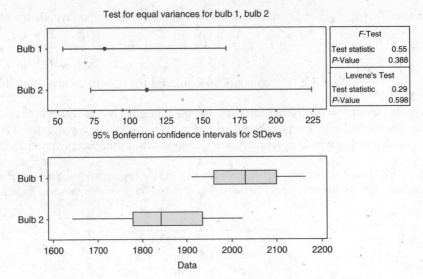

Figure 6.26   MINITAB output for the equal variances test with graphs.

Test for Equal Variances: bulb 1, bulb 2

95% Bonferroni confidence intervals for standard deviations

	N	Lower	StDev	Upper
bulb 1	10	53.9591	82.491	166.108
bulb 2	10	72.6845	111.117	223.752

F-Test (normal distribution)

Test statistic = 0.55, p-value = 0.388

Levene's Test (any continuous distribution)

Test statistic = 0.29, p-value = 0.598

The assumptions are reasonable for the test of equal means that assumes normality and equal variances. We proceed with the test $H_o: (\mu_1 - \mu_2) = 0$ versus $H_a: (\mu_1 - \mu_2) \neq 0$. The MINITAB pulldown menu to use is **Stat ⟹ Basic Statistics ⟹ 2-sample t**. The dialog box for the two-sample *t*-test is shown in Figure 6.27.

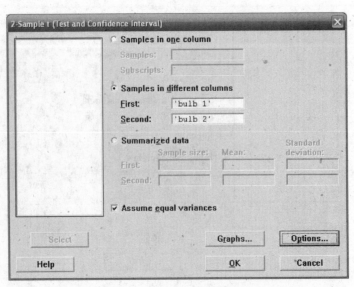

Figure 6.27   Dialog box for the 2-sample *t*-test.

```
Two-Sample t-Test and CI: bulb 1, bulb 2
Two-sample T for bulb 1 vs bulb 2

                                     SE
           N     Mean     StDev    Mean
bulb 1     10    2026.9   82.5     26
bulb 2     10    1845     111      35

Difference = mu (bulb 1) - mu (bulb 2)
Estimate for difference: 181.899
95% CI for difference: (89.957, 273.841)
T-Test of difference = 0 (vs not =): T-Value = 4.16 P-Value = 0.001 DF = 18
Both use Pooled StDev = 97.8565
```

The point estimate for the difference in means is that bulb type 1 lasts about 182 hours on the average longer than bulb type 2. The hypothesis of no difference in mean lifetimes is rejected at $\alpha = 0.05$. Again, any one of the three methods of testing may be used. The three summaries follow:

1. The two-tailed rejection region is $t < -2.100922$ or $t > 2.100922$. The computed test statistic is $t = 4.16$. Since this computed test statistic is greater than 2.100922 and is, therefore, in the rejection region, the null is rejected for $\alpha = 0.05$. This t-value may be found using any of the software packages.

2. Using the p-value approach, compare the p-value = 0.001 with the pre-set $\alpha = 0.05$. Since the p-value is less than α, reject the null.

3. The confidence interval approach calculates the 95% two-sided confidence interval and if it does not contain $D_o = 0$, reject the null. The 95% confidence interval is (89.957, 273.841). Since 0 is not in this interval, reject the null.

Note: once again, any one of the three methods of testing may be used. You will always reach the same conclusion no matter which method you use.

Now consider the output from other packages.

EXCEL The pulldown **Tools** $\Rightarrow$ **Data Analysis** leads to a dialog box where t-Test: Two-Sample Assuming Equal Variances may be chosen. This produces another dialog box which gives the analysis shown in Figure 6.28. This is the same analysis given by MINITAB.

t-Test: Two-sample assuming equal variances		
	Variable 1	Variable 2
Mean	2027	1844.9
Variance	6774.444	12309.21
Observations	10	10
Pooled variance	9541.828	
Hypothesized mean Difference	0	
df	18	
t Stat	4.168494	
$P(T<=t)$one-tail	0.000289	
t Criticalone-tail	1.734064	
$P(T<=t)$two-tail	0.000577	
t Criticaltwo-tail	2.100922	

Figure 6.28   EXCEL analysis assuming equal variances.

STATISTIX

```
Statistix 8.0
Two-Sample T Tests for lifetime by type
type            Mean       N    SD        SE
1               2027.0     10   82.307    26.028
2               1844.9     10   110.95    35.084

Difference      182.10

Null Hypothesis: difference = 0
Alternative Hyp: difference <> 0
```

```
                                              95% CI for Difference
Assumption             T          DF       P        Lower      Upper
Equal Variances        4.17       18       0.0006   90.322     273.88
Unequal Variances      4.17       16.6     0.0007   89.765     274.43

Test for Equality      F          DF       P
   of Variances        1.82       9,9      0.1935
```

The STATISTIX output provides a test for the equality of variances. The large p-value indicates that the equal variances model should be chosen. When the equal variances portion of the output is compared with the EXCEL and MINITAB output it is seen to be the same.

SPSS   The SPSS output is

Independent Samples Test

		Levene's Test for Equality of Variances		t-test for Equality of Means						
		F	Sig.	t	df	Sig. (2-tailed)	Mean Difference	Std. Error Difference	95% Confidence Interval of the Difference Lower	Upper
Lifetime	Equal variances assumed	.293	.595	4.168	18	.001	182.10000	43.68484	90.32155	273.87845
	Equal variances not assumed			4.168	16.603	.001	182.10000	43.68484	89.76504	274.43496

Note that SPSS also gives both the equal and unequal variances model.

SAS

```
Two Sample t-test for the Means of lifetime within type
Sample Statistics
   Group    N    Mean      Std. Dev.    Std. Error
   --------------------------------------------------------
   1        10   2027      82.307       26.028
   2        10   1844.9    110.95       35.084

Hypothesis Test
Null hypothesis:      Mean 1 - Mean 2 = 0
Alternative:          Mean 1 - Mean 2 ^= 0
   If Variances Are    t statistic    Df      Pr > t
   --------------------------------------------------------
   Equal               4.168          18      0.0006
   Not Equal           4.168          16.60   0.0007
```

SAS gives the choice just as SPSS. Once you decide on the equal or unequal variances model, select the output for that model.

Unequal Variances Model

The solution to this case is called the **Smith–Satterthwaite test**. I will give the equations, but the engineer will not have to be burdened with them. He or she can thank the software manufacturers.

To test $H_o: (\mu_1 - \mu_2) = D_o$ versus any one of the three alternative hypotheses $H_a: (\mu_1 - \mu_2) < D_o$, $H_a: (\mu_1 - \mu_2) > D_o$, or $H_a: (\mu_1 - \mu_2) \neq D_o$, use the test statistic $T = \dfrac{\bar{X}_1 - \bar{X}_2 - D_o}{\sqrt{S_1^2/n_1 + S_2^2/n_2}}$. The degrees of freedom are

$$\nu = \frac{(s_1^2/n_1 + s_2^2/n_2)^2}{(s_1^2/n_1)^2/n_1 - 1 + (s_2^2/n_2)^2/n_2 - 1}.$$ The interval $(\bar{X}_1 - \bar{X}_2) \pm t_{\alpha/2}\sqrt{\dfrac{S_1^2}{n_1} + \dfrac{S_2^2}{n_2}}$ is a $(1 - \alpha)$ confidence

interval for $\mu_1 - \mu_2$ and $\nu = \dfrac{(s_1^2/n_1 + s_2^2/n_2)^2}{(s_1^2/n_1)^2/n_1 - 1 + (s_2^2/n_2)^2/n_2 - 1}$.

EXAMPLE 6.19 An engineer wishes to compare the mean lifetimes of two bulb types. She selects samples of size 10 from each and determines the lifetimes of the 20 bulbs. The lifetimes (in hours) are shown in Figure 6.29.

Bulb type 1	Bulb type 2
1726	1787
1779	1973
1824	1803
1718	1903
1898	1971
1827	2378
1813	2226
1806	2040
1896	2200
1803	1675

Figure 6.29   Bulb lifetimes, unequal variability.

The Kolmogorov–Smirnov test for normality for bulb type 1 gives a p-value > 0.150 and for bulb type 2 gives a p-value > 0.15. The normality assumption is reasonable for both populations.

The output for the test of equal variances is shown in Figure 6.30. The F-test is valid if both sets of data come from normal populations. We have already tested and accepted that both sets of data came from normal populations. The null hypothesis is that the variances are equal. This hypothesis is rejected since the p-value is much smaller than 0.05.

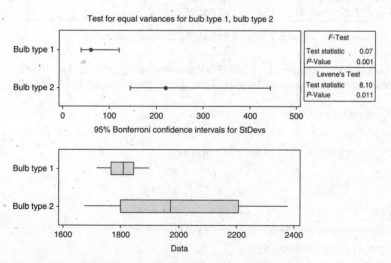

Figure 6.30   MINITAB output for the equal variances test.

We have two small samples and we have concluded that the two populations do not have equal variances. We will test the equality of means first using MINITAB.

```
Two-Sample t-Test and CI: Bulb type 1, Bulb type 2
Two-sample T for Bulb type 1 vs Bulb type 2

                                              SE
                 N      Mean     StDev       Mean
Bulb type 1     10     1809.0    59.8         19
Bulb type 2     10     1996      220          70

Difference = mu (Bulb type 1) - mu (Bulb type 2)
Estimate for difference: -186.600
95% CI for difference: (-347.328, -25.872)
T-Test of difference = 0 (vs not =): T-Value = -2.59 P-Value = 0.027 DF = 10
```

The point estimate for the difference in means is that bulb type 2 lasts about 187 hours on the average longer than bulb type 1. The hypothesis of no difference in mean lifetimes is rejected at $\alpha = 0.05$. Again, any one of the three methods of testing may be used. Please see Figure 6.31 (see the three summaries below).

1. The two-tailed rejection region is $t < -2.228$ and $t > 2.228$. The computed test statistic is $t = -2.59$. Since this computed test statistic is in the rejection region, the null is rejected for $\alpha = 0.05$. This t-value may be found using any of the software packages.

2. Using the p-value approach, compare the p-value $= 0.027$ with the pre-set $\alpha = 0.05$. Since the p-value is less than α, reject the null.

3. The confidence interval approach calculates the 95% two-sided confidence interval and if it does not contain $D_o = 0$, reject the null. The 95% confidence interval is $(-347.328, -25.872)$. Since 0 is not in this interval, reject the null.

EXCEL

t-Test: Two-sample assuming unequal variances		
	Type 1	*Type 2*
Mean	1809	1995.6
Variance	3574.444	48460.93
Observations	10	10
Hypothesized mean		
Difference	0	
Df	10	
t Stat	-2.5868	
P (*T*<=*t*) one-tail	0.013549	
t Critical one-tail	1.812461	
P (*T*<=*t*) two-tail	0.027098	
t Critical two-tail	2.228139	

Figure 6.31   EXCEL analysis assuming unequal variances.

STATISTIX   The data file is structured for STATISTIX as shown in Figure 6.32. The lifetimes are referred to as the dependent variable and the bulb type as a categorical variable by the package STATISTIX. The dialog box is shown in Figure 6.33.

The STATISTIX output is as follows:

```
Statistix 8.0

Two-Sample T Tests for lifetime by type

type     Mean        N         SD         SE
1        1809.0      10        59.787     18.906
2        1995.6      10        220.14     69.614

Difference            -186.60

Null Hypothesis: difference = 0
Alternative Hyp: difference <> 0
```

Lifetime	Type
1726	1
1779	1
1824	1
1718	1
1898	1
1827	1
1813	1
1806	1
1896	1
1803	1
1787	2
1973	2
1803	2
1903	2
1971	2
2378	2
2226	2
2040	2
2200	2
1675	2

Figure 6.32 Data file with lifetimes in one column and bulb type in the other.

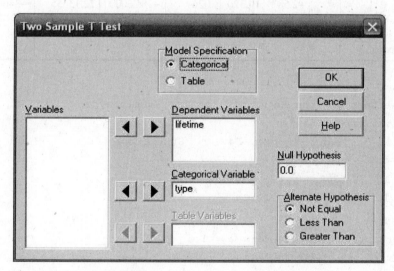

Figure 6.33 STATISTIX dialog box for two-sample *t*-test.

				95% CI for Difference	
Assumption	T	DF	P	Lower	Upper
Equal Variances	−2.59	18	0.0186	−338.15	−35.049
Unequal Variances	−2.59	10.3	0.0265	−346.65	−26.546
Test for Equality	F	DF	P		
of Variances	13.56	9,9	0.0003		

The STATISTIX output gives the information needed to make a decision regarding equality of variance. Since the p-value equals 0.0003, I would reject the hypothesis of equal variances. I would then go to the assumption of unequal variances. The t-value is computed using the formula $T = \dfrac{\bar{X}_1 - \bar{X}_2 - D_o}{\sqrt{S_1^2/n_1 + S_2^2/n_2}}$. The degrees of freedom is computed

using $v = \dfrac{(s_1^2/n_1 + s_2^2/n_2)^2}{(s_1^2/n_1)^2/n_1 - 1 + (s_2^2/n_2)^2/n_2 - 1}$. It is recommended that the degrees of freedom be rounded down when using this formula. The degrees of freedom would be 10. The information needed to make a decision regarding the null hypothesis of no difference in the means, using any one of the three methods, is contained in the output. Note that you can get the rejection region by using 10 degrees of freedom and the Student's t distribution or you can compare the p-value = 0.0265 with $\alpha = 0.05$ or you can see that the confidence interval $(-346.65, -26.546)$ does not contain $D_o = 0$.

SPSS SPSS expects the data file as shown in Figure 6.32. The test for equality of variances uses Levene's test which gives a different p-value than does STATISTIX. Otherwise, the output is the same as STATISTIX.

Independent Samples Test

		Levene's Test for Equality of Variances		t-test for Equality of Means					95% Confidence Interval of the Difference	
		F	Sig.	t	df	Sig. (2-tailed)	Mean Difference	Std. Error Difference	Lower	Upper
Lifetime	Equal variances assumed	9.884	.006	−2.587	18	.019	−186.60000	72.13555	−338.151	−35.04883
	Equal variances not assumed			−2.587	10.320	.026	−186.60000	72.13555	−346654	−26.54595

SAS

```
Two Sample t-test for the Means of lifetime within type
Sample Statistics
   Group    N     Mean      Std. Dev.    Std. Error
   --------------------------------------------------------
   1        10    1809      59.787       18.906
   2        10    1995.6    220.14       69.614

Hypothesis Test
   Null hypothesis:    Mean 1 - Mean 2 = 0
   Alternative:        Mean 1 - Mean 2 ^= 0
   If Variances Are    t statistic    Df        Pr > t
   --------------------------------------------------------
   Equal               -2.587         18        0.0186
   Not Equal           -2.587         10.32     0.0265
95% Confidence Interval for the Difference between Two Means
   Lower Limit         Upper Limit
   -338.15             -35.05
```

SAS expects the data file to be in the form shown in Figure 6.32. The output is basically the same as SPSS and STATISTIX.

Inferences Concerning Two Means—Dependent Samples

In the previous two sections the two samples were purposely selected independently of one another. If a group of n objects are available for a study that was to have treatment 1 applied to n_1 units and treatment 2 applied to n_2 units and $n = n_1 + n_2$, we were interested in whether the mean for treatment 1 differs from the mean for treatment 2. The n objects are numbered 1 through n. For example, suppose 50 rats are available and they are numbered 1 through 50. Twenty-five numbers between 1 and 50 are randomly selected. Those rats with the numbers selected go into one group and receive diet 1. The remaining rats go into group 2 and receive diet 2. The lifetime of each rat is recorded and at the end of the experiment, the mean age of each group is determined. It is of interest to know if the mean lifetimes of the two groups are equal or not. The analysis given in the previous section would be applied to the data.

Now, imagine a different scenario. Rats are paired up according to age, sex, and other factors that are important to longevity. For each pair, one of the rats is selected and a coin is flipped. If a head turns up, that rat is put on diet 1. If a tail turns up, that rat is placed on diet 2. Thus, the other member of the pair is placed on the other diet. In this way the treatments are randomly assigned to the pairs. The age at death is recorded for each member of each pair. It is of interest if the diets differ with respect to mean age reached by the rats. The two samples are dependent or paired. The analysis of such an experiment is what concerns us in this section.

EXAMPLE 6.20   Suppose 20 rats have been paired according to certain salient characteristics. The members of each pair have been randomly assigned to diet 1 or diet 2. There are 10 pairs. The experiment lasts until all 20 rats are dead. The age of each rat is recorded at death. The data is given in Figure 6.34.

Diet 1	Diet 2	Difference
1271	1276	−5
1311	1359	−48
1221	1312	−91
1275	1247	28
1221	1279	−58
1289	1388	−99
1291	1296	−5
1393	1343	50
1224	1402	−178
1223	1365	−142

Figure 6.34   Ages at death (in days) of each rat for each diet.

The null hypothesis is that the mean difference is 0 or $H_o: \mu_d = 0$ versus one of three alternatives: $H_a: \mu_d < 0, H_a: \mu_d > 0$, or $H_a: \mu_d \neq 0$. The analysis is the same as for testing a single mean in the section on testing hypotheses about a single mean. The small samples test is the t-test and the large samples test is the z-test. The differences are analyzed. The test statistic is $t = \dfrac{\bar{d} - 0}{s_d / \sqrt{n}}$ where n is the number of pairs. The Kolmogorov–Smirnov test confirms that normality is not rejected. Using MINITAB, the results are

```
One-Sample T: C3
Test of mu = 0 vs not = 0

Variable   N     Mean      StDev    SE Mean   95% CI                 T       P
C3         10    -54.6759  73.7114  23.3096   (-107.4059, -1.9460)   -2.35   0.044
```

The mean difference is $\bar{d} = -54.7$. The differences were formed by taking diet 1 lifetime minus diet 2 lifetime. The minus means diet 2 lifetime is larger on the average by 54.7 days than diet 1 lifetime. The difference is significant since the p-value < 0.05.

EXAMPLE 6.21　Traffic engineers installed traffic control devices at 35 different intersections around the city. They recorded the average number of accidents per week before and after installing the devices. Do the following summary data support the hypothesis that the devices reduced the number of accidents? $n = 35$, $\bar{d} = 10.5$, $s_{\bar{d}} = 3.7$. The differences are formed by taking before and subtracting after.

The test statistic is $z = \dfrac{\bar{d} - 0}{s_{\bar{d}}/\sqrt{n}} = \dfrac{10.5}{3.7} = 2.84$. Positive differences tend to support the hypothesis that the devices reduced accidents.

MINITAB allows for the case where your data has been summarized. Figure 6.35 is the dialog box for testing a single mean for large samples. The dialog box asks for the sample size, the mean, the standard deviation, and the test mean. The sample size is 35, the mean difference is 10.5, and the test mean is 0. Even though the standard deviation is not given, it can be figured. The formula for $s_{\bar{d}}$ is $s_{\bar{d}} = \dfrac{s_d}{\sqrt{n}}$. Solving for the standard deviation gives $s_d = \sqrt{n}(s_{\bar{d}})$. The output is

```
One-Sample Z

Test of mu = 0 vs > 0
The assumed standard deviation = 21.9

                            95%
                            Lower
N      Mean     SE Mean     Bound     Z        P
35     10.5000  3.7018      4.4111    2.84     0.002
```

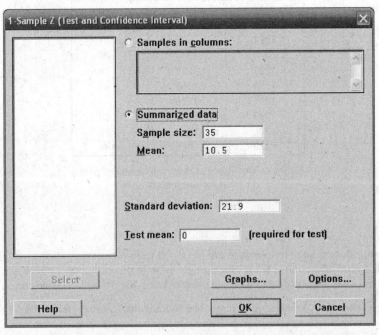

Figure 6.35　MINITAB dialog box for z-test when data is summarized.

The null may be rejected because the p-value $= 0.002 < \alpha = 0.05$. Or the one-sided confidence interval is $(4.4111, \infty)$ and it does not contain the null value which is 0.

SOLVED PROBLEMS

Point estimation

6.1. During the filling process for 8-ounce milk cartons, the weight was determined for 5 cartons. The weights were 263.9, 266.2, 266.3, 266.8, and 265.0 grams for a given day. Set a 95% confidence interval on the mean weight. Historically the standard deviation for these weights is 1.65 grams. Use MINITAB to set this confidence interval.

SOLUTION

The output produced by MINITAB is

One-Sample Z: weight

The assumed standard deviation = 1.65

Variable	N	Mean	StDev	SE Mean	95% CI
weight	5	265.640	1.176	0.738	(264.194, 267.086)

Note that 1.65 is used in the confidence interval for σ and not 1.176. $E = z_{\alpha/2} \dfrac{\sigma}{\sqrt{n}}$ is computed and the formula $\bar{X} \pm E$ is used to find the confidence interval.

6.2. In Problem 6.1, suppose a government inspector requires this company to generate a 95% confidence interval with a width of ± 0.5 grams. What size of sample will give this interval?

SOLUTION

We need $1.96 \dfrac{\sigma}{\sqrt{n}}$ to equal ± 0.5. Setting these equal and solving for n we find n equal to 41.83 and rounding up we have $n = 42$.

Interval estimation

6.3. An engineer studied the concentricity of an oil seal groove and a base cylinder in the interior of the groove. He measured the concentricity as a positive deviation using a dial indicator gauge. Historically, the standard deviation for the concentricity is 0.9. To monitor the process, the engineer takes a random sample of five measurements. A recent sample gave a sample mean equal to 6.2. (a) Construct a 99% confidence interval for the true mean concentricity using MINITAB. (b) What assumption must be checked out for this confidence interval to be valid?

SOLUTION

(a) The MINITAB dialog box is and 99% confidence is chosen under options.

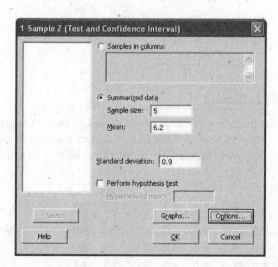

The MINITAB output is

One-Sample Z

The assumed standard deviation = 0.9

```
N    Mean    SE Mean    99% CI
5    6.200   0.402      (5.163, 7.237)
```

(b) Normality should be checked using the Anderson–Darling test, the Ryan–Joiner test, or the Kolmogorov–Smirnov test.

6.4. A process for the manufacture of steel bolts that continuously feed an assembly line downstream is sampled. Historically, the thicknesses of these bolts follow a normal distribution with a standard deviation of 1.6 mm. A recent sample yielded the following thicknesses: 9.7, 9.4, 9.9, 9.9, 10.3, 10.1, 10.1, 9.7, 10.5, and 10.3 mm. (a) Test that the data came from a normal distribution. (b) Construct a 95% confidence interval for the true mean thickness.

SOLUTION

(a) The Kolmogorov–Smirnov test for normality is applied and when the test for normality of MINITAB is applied, the p-value of the test is found to be > 0.15. The null hypothesis that the data come from a normal population is not rejected.

(b) The 95% confidence interval given by MINITAB is as follows:

One-Sample Z: thickness

The assumed standard deviation = 1.6

```
Variable     N    Mean     StDev    SE Mean    95% CI
thickness    10   9.990    0.335    0.506      (8.998, 10.982)
```

Basic concepts of testing hypotheses

6.5. Alpha represents the probability of rejecting the null when the null is true. What event has probability $(1 - \alpha)$? Beta represents the probability of not rejecting the null when it is false. What event has probability $(1 - \beta)$?

SOLUTION

The event that null hypothesis is true and it is not rejected has probability $(1 - \alpha)$. The event that the null is not true and it is rejected has probability $(1 - \beta)$.

6.6. True or false: The only way that α and β can both be made arbitrarily small is by taking the sample size large enough.

SOLUTION

True

Testing hypotheses about a single mean

6.7. An engineer studied the thickness of metal wires produced in a chip-manufacturing process. These wires should have a target thickness of 8 microns. Conduct the hypothesis that the mean thickness is 8 microns versus it is not 8 microns at a 0.05 significance level. The sample data is:

Thickness values

```
8.4  7.9  8.1  7.8  7.8  8.0  8.2  8.1  7.9  8.2  7.8  7.9  8.0
8.4  7.7  8.0  7.8  8.0  7.7  8.3  7.9  7.9  8.3  8.0  7.8  7.7
7.9  7.8  7.9  8.3  8.0  8.0  8.2  8.0  7.8  7.9  8.0  8.3  7.7
8.0  8.2  7.6  8.0  7.7  8.2  7.9  8.2  8.0  7.8  7.8
```

SOLUTION

The standard deviation of the sample values is 0.2006. The MINITAB output for testing the hypothesis is:

```
Test of mu = 8 vs not = 8
The assumed standard deviation = 0.2006
Variable    N    Mean    StDev    SE Mean    95% CI              Z       P
thickness   50   7.9760  0.2006   0.0284     (7.9204, 8.0316)  -0.85   0.398
```

The *p*-value is 0.398 which exceeds 0.05. The sample data does not cause us to reject that μ equals 8 microns.

6.8. Engineers studied a batch operation at a chemical plant where an important quality characteristic is the product viscosity, which has a target value of 14.90. Production personnel use a viscosity measurement to monitor this process. The viscosity measurements for the past 10 batches are: 13.3, 14.8, 14.5, 15.2, 15.3, 14.9, 15.3, 14.6, 14.3, and 14.1. Conduct the hypothesis that $\mu = 14.90$ versus it is not 14.90 at significance level 0.05.

SOLUTION

The following MINITAB output is produced:

```
One-Sample T: viscosity
Test of mu = 14.9 vs not = 14.9
Variable     N    Mean     StDev    SE Mean    95% CI                T       P
viscosity    10   14.630   0.624    0.197      (14.184, 15.076)    -1.37   0.204
```

Do not reject the null hypothesis because the *p*-value = 0.204 > $\alpha = 0.05$.

Relationship between tests of hypotheses and confidence intervals

6.9. Refer to Problem 6.7 above. Test the hypothesis by using the confidence interval technique.

SOLUTION

If a null hypothesis is being tested at α, and a $(1 - \alpha)$ confidence interval for μ is formed, then the *p*-value < α **if and only if** the null hypothesis value μ_o is not contained in the $(1 - \alpha)$ confidence interval for μ. The 95% confidence interval for μ is (7.9204, 8.0316). Since μ_o is contained in the confidence interval, the null hypothesis would not be rejected at the 0.05 significance level.

6.10. Refer to Problem 6.8 above. Test the hypothesis by using the confidence interval technique.

SOLUTION

If a null hypothesis is being tested at α, and a $(1 - \alpha)$ confidence interval for μ is formed, then the *p*-value < α **if and only if** the null hypothesis value μ_o is not contained in the $(1 - \alpha)$ confidence interval for μ. The 95% confidence interval for μ is (14.184, 15.076). Since μ_o is contained in the confidence interval, the null hypothesis would not be rejected at the 0.05 significance level.

Inferences concerning two means—independent and large samples

6.11. Engineers studied two modern high strength micro-alloyed structural steels. Thirty Charpy V notched specimens were machined from a 50-mm-thick plate of steel A. Similarly, a set of 30 specimens were machined form a piece of pipe of steel B. All specimens were tested in four-point bending tests at a temperature of 77 K and the values of fracture load recorded. The data is

```
                          steel A
    24700  30200  40400  31300  32700  31300  29400  30900  32600
    31500  42000  34700  28700  22800  33100  31900  36000  23200
    28400  28000  24300  27200  27700  35500  32800  30500  25300
    30200  28500  24600

                          steel B
    34510  21010  34110  28730  28880  26750  36380  35150  23540
    32060  28780  32300  25810  35970  34140  21900  37960  33140
    28050  28840  28330  37510  26400  32220  29960  24770  22040
    39730  26580  28610
```

Use EXCEL to perform the test at $\alpha = 0.05$.

SOLUTION

The fracture load for steel A is entered into column A and for steel B is entered into column B.
The EXCEL output is as follows:

t-Test: Two-Sample Assuming Equal Variances

	steel A	*steel B*
Mean	30346.67	30138.67
Variance	21126713	25561950
Observations	30	30
Pooled Variance	23344331	
Hypothesized Mean Difference	0	
Df	58	
t Stat	0.166732	
P(T<=t) one-tail	0.434081	
t Critical one-tail	1.671553	
P(T<=t) two-tail	0.868161	
t Critical two-tail	2.001717	

The computed z-value is 0.167 and p-value is 0.868. The null hypothesis of equal means is not rejected.

6.12. Do Problem 6.11 using MINITAB.

SOLUTION

The 30 fracture loads for A and B are entered into C1. A 1 is entered 30 times and a 2 is entered 30 times into C2. The following output is obtained. This is the same output as given in EXCEL in Problem 6.11.

```
Two-Sample t-Test and CI: fracture load, steel
Two-sample T for fracture load

steel     N     Mean     StDev     SE Mean
1        30     30347     4596      839
2        30     30139     5056      923

Difference = mu (1) - mu (2)
Estimate for difference:  208
95% CI for difference:  (-2290, 2706)
T-Test of difference = 0 (vs not =): T-Value = 0.17  P-Value = 0.868  DF = 57
```

The null is not rejected since the p-value is > alpha.

Inferences concerning two means—independent and small samples

6.13. The following data shows the yields for the last 8 hours of production from two ethanol-water distillation columns. It is desired to test the null hypothesis that $\mu_1 = \mu_2$. Test the three assumptions that are needed on the data to do the test of hypothesis.

```
                column1
70      74      73      72      72      73      72      73

                column2
71      74      72      71      72      70      72      72
```

SOLUTION

The assumption of equal variances is tested using MINITAB. The output of the test is as follows:

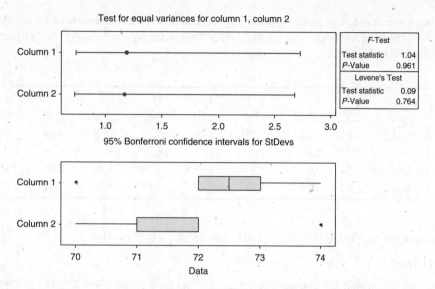

Test for equal variances for column 1, column 2

	F-Test	
Test statistic	1.04
P-Value	0.961
	Levene's Test	
Test statistic	0.09
P-Value	0.764

95% Bonferroni confidence intervals for StDevs

The p-value for testing the equality of variances is 0.961 for the F-test and 0.764 for Levene's test. Because of the large size of the p-values, the null hypothesis for testing equal variances is not rejected.

The data for column 1 and column 2 both pass the normality tests of MINITAB.

Engineers need to make sure that an effort has been made to insure that the two samples are independent samples.

6.14. An engineer compared two brands of ultrasonic humidifiers with respect to the rate at which they output moisture. The following data shows the maximum outputs (in fluid ounces) per hour as measured in a chamber controlled at a temperature of 70 degrees and a relative humidity of 30%.

Brand 1: 14.0, 14.3, 12.2, 15.1   Brand 2: 12.1, 13.6, 11.9, 11.2

Use SPSS to test for equal means at $\alpha = 0.05$.

SOLUTION

The output for SPSS is given below.

Group Statistics

Number		N	Mean	Std. Deviation	Std. Error Mean
Brand	1.0	4	13.900	1.2247	.6124
	2.0	4	12.200	1.0100	.5050

Independent Samples Test

		Levene's Test for Equality of Variances		t-test for Equality of Means		
		F	Sig.	t	df	Sig. (2-tailed)
brand	Equal variances assumed	.100	.763	2.142	6	.076
	Equal variances not assumed			2.142	5.790	.078

The computed value of t is 2.142 and the computed two-tailed p-value is 0.076.

The null hypothesis is not rejected for $\alpha = 0.05$.

Inferences concerning two means—dependent samples

6.15. An engineer carried out an experiment on scale models in a wave tank to investigate how the choice of mooring method affected the bending stress produced in a device used to generate electricity from wave power at sea. The model system was subjected to the same sample of 18 sea states with each of the two mooring methods. The resulting data (root mean square bending moment in Newton-meters) are entered in MINITAB.

Sea state	1	2	3	4	5	6	7	8	9
Method 1	2.23	2.55	7.99	4.09	9.62	1.59	8.98	0.82	10.83
Method 2	1.82	2.42	8.26	3.46	9.77	1.40	8.88	0.87	11.20
Sea state	10	11	12	13	14	15	16	17	18
Method 1	1.54	10.75	5.79	5.91	5.79	5.50	9.96	1.92	7.38
Method 2	1.33	10.32	5.87	6.44	5.87	5.30	9.82	1.69	7.41

Conduct the hypothesis that $H_0: \mu_d = 0$ versus $H_a: \mu_d \neq 0$ at $\alpha = 0.05$.

SOLUTION

The data for Method 1 is entered into column C1 and for Method 2 into C2. The pulldown **Stat ⇒ Basic statistics ⇒ Paired t** is given. The following output results:

Paired t-Test and CI: Method 1, Method 2

Paired T for Method 1 - Method 2

	N	Mean	StDev	SE Mean
Method 1	18	5.736	3.439	0.811
Method 2	18	5.674	3.542	0.835
Difference	18	0.0617	0.2901	0.0684

95% CI for mean difference: (−0.0826, 0.2059)
T-Test of mean difference = 0 (vs not = 0): T-Value = 0.90 P-Value = 0.380

The p-value = 0.38 is not significant. There is no significant difference in the two methods. The null hypothesis is not rejected.

6.16. The value of the turbine prediction coefficient was calculated for 14 pumps using two different methods. The data for the study is

Pump	1	2	3	4	5	6	7
Childs	0.78	1.64	2.25	0.80	1.09	1.39	0.47
Stepanoff	0.60	1.82	2.25	0.80	1.17	0.91	1.12
Difference	0.18	−0.18	0	0	−0.08	0.48	−0.65
Pump	8	9	10	11	12	13	14
Childs	2.66	0.44	1.23	0.37	0.41	0.56	2.45
Stepanoff	2.88	0.23	0.89	0.86	0.67	1.23	1.89
Difference	−0.22	0.21	0.34	−0.49	−0.26	−0.67	0.56

In order to test $H_0: \mu_d = 0$ versus $H_a: \mu_d \neq 0$ the one sample paired t-test is used. The differences must be normally distributed. Test to see if this is a reasonable assumption.

SOLUTION

The data passes all three of the normality tests in MINITAB. The p-value for the Anderson–Darling test is 0.909. The p-value for the Ryan–Joiner is > 0.1. The p-value for the Kolmorgorov–Smirnoff is > 0.150. The normality assumption is reasonable.

SUPPLEMENTARY PROBLEMS

6.17. A nuclear power plant is adjacent to a large reservoir and utilizes the water from the reservoir in the production of power. The mean temperature of the reservoir water is of interest to a nuclear engineer. Twenty random selections of water temperatures are made and the results are given in Figure 6.36. The data passes the normality test. Set a 95% confidence interval on μ using EXCEL, MINITAB, SAS, SPSS, and STATISTIX.

65.7	68.4
64.9	66.4
69.8	69.2
70.3	66.6
67.3	69.9
64.9	70.0
67.7	68.5
71.5	65.7
68.7	69.2
67.5	65.7

Figure 6.36   Twenty Reservoir Water Temperatures.

6.18. The balance of a die is in question. In particular, the probability of a 6 appearing on a given roll is believed not to be $1/6 = 0.167$. The null hypothesis is $H_o: p = 0.167$ and the alternative is $H_a: p \neq 0.167$. The rejection region is chosen to be $X \leq 2$ or $X \geq 10$, where $X =$ the number of times a 6 appears in 36 rolls of the die. Answer the following questions using EXCEL.
(a) Find α.
(b) Construct a graph of type II errors and a power curve.
(c) If the experiment is conducted and $X = 9$, find the p-value.

6.19. A large sample is used to test $H_o: \mu = 3$ milligrams per liter versus $H_a: \mu \neq 3$, where μ is the mean fluoride in drinking water for a region. A sample of size 50 has a sample mean equal to 3.4 and a standard deviation equal to 1.5. Alpha is set equal to 0.05. Use EXCEL and test using (a) the classical method, (b) the confidence interval method, and (c) the p-value method.

6.20. A small sample is used to test $H_o: \mu = 3$ milligrams per liter versus $H_a: \mu \neq 3$, where μ is the mean fluoride in drinking water for a region. A sample of size 13 has a sample mean equal to 3.9 and a standard deviation equal to 1.5. Alpha is set equal to 0.05. Use EXCEL and test the hypothesis using (a) the rejection region method, (b) the confidence interval method, and (c) the p-value method.

6.21. An industrial engineer wishes to test if two processes are producing products of the same weight on the average. She wishes to test $H_o: (\mu_1 - \mu_2) = 0$ versus the alternative hypothesis $H_a: (\mu_1 - \mu_2) \neq 0$ where μ_1 represents the mean weight of products from process 1 and μ_2 represents the mean weight of products from process 2. She takes a sample of size 65 from process 1 and finds that the mean equals 16.10 ounces and the standard deviation equals 0.25 ounces. The sample of size 65 from process 2 has a mean equal to 16.35 ounces and a standard deviation equal to 0.55. Use MINITAB and test the hypothesis using (a) the classical method, (b) the confidence interval method, and (c) the p-value method. ($\alpha = 0.05$)

6.22. An industrial engineer wishes to test if two processes are producing products of the same weight on the average. She wishes to test $H_o: (\mu_1 - \mu_2) = 0$ versus the alternative hypothesis $H_a: (\mu_1 - \mu_2) \neq 0$ where μ_1 represents the mean weight of products from process 1 and μ_2 represents the mean weight of products from process 2. She takes a sample of size 10 from process 1 and finds that the mean equals 16.10 ounces and the standard deviation equals 0.25 ounces. The sample of size 15 from process 2 has a mean equal to 16.35 ounces and a standard deviation equal to 0.55. Use MINITAB and test the hypothesis using (a) the rejection region method, (b) the confidence interval method, and (c) the p-value method. ($\alpha = 0.05$). Assume normality of both populations and equal variances. Solve using MINITAB.

6.23. Nine automobiles were tested under stop-and-go as well as highway driving. The miles per gallon were measured and the results in Figure 6.37 were obtained.

Vehicle	Stop-and-go	Highway	Difference
1	25	29	4
2	30	33	3
3	27	35	8
4	25	30	5
5	23	26	3
6	27	33	6
7	20	28	8
8	31	33	2
9	29	34	5

Figure 6.37   Comparison of 9 automobiles mpg values under stop-and-go and highway.

Test the null hypothesis $H_o: (\mu_1 - \mu_2) = 5$ versus $H_a: (\mu_1 - \mu_2) < 5$ at $\alpha = 0.05$ using EXCEL, MINITAB, and STATISTIX.

6.24. It is desired to compare the mean pressures in two processes. The engineer making the comparison is certain that the variability in pressure is not the same for the two processes. The sample data collected is shown in Figure 6.38.

Process 1	Process 2
84.3	77.0
75.3	72.3
76.0	90.4
75.2	101.9
78.1	60.8
74.3	89.1
78.5	70.0
70.3	20.8
75.0	26.0
77.4	91.4

Figure 6.38   Pressures from two processes.

Calculate the degrees of freedom using EXCEL for the t-test that is used to compare means when equal variability is not assumed. Compare this computed degrees of freedom with the degrees of freedom given by MINITAB.

6.25. The mean lifetimes of two bulbs are compared. Normality and equal variances are assumed. When the variances are equal, the two sample variances are pooled to form an estimate of the common population variance as follows: $S_p^2 = \dfrac{(n_1 - 1)S_1^2 + (n_2 - 1)S_2^2}{n_1 + n_2 - 2}$ and the **pooled standard deviation** is the square root of this quantity. This pooled variance is part of the t variable as follows: $t = \dfrac{\bar{X}_1 - \bar{X}_2 - D_o}{\sqrt{S_p^2(1/n_1 + 1/n_2)}}$. Use EXCEL to compute the pooled standard deviation for the following sample data and compare your results with the MINITAB output.

Bulb type 1	Bulb type 2
2164	1641
2040	1827
1926	1847
1911	1800
1973	1932
2064	1937
2018	2023
2097	1836
2105	1891
1972	1715

CHAPTER 7

Inferences Concerning Variances

Estimation of Variances

In this chapter we shall investigate how to make inferences about the variance of a single population and about the ratio of the variances of two populations. In particular, in this section we will see how to set a confidence interval on the variance of a single population or the ratio of two population variances. In the section on hypothesis concerning one variance we shall investigate testing hypothesis about one population variance. In the section on hypothesis concerning two variance we shall investigate testing hypothesis about two population variances.

The distribution of the variable $\dfrac{(n-1)S^2}{\sigma^2}$ is chi-square with $(n-1)$ degrees of freedom, σ^2 is the population variance and S^2 is the sample variance. The population from which the sample is taken has a normal distribution. The symbol χ_α^2 represents the value on the horizontal axis of the chi-square curve with area α to its right. Recall that area represents probability.

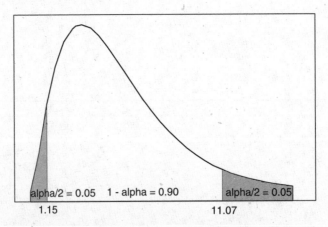

Figure 7.1   Chi-square distribution with 5 degrees of freedom
and $\chi_{0.05}^2 = 11.07$ and $\chi_{0.95}^2 = 1.15$.

Figure 7.1 shows a chi-square distribution with $\nu = 5$ degrees of freedom. The points $\chi_{0.05}^2 = 11.07$ and $\chi_{0.95}^2 = 1.15$ are also shown. They are found using EXCEL as follows: =CHIINV(0.05,5) gives 11.0705 and =CHIINV(0.95,5) gives 1.1455.

EXAMPLE 7.1 Take a random sample of size 10 from a normal population with unknown variance σ^2. Suppose we wish to construct a 95% confidence interval for σ^2. The variable $\frac{9S^2}{\sigma^2}$ will have a chi-square distribution with 9 degrees of freedom. Using EXCEL, the function =CHIINV(0.025,9) gives 19.02277. There is 0.025 area to the right of 19.02277 on the horizontal axis of the chi-square curve. There is 0.975 area to the right of =CHIINV(0.975,9) which equals 2.70039. Now, we may write the following concerning the area under the curve: $P(2.700 < \frac{(n-1)S^2}{\sigma^2} < 19.023) = 0.95$.

Solving the inequality inside the probability statement for σ^2 we find $\frac{(10-1)S^2}{19.023} < \sigma^2 < \frac{(10-1)S^2}{2.0700}$. Once the sample of size 10 is taken and S^2 is replaced by its value we will have a 95% confidence interval for σ^2.

Generalizing from the example, we have the following $(1-\alpha)$ confidence interval for σ^2.

$$\frac{(n-1)S^2}{\chi^2_{\alpha/2}} < \sigma^2 < \frac{(n-1)S^2}{\chi^2_{1-\alpha/2}}$$

It is interesting to contrast EXCEL and MINITAB as to how they find the values $\chi^2_{.025}$ and $\chi^2_{0.975}$. The value for $\chi^2_{.025}$ is 19.023. Using MINITAB, the area to the left of $\chi^2_{.025}$ is 0.975. Using the pulldown **Calc ⇒ Probability Distribution ⇒ Chi-square** and choosing inverse cumulative probability, gives the following:

```
Inverse Cumulative Distribution Function

Chi-Square with 9 DF

P( X <= x )   x

0.975       19.0228
```

The area to the right of $\chi^2_{.025}$ is 0.025 and to the left of $\chi^2_{.025}$ is 0.975.

EXAMPLE 7.2 An industrial engineer is interested in the variation of content in containers labeled as 180 fluid ounces. Twenty are selected from production and the volumes in fluid ounces are shown in Figure 7.2. Set a 99% confidence interval on the variance in volume for this process. First, we need to confirm that the sample data came from a normal population. The Kolmogorov–Smirnov test gives a p-value > 0.15. Normality is therefore a reasonable assumption.

Figure 7.3 shows the EXCEL computation of a 99% confidence interval for the population variance. The data is contained in A1:B10. The S^2 computation is in C1. The computation $19S^2$ is in C2. The chi-square values are in C3:C4. The lower limit of the 99% confidence interval is in C5 and the upper limit is in C6. The commands in column C are shown in column D. The population variance is between 1.061 and 5.980 with 99% confidence. The 99% confidence interval for the population standard deviation is found by taking square roots. The 99% confidence interval for σ is (1.030, 2.445).

181.5	180.8
179.1	182.4
178.7	178.5
183.9	182.2
179.7	180.9
180.6	181.4
180.4	181.4
178.5	180.6
178.8	180.1
181.3	182.2

Figure 7.2 Volume in fluid ounces.

A	B	C	D
181.5	180.8	2.154211	=VAR(A1:B10)
179.1	182.4	40.93	=19*C1
178.7	178.5	38.58226	=CHIINV(0.005,19)
183.9	182.2	6.843971	=CHIINV(0.995,19)
179.7	180.9		
180.6	181.4	1.06085	=C2/C3
180.4	181.4	5.980446	=C2/C4
178.5	180.6		
178.8	180.1		
181.3	182.2		

Figure 7.3 EXCEL worksheet computation of 99% confidence interval for σ^2.

If SAS is used to set a 99% confidence interval on the variance, the following is found:

```
One Sample Chi-square Test for a Variance

Sample Statistics for volume

   N      Mean     Std. Dev.    Variance
------------------------------------------
   20    180.65    1.4677        2.1542

99% Confidence Interval for the Variance

   Lower Limit     Upper Limit
   --------------------------
   1.06085         5.98045
```

When sample variances, S_1^2 and S_2^2, based on independent sample sizes n_1 and n_2, are taken from two normal populations, it can be shown that $\frac{S_1^2}{S_2^2}\frac{\sigma_2^2}{\sigma_1^2}$ has an F distribution with $v_1 = n_1 - 1$ and $v_2 = n_2 - 1$ degrees of freedom. Let $F_L = F_{1-\alpha/2}$ and $F_U = F_{\alpha/2}$, then we may write $P(F_L < \frac{S_1^2}{S_2^2}\frac{\sigma_2^2}{\sigma_1^2} < F_U) = 1 - \alpha$. Now, we solve the inequality $F_L < \frac{S_1^2}{S_2^2}\frac{\sigma_2^2}{\sigma_1^2} < F_U$ for $\frac{\sigma_1^2}{\sigma_2^2}$ and we will have a $(1 - \alpha)$ confidence interval for the ratio of population variances. After a small bit of algebra, we have the following $(1 - \alpha)$ confidence interval for the ratio of population variances.

$$\frac{S_1^2}{S_2^2}\frac{1}{F_{\alpha/2(v_1,v_2)}} < \frac{\sigma_1^2}{\sigma_2^2} < \frac{S_1^2}{S_2^2}F_{\alpha/2(v_2,v_1)}$$

EXAMPLE 7.3   Two assembly lines using different methods of assembling video game consoles were compared. The number of consoles assembled per day was recorded for assembly line 1 for 13 days and for assembly line 2 for 15 days. The engineer in charge of the process wanted a 95% confidence interval for the ratio of the variance of line 1 to the variance of line 2. The data for the two assembly lines is shown in Figure 7.4.

Both sets of sample data pass the Kolmogorov–Smirnov normality test with *p*-values > 0.15.

Line 1	Line 2
81	99
104	100
115	104
111	98
85	103
121	113
95	95
112	107
100	98
117	95
113	101
109	109
101	99
	93
	105

Figure 7.4   Number of video game consoles assembled for 2 assembly lines.

Figure 7.5 gives the EXCEL computation of $\frac{S_1^2}{S_2^2}\frac{1}{F_{\alpha/2(v_1,v2)}} < \frac{\sigma_1^2}{\sigma_2^2} < \frac{S_1^2}{S_2^2}F_{\alpha/2(v_2,v_1)}$, the 95% confidence interval for

$\frac{\sigma_1^2}{\sigma_2^2}$. The data is shown in A2:B16. The sample variance of assembly line 1 is shown in C1 as 148.58 and the sample variance of assembly line 2 is shown in C2 as 31.07. The commands given in column C are shown in column D. The

95% confidence interval for $\frac{\sigma_1^2}{\sigma_2^2}$ is (1.568, 15.334) and the 95% confidence interval for $\frac{\sigma_1}{\sigma_2}$ is (1.252, 3.916).

A	B	C	D
Line 1	Line 2	148.5769231	=VAR(A2:A14)
81	99	31.06666667	=VAR(B2:B16)
104	100	3.050154789	=FINV(0.025,12,14)
115	104	3.2062117	=FINV(0.025,14,12)
111	98	1.567959436	=(C1/C2)/C3
85	103	15.33376832	=(C1/C2)*C4
121	113		
95	95	1.25218187	=SQRT(C5)
112	107	3.915835584	=SQRT(C6)
100	98		
117	95		
113	101		
109	109		
101	99		
	93		
	105		

Figure 7.5   EXCEL computation of 95% confidence interval for σ_1/σ_2.

The SAS solution to the above example is as follows:

```
Two Sample Test for Variances of units within line
Sample Statistics

    line

    Group    N      Mean       Std. Dev.   Variance
    ------------------------------------------------
    1        13     104.9231   12.189      148.5769
    2        15     101.2667   5.5737      31.06667

95% Confidence Interval of the Ratio of Two Variances

    Lower Limit    Upper Limit
    ---------------------------
    1.568          15.334
```

EXAMPLE 7.4   Construct the F curve for $v_1 = 12$ and $v_2 = 14$ degrees of freedom and locate the points that give 0.025 area in the left-hand tail and 0.025 in the right-hand tail.

The two points shown on the graph in Figure 7.6 may be confirmed using MINITAB as follows:

Inverse Cumulative Distribution Function
```
F distribution with 12 DF in numerator and 14 DF in denominator
P( X <= x )   x
0.025         0.311895
```

Inverse Cumulative Distribution Function
```
F distribution with 12 DF in numerator and 14 DF in denominator
P( X <= x )   x
0.975         3.05015
```

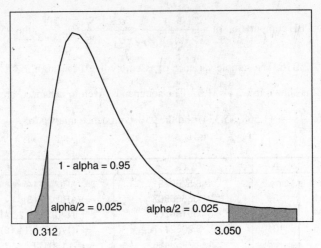

Figure 7.6 *F* distribution with $v_1 = 12$ and $v_2 = 14$ degrees of freedom.

Hypotheses Concerning One Variance

Testing hypotheses about the variance of one population is the topic for discussion in this section. First, we shall outline the technique and then apply the technique to an example. We shall call upon some of the theory we learned in the previous section when estimating population variances. The test of hypothesis follows the following outline.

One-tailed test   Two-tailed test

$H_o: \sigma^2 = \sigma_o^2$ $H_o: \sigma^2 = \sigma_o^2$

$H_a: \sigma^2 < \sigma_o^2$ $H_a: \sigma^2 \neq \sigma_o^2$

$\quad\ \sigma^2 > \sigma_o^2$

Test statistic: $\chi^2 = \dfrac{(n-1)S^2}{\sigma_o^2}$ has a chi-square distribution with $v = n - 1$ degrees of freedom.

Rejection region   Rejection region

$\chi^2 < \chi^2_{1-\alpha}$ $\chi^2 < \chi^2_{1-\alpha/2}$ or $\chi^2 > \chi^2_{\alpha/2}$

$\chi^2 > \chi^2_{\alpha}$

Assumption: The population from which the sample is selected is normal.

EXAMPLE 7.5 An industrial engineer is interested in the variation of content in containers labeled as 180 fluid ounces. Twenty are selected from production and the volumes in fluid ounces are shown in Figure 7.7. Test the null

181.5	180.8
179.1	182.4
178.7	178.5
183.9	182.2
179.7	180.9
180.6	181.4
180.4	181.4
178.5	180.6
178.8	180.1
181.3	182.2

Figure 7.7   Volume in fluid ounces.

hypothesis that the standard deviation of the population is 1.5 versus the alternative that the standard deviation is not 1.5 at $\alpha = 0.05$. Test using EXCEL and the three methods that have been discussed in the previous section and then follow this by SAS output that gives the solution.

First, we need to confirm that the sample data came from a normal population. The Kolmogorov–Smirnov test gives a p-value > 0.15. Normality is therefore a reasonable assumption.

METHOD 1   The rejection region will be found first using EXCEL. The EXCEL solution is shown in Figure 7.8. The data are shown in A2:A21. The critical values are shown in B1:B2. The computed test statistic is given in B4. The EXCEL commands are given in C1, C2, and C4. The figure tells us that the rejection region is $\chi_2 < 8.906$ or $\chi_2 > 32.852$. The computed test statistic is 18.191. We are unable to reject the null hypothesis at $\alpha = 0.05$.

A	B	C
Contents	32.85233	=CHIINV(0.025,19)
181.5	8.906517	=CHIINV(0.975,19)
179.1		
178.7	18.19111	=19*VAR(A2:A21)/2.25
183.9		
179.7		
180.6		
180.4		
178.5		
178.8		
181.3		
180.8		
182.4		
178.5		
182.2		
180.9		
181.4		
181.4		
180.6		
180.1		
182.2		

Figure 7.8   EXCEL solution to H_o: $\sigma^2 = 2.25$ versus H_a: $\sigma^2 \neq 2.25$.

Figure 7.9 shows the chi-square distribution, the rejection region, and the computed test statistic.

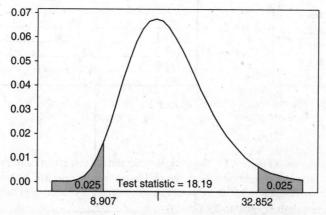

Figure 7.9   Chi-square distribution curve showing rejection region and test statistic for a two-tailed test.

METHOD 2 Compute the *p*-value, and then follow the rule that if the *p*-value is less than the pre-set alpha, reject the null. The *p*-value is given by 2*(1-CHIDIST(18.19,19)) which equals 0.98. Since 0.98 is not less than 0.05 the null is not rejected.

METHOD 3 The 95% confidence interval is computed using $\frac{(n-1)S^2}{\chi^2_{0.025}} < \sigma^2 < \frac{(n-1)S^2}{\chi^2_{0.975}}$. 19*VAR(A2:A21)/

CHIINV(0.025,19) = 1.246 and 19*VAR(A2:A21)/CHIINV(0.975,19) = 4.596. Since $\sigma_o^2 = 2.25$ is contained in the interval (1.246, 4.596) we do not reject the null hypothesis.

The SAS solution to the test of hypothesis is as follows:

```
One Sample Chi-square Test for a Variance
Sample Statistics for volume
    N      Mean      Std. Dev.     Variance
    ------------------------------------------
    20     180.65    1.4677        2.1542

Hypothesis Test
    Null hypothesis: Variance of volume = 2.25
    Alternative: Variance of volume ^= 2.25

    Chi-square   Df   Prob
    --------------------------
    18.191       19   0.9806

95% Confidence Interval for the Variance
    Lower Limit    Upper Limit
    --------------------------
    1.2459         4.5955
```

The SAS output above contains the same information as the EXCEL output.

EXAMPLE 7.6 An industrial engineer is interested in the variation of content in containers labeled as 180 fluid ounces. Ten are selected from production and the volumes in fluid ounces are shown in Figure 7.10. Test the null hypothesis that the standard deviation of the population is 1.0 versus the alternative that the standard deviation is greater than 1.0 at $\alpha = 0.05$. Test using EXCEL.

	A	B	C
	Contents	16.91898	CHIINV(0.05,9)
181.5	181.5		
179.1	179.1	25.325	9*VAR(A2:A11)/1
178.7	178.7		
183.9	183.9		
179.7	179.7		
180.6	180.6		
180.4	180.4		
178.5	178.5		
178.8	178.8		
181.3	181.3		

Figure 7.10　Volume in fluid ounces.

Figure 7.11　EXCEL solution to $H_o: \sigma^2 = 1.0$ versus $H_a: \sigma^2 > 1.0$.

The EXCEL solution is shown in Figure 7.11. The critical value and the test statistic are shown in Figure 7.12. Since the computed test statistic falls in the rejection region, the null hypothesis is rejected. If the null hypothesis were true and the population variance were equal to 1.0, there would be less than a 5% chance of getting a sample variance equal to 2.814 or equivalently, a test statistic equal to 25.325.

The conclusion is that the population variance is most likely greater than 1.

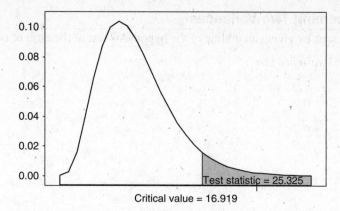

Figure 7.12 Chi-square distribution curve showing rejection region and test statistic for a one-tailed test (upper-tailed).

EXAMPLE 7.7 An industrial engineer is interested in the variation of content in containers labeled as 180 fluid ounces. Ten are selected from production and the volumes in fluid ounces are shown in Figure 7.13. Test the null hypothesis that the standard deviation of the population is 1.5 versus the alternative that the standard deviation is less than 1.5 at $\alpha = 0.05$. Test using EXCEL.

180.8
182.4
178.5
182.2
180.9
181.4
181.4
180.6
180.1
182.2

A	B	C
Contents	3.325113	=CHIINV(0.95,9)
180.8		
182.4	5.513333	=9*VAR(A2:A11)/2.25
178.5		
182.2		
180.9		
181.4		
181.4		
180.6		
180.1		
182.2		

Figure 7.13 Volume in fluid ounces.

Figure 7.14 EXCEL solution to H_o: $\sigma^2 = 1.5$ versus H_a: $\sigma^2 < 1.5$.

The EXCEL solution is shown in Figure 7.14. The critical value and the test statistic are shown in Figure 7.15. Since the computed test statistic does not fall in the rejection region, the null hypothesis is not rejected. It would not be considered unusual to get a sample variance equal to 1.378 if the population variance equaled 2.25.

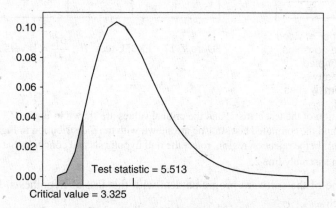

Figure 7.15 Chi-square distribution curve showing rejection region and test statistic for a one-tailed test (lower-tailed).

Hypotheses Concerning Two Variances

In this section we will start by giving an outline of the hypothesis test of the ratio of two variances.

One-tailed test   Two-tailed test

$$H_o: \frac{\sigma_1^2}{\sigma_2^2} = 1 \qquad H_o: \frac{\sigma_1^2}{\sigma_2^2} = 1$$

$$H_a: \frac{\sigma_1^2}{\sigma_2^2} < 1 \qquad H_a: \frac{\sigma_1^2}{\sigma_2^2} \neq 1$$

$$H_a: \frac{\sigma_1^2}{\sigma_2^2} > 1$$

TEST STATISTIC: $F = \frac{S_1^2}{S_2^2}$ has a F distribution with $v_1 = n_1 - 1$ and $v_2 = n_2 - 1$. v_1 is called the numerator degrees of freedom and v_2 the denominator degrees of freedom.

Rejection region   Rejection region

$$F < F_\alpha \qquad\qquad F < F_{\alpha/2} \quad \text{or} \quad F > F_{\alpha/2}$$

$$F > F_\alpha$$

Assumptions:

1. The samples are from normal populations.

2. The samples are selected independently of one another.

EXAMPLE 7.8 Two assembly lines using two different methods of assembling video game consoles were compared. The number of consoles assembled per day was recorded for assembly line 1 for 5 days and for assembly line 2 for 7 days. The engineer in charge of the process wanted to test if the ratio of variances was equal to 1 versus not equal to 1. The data for the two assembly lines is shown in Figure 7.16.

Line 1	Line 2
81	99
104	100
115	104
111	98
85	103
	113
	95

A	B	C	D
Line 1	Line 2	6.227161	=FINV(0.025,4,6)
81	99	0.108727	=FINV(0.975,4,6)
104	100	6.966573	=VAR(A2:A6)/VAR(B2:B8)
115	104		
111	98		
85	103		
	113		
	95		

Figure 7.16   Number of video game consoles assembled for the two assembly lines.

Figure 7.17   EXCEL test $H_o: \frac{\sigma_1^2}{\sigma_2^2} = 1$ versus $H_a: \frac{\sigma_1^2}{\sigma_2^2} \neq 1$.

The EXCEL computation of the test statistic and the critical values are shown in Figure 7.17.

The rejection regions and the computed test statistic are shown with the F distribution in Figure 7.18. Since the computed test statistic falls within the rejection region, reject the null hypothesis and conclude that there is more variability in assembly line 1 than in assembly line 2.

Let us see how some of the software programs deal with this test of hypothesis. EXCEL has a built-in test of $H_o: \frac{\sigma_1^2}{\sigma_2^2} = 1$ versus $H_a: \frac{\sigma_1^2}{\sigma_2^2} \neq 1$. Use the EXCEL pulldown **Tools $\Rightarrow$ Data Analysis $\Rightarrow$ F-test Two sample for variances** to perform the test of hypothesis in Figure 7.17. The output for the test is shown in Figure 7.19.

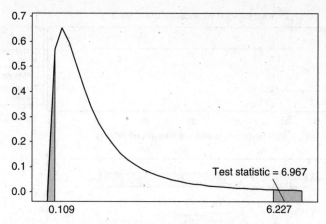

Figure 7.18 *F* distribution curve showing rejection region and test statistic for a two-tailed test.

F-Test two-sample for variances		
	Line 1	Line 2
Mean	99.2	101.7143
Variance	236.2	33.90476
Observations	5	7
df	4	6
F	6.966573	
P (F< = f) one-tail	0.019288	
F Critical one-tail	4.533677	

Figure 7.19 EXCEL pulldown test $H_o: \dfrac{\sigma_1^2}{\sigma_2^2} = 1$

versus $H_a: \dfrac{\sigma_1^2}{\sigma_2^2} \neq 1$.

The following are shown in Figure 7.19: $S_1^2 = 236.2, S_2^2 = 33.9, F = \dfrac{S_1^2}{S_2^2} = 6.97$. The one-tailed *p*-value is $P(F > 6.967) = 0.019288$. Since the test is two-tailed, *p*-value $= 2(0.019288) = 0.039$. The $\alpha = 0.05$ right-hand critical value is also given. A little work is saved if the built-in EXCEL routine in Figure 7.19 is used rather than the functions in Figure 7.17.

The MINITAB solution to the example uses the pulldown **Stat $\Rightarrow$ Basic Statistics $\Rightarrow$ 2 Variances**. The following output graphic results. (see Figure 7.20)

The *F* test results are shown along with the two-tailed *p*-value in the upper right corner of Figure 7.20.

EXAMPLE 7.9 Consider some of the changes that occur when the past example is one-tailed rather than two-tailed. Two assembly lines using two different methods of assembling video game consoles were compared. The number of consoles assembled per day was recorded for assembly line 1 for 5 days and for assembly line 2 for 7 days. The engineer in charge of the process wanted to test if the ratio of variances was equal to 1 versus the ratio was greater than 1 at $\alpha = 0.05$. The data for the two assembly lines are shown in Figure 7.16. The computed test statistic $F = 6.97$ remains the same. Consider how the rejection region changes. It is now on the upper side of the *F* distribution curve (See Figure 7.21).

The SAS output for the analysis of the hypothesis $H_o: \dfrac{\sigma_1^2}{\sigma_2^2} = 1$ versus $H_a: \dfrac{\sigma_1^2}{\sigma_2^2} > 1$ follows. Note the computed value of the test statistic under *F*. The one-tailed *p*-value, listed as $Pr > F$, is given as 0.0193. Because the *p*-value < the pre-set α, the null would be rejected. Also, note the 95% confidence interval for the ratio of variances, $1.1187 < \dfrac{\sigma_1^2}{\sigma_2^2} < 64.074$. Since the interval does not contain 1, the value specified in the null hypothesis, the null would be rejected using the confidence interval method of testing.

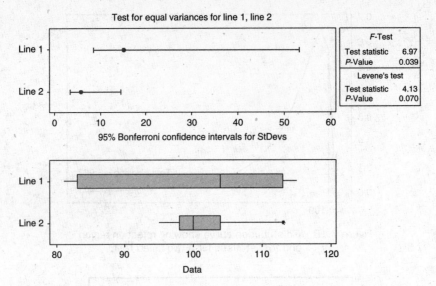

Test for equal variances for line 1, line 2

	F-Test
Test statistic	6.97
P-Value	0.039
Levene's test	
Test statistic	4.13
P-Value	0.070

Figure 7.20   MINITAB pulldown test $H_o: \dfrac{\sigma_1^2}{\sigma_2^2} = 1$ versus $H_a: \dfrac{\sigma_1^2}{\sigma_2^2} \neq 1$.

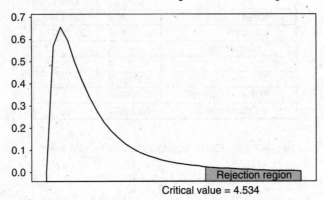

Critical value = 4.534

Figure 7.21   *F* distribution curve showing upper-tailed
rejection region.

```
Two Sample Test for Variances of units within line
Sample Statistics
line

    Group      N       Mean        Std. Dev.        Variance
    ----------------------------------------------------------
    1          5       99.2        15.369           236.2
    2          7       101.7143    5.8228           33.90476

Hypothesis Test
    Null hypothesis: Variance 1 / Variance 2 <= 1
    Alternative: Variance 1 / Variance 2 > 1

                   - Degrees of Freedom -

    F          Numer.      Denom.      Pr > F
    -------------------------------------------------
    6.97       4           6           0.0193

95% Confidence Interval of the Ratio of Two Variances
    Lower Limit              Upper Limit
    -----------              -----------
    1.1187                   64.074
```

EXAMPLE 7.10   It is believed that variability in female engineers' salaries is larger than the variability in male engineers' salaries. The data on salaries in Figure 7.22 was collected in a random sampling of males and females who had been working for 10 years or less.

Male	Female
61	65
73	67
70	70
71	70
71	51
71	75
65	72
65	58
71	67
71	83
67	73
68	53
65	70
77	47

Figure 7.22   Engineers salaries in thousands.

Identifying males as group 1 and females as group 2, we wish to test $H_o: \frac{\sigma_1^2}{\sigma_2^2} = 1$ versus $H_a: \frac{\sigma_1^2}{\sigma_2^2} < 1$ at pre-set $\alpha = 0.05$. The critical value is =FINV(0.95,13,13) or 0.388. The rejection region is shown in Fig. 7.23.

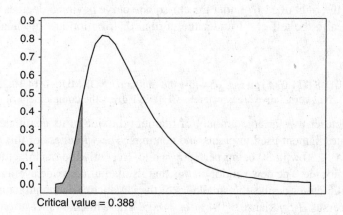

Critical value = 0.388

Figure 7.23   *F* distribution curve showing lower-tail rejection region.

The variance of the 14 male salaries is $S_1^2 = 16.769$ and the variance of the 14 female salaries is $S_2^2 = 101.874$ and the computed test statistic is $F = 16.769/101.874 = 0.165$. This computed test statistic falls in the rejection region, and we would conclude that females' salaries have a greater variance than males' salaries.

The SAS output is as follows:

```
Two Sample Test for Variances of salary within sex
Sample Statistics
sex

  Group     N      Mean       Std. Dev.     Variance
  -------------------------------------------------------
  1        14      69         4.095         16.76923
  2        14      65.78571   10.093        101.8736
```

```
Hypothesis Test
    Null hypothesis: Variance 1 / Variance 2 => 1
    Alternative: Variance 1 / Variance 2 < 1

            - Degrees of Freedom -
    F       Numer.    Denom.     Pr < F
    -------------------------------------
    0.16     13        13        0.0013
95% Confidence Interval of the Ratio of Two Variances
    Lower Limit            Upper Limit
    -----------            -----------
    0.0528                 0.5128
```

The 95% confidence interval for $\frac{\sigma_1^2}{\sigma_2^2}$ is $0.0528 < \frac{\sigma_1^2}{\sigma_2^2} < 0.5128$. Note that this confidence interval does not include the null value for $\frac{\sigma_1^2}{\sigma_2^2}$ which is 1 and, therefore, the null hypothesis is rejected.

SOLVED PROBLEMS

Estimation of variances

7.1. The EXCEL function =CHIDIST is used to find areas under the chi-square curve. Consider the chi-square curve with 5 degrees of freedom. Find the area **to the right** of 11.07. The function =CHIDIST(11.07,5) gives the area to the right of 11.07 under the chi-square curve having 5 degrees of freedom. Contrast this with the function =NORMDIST. The function =NORMDIST(14,10,2,1) gives the area under the normal curve with mean = 10 and standard deviation = 2 **to the left** of 14. Find the area to the right of 11.07 under the chi-square curve having 5 degrees of freedom. Contrast this with the area to the left of 14 under the normal distribution having mean = 10 and standard deviation = 2.

SOLUTION

The function =CHIDIST(11.07,5) gives 0.05 while the function =NORMDIST(14,10,2,1) gives 0.97725. Note that NORMDIST gives cumulative areas whereas CHIDIST gives the complement of cumulative areas.

7.2. A paint manufacturer uses a large amount of titanium dioxide in its coatings. Titanium dioxide is the primary white pigment used in paints and coatings. The whiteness of this pigment is measured using a scale of 0 – 30 with 30 being perfectly white. Recently the manufacturer switched vendors for its titanium dioxide. The new vendor claims that its titanium dioxide averages 24 with a variance of 0.4. The manufacturer doubts the smallness of the variance. The manufacturer wishes to test H_o: variance = 0.4 versus H_a: variance > 0.4 at $\alpha = 0.05$. The whiteness measurements for a sample of 10 shipments were: 24, 26, 25, 24, 27, 25, 25, 26, 26, and 25. Calculate the test statistic for this test. Give the rejection region for the test. What is your conclusion?

SOLUTION

```
Sample Statistics for whiteness
  N    Mean   Std. Dev.  Variance
  ------------------------------
  10   25.3   0.9487     0.9

Hypothesis Test
    Null hypothesis: Variance of whiteness <= 0.4
    Alternative: Variance of whiteness > 0.4
    Chi-square    Df     Prob
    ---------------------------
    20.250        9      0.0164
```

The above SAS output gives the computed test statistic as chi-square $= (10-1)S^2/0.4 = 20.25$ with a p-value $= 0.0164$. The critical value is 16.919 and the test statistic exceeds the critical value. We reject the null hypothesis.

7.3. Give a 95% confidence interval for the population variance in Problem 7.2.

SOLUTION

$$\frac{(n-1)S^2}{\chi^2_{0.025}} < \sigma^2 < \frac{(n-1)S^2}{\chi^2_{0.975}} \quad \frac{(10-1)S^2}{19.028} < \sigma^2 < \frac{(n-1)S^2}{2.7004} \text{ or } (0.42, 3.00)$$

The 95% confidence interval for variance ranges from 0.42 to 3.00. Part of the MINITAB output is as follows:

```
95% Confidence Intervals

                                        CI for
Variable     Method       CI for StDev  Variance
whiteness    Chi-Square   (0.653, 1.732)  (0.426, 3.000)
             Bonett       (0.639, 1.751)  (0.409, 3.067)
```

The MINITAB output is obtained from the pulldown **Stat** $\Rightarrow$ **Basic statistics** $\Rightarrow$ **1-variance**.

Hypotheses concerning one variance

7.4. A study involving aluminum contamination in recycled PET plastic was conducted by an engineer. She collected 26 samples and measured in parts per million (ppm) the amount of aluminum contamination. The designer claimed that the standard deviation of these concentrations is 60. The data was as follows:

291	63	119	222	30	511	125	140	120	79	101	172	145
102	70	119	87	30	244	183	90	118	60	115	182	191

Normality is rejected using the Ryan–Joiner, the Anderson–Darling, and the Kolmogorov–Smirnov tests for normality.

Test the claim concerning variability at a 0.05 significance level.

SOLUTION

The MINITAB output is obtained from the pulldown **Stat** $\Rightarrow$ **Basic statistics** $\Rightarrow$ **1-variance**.

```
Null hypothesis          Sigma = 60
Alternative hypothesis   Sigma not = 60

The chi-square method is only for the normal distribution.
The Bonett method is for any continuous distribution.

Statistics
Variable    N      StDev    Variance
ppm         26     98.2     9644

95% Confidence Intervals
                         CI for         CI for
Variable    Method       StDev          Variance
ppm         Chi-Square   (77.0, 135.6)  (5932, 18377)
            Bonett       (53.4, 195.1)  (2857, 38081)

Tests
                         Test
Variable    Method       Statistic  DF   P-Value
ppm         Chi-Square   66.97      25   0.000
            Bonett       –          –    0.107
```

The Bonett method for the test that the standard deviation is 60 versus it is not, has a *p*-value equal to 0.107. The null hypothesis is not rejected.

7.5. A study looked at the average particle size of a product. Specifications suggest that the variance of particle sizes is 144. Production personnel measured the particle size distribution using a set of screening sieves. They tested one sample a day and found the average particle sizes for the past 25 days to be:

particle size

99.6	102.8	101.5	103.2	102.3	92.1	100.9	96.7	97.5
93.8	103.8	100.5	96.8	98.3	102.7	95.3	102.7	97.8
105.8	94.9	101.6	96.9	104.7	100.6	94.9		

Test the data for normality using MINITAB. Test the claim about variability at the 0.05 significance level. Construct a 95% confidence interval for the true variance of the particle sizes.

SOLUTION

Using MINITAB, we find that normality cannot be rejected using the Anderson–Darling test, the Ryan–Joiner test, and the Kolmogorov–Smirnoff test.

The MINITAB output from the test that the standard deviation is 12 is as follows:

```
Null hypothesis          Sigma = 12
Alternative hypothesis   Sigma not = 12
The chi-square method is only for the normal distribution.
The Bonett method is for any continuous distribution.

Statistics
Variable          N          StDev          Variance
particlesize      25         3.70           13.7

95% Confidence Intervals

                             CI for         CI for
Variable          Method     StDev          Variance
particlesize      Chi-Square (2.89, 5.14)   (8.3, 26.5)
                  Bonett     (3.04, 4.87)   (9.3, 23.7)

Tests
Test
Variable          Method     Statistic      DF    P-Value
particlesize      Chi-Square 2.28           24    0.000
                  Bonett     —              —     0.000
```

Since the normality assumption is not rejected, use the analysis for the chi-square method. Because of the *p*-value (0.000), reject the null hypothesis. The 95% confidence interval for the variance is (8.3, 26.5).

7.6. Highway engineers analyzed a 957-m stretch of a recently constructed two-lane highway. A borehole test for silt was conducted. The goal of the study is to show that the variance in thickness is less than 0.05 m. The data for the study is: 0.6, 0.8, 0.7, 1.1, 0.8, 0.8, 0.6, 0.5, 0.4, 0.4, 0.8, 0.6. Use SAS to test this criterion about variability using a 0.10 significance level.

SOLUTION

```
One Sample Chi-square Test for a Variance          1

   Sample Statistics for thickness
   N     Mean     Std. Dev.    Variance
   ------------------------------------
   12    0.675    0.2006       0.0402
```

```
Hypothesis Test
    Null hypothesis: Variance of thickness => 0.05
    Alternative: Variance of thickness < 0.05

        Chi-square    Df    Prob
        ---------------------------
        8.850         11    0.3643
```

Since the *p*-value = 0.3643 is greater than 0.10, the null hypothesis cannot be rejected.

Hypotheses concerning two variances

7.7. A chemical engineer describes materials prepared by selective functionalization of olefin-containing polymers to produce novel EPDM and polybutadiene polyols. The properties of the final urethane can be modified by the addition of other agents, such as 1,4-butanediol. One question of interest is how the variation is affected by increasing the ratio of 1,4-butanediol to polyol. The ultimate strength in psi is:

Ratio-1:1	513	1415	619	1699
Ratio-1:2	1278	2528	758	2332

Test for equal variability using MINITAB.

SOLUTION

Both samples pass the three normality tests that are in MINITAB. This is an assumption of the equal variability test.

The MINITAB output for testing equal variances is shown below.

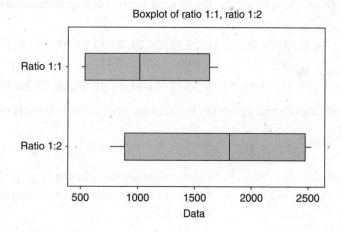

Boxplot of ratio 1:1, ratio 1:2

The *F* test gives $F = \dfrac{S_1^2}{S_2^2} = 0.48$ and the *p*-value is 0.560. Using the *p*-value approach you do not reject the null hypothesis of equal variances.

7.8. Engineers investigated the production of cyclo-dextrin glycosyl-transferase enzymes by bacterial cultures. Enzyme production was done in shaken and surface cultures. An important question is whether the two methods have equal variances. The protein content of the cultures in milligrams per milliliter is:

Shaken culture	1.91	1.66	2.64	2.62	2.57	1.85
Surface culture	1.71	1.57	2.51	2.30	2.25	1.15

Test the hypothesis of equal variability for the two methods versus unequal variability for the two methods at a level of significance equal to 0.10. Give the SAS output.

SOLUTION

```
Sample Statistics
   group
   Group    N      Mean      Std. Dev.    Variance
   -----------------------------------------------------
   1        6      2.208333   0.4483      0.200937
   2        6      1.915      0.5217      0.27215
```

Hypothesis Test

```
    Null hypothesis: Variance 1 / Variance 2 = 1
    Alternative: Variance 1 / Variance 2 ^= 1

               - Degrees of Freedom -
    F         Numer.       Denom.       Pr > F
    -----------------------------------------
    0.74      5            5            0.7473
```

95% Confidence Interval of the Ratio of Two Variances

```
    Lower Limit      Upper Limit
    -----------      -----------
    0.1033           5.2764
```

The *p*-value (0.7473) is not less than 0.10. The hypothesis of equal variability is not rejected.

7.9. Chemical engineers made spaghetti from durum wheats with strong Vic gluten. The spaghetti was dried at a high temperature of 80°C and a low temperature of 40°C. Of interest was whether the variability in the retention times was the same at the two drying temperatures.

Retention times at 40°C:
     28.68, 19.76, 20.40, 22.96, 22.31, 25.27, 24.29, 23.75, 26.68, 30.69, 19.13, 27.66

Retention times at 80°C:
     25.97, 28.44, 30.37, 23.55, 22.75, 22.11, 24.09, 18.97, 26.48, 27.39, 19.62, 20.22

   Test the hypothesis of equal variability in retention times at the two temperatures using MINITAB.

SOLUTION

```
Method
Null hypothesis           Sigma(40degrees)/Sigma(80degrees) = 1
Alternative hypothesis    Sigma(40degrees)/Sigma(80degrees) not = 1
Significance level        Alpha = 0.05

Statistics
Variable        N     StDev  Variance
40degrees       12    3.650  13.320
80degrees       12    3.638  13.238

Continuous     (0.524, 1.808) (0.274, 3.270)

Tests
Test
Method                            DF1   DF2   Statistic   P-Value
F Test (normal)                   11    11    1.01        0.992
Levene's Test (any continuous)    1     22    0.00        0.955
```

Do not reject the null hypothesis.

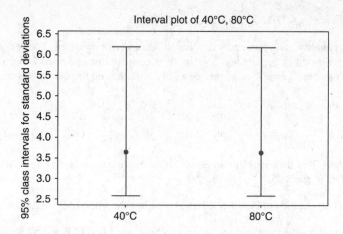

SUPPLEMENTARY PROBLEMS

7.10. Find $\chi^2_{.99}$ and $\chi^2_{.01}$ for a chi-square distribution having 10 degrees of freedom using EXCEL and MINITAB.

7.11. The diameter of a manufactured piston is of interest and the null hypothesis is that the variance is 0.01 centimeters. If the variance exceeds 0.01, the process must be stopped and repaired. The data is: 1.6, 1.5, 1.8, 1.5, 1.4, 1.5, 1.6, 1.7, 1.9, and 1.5. Use EXCEL to test the null hypothesis that the variance equals 0.01 versus the alternative hypothesis that the variance is greater than 0.01 at level of significance 0.1. Give the output for the computed value of the test statistic. Give the computed *p*-value. What is your conclusion? Why?

7.12. Two types of drill bits were compared with respect to the variation in their lifetimes. Ten from Company 1 were tested with the result that the sample variance equaled 3.78 and ten from Company 2 were tested with the result that the sample variance equaled 1.76. Test $H_o: \dfrac{\sigma_1^2}{\sigma_2^2} = 1$ against $H_a: \dfrac{\sigma_1^2}{\sigma_2^2} \neq 1$ at $\alpha = 0.10$. Give your conclusion.

7.13. The number of accidents was measured at busy intersections being controlled by two different devices. Type 1 device is used at 10 intersections and Type 2 device is used at 12 other intersections. The number of accidents at the intersections over a one-week period is shown in the table below. Test $H_o: \dfrac{\sigma_1^2}{\sigma_2^2} = 1$ versus $H_a: \dfrac{\sigma_1^2}{\sigma_2^2} \neq 1$ at $\alpha = 0.05$. Use SAS to test the hypothesis. Give the SAS output, the calculated test statistic, the *p*-value, and your conclusion.

Comparison of Two Devices Used to Control Traffic

Type 1	Type 2
43	64
52	36
46	50
53	42
62	52
62	52
54	57
41	64
51	52
42	70
	68
	67

7.14. Find $F_{0.01}(4,7)$ and $F_{0.99}(4,7)$ using EXCEL and MINITAB.

7.15. Tests of product quality using inspectors can lead to serious error problems. The performance of inspectors using novice and experienced inspectors was compared. Each inspector classified 200 products as defective or non-defective. The number of errors for 12 novice and 12 experienced inspectors was recorded. The results were as below.

Novice	25	35	26	40	46	20	45	31	33	29	21	49
Experienced	31	15	25	19	28	17	19	18	24	10	20	21

Test the hypothesis that the variability in errors for novice and experienced inspectors are the same versus they are different. Test at $\alpha = 0.05$.

7.16. Give the MINITAB solution to Problem 7.11 and compare the MINITAB solution with the EXCEL solution.

7.17. Give the MINITAB solution to Problem 7.13 and compare the MINITAB and SAS solutions.

CHAPTER 8

Inferences Concerning Proportions

Estimation of Proportions

Engineers are often concerned with proportions or percentages. As usual we can estimate the percentage of defectives that a process produces when it is mass producing items such as cell phones, video game consoles, automobiles, etc. Or we may be interested in testing hypotheses about the percentage of defectives that are produced when we are mass producing items. Suppose we are producing a product and are interested in the percentage of defectives that we are manufacturing, p. We need to keep up with the value of p so that we can remain competitive. The way we do this is to select a sample of size n and determine X, the number of defectives in the sample. The **sample proportion** $\hat{p}$ is the proportion in the sample that have the characteristic of interest, such as being defective. Before the sample is taken, the sample proportion $\hat{p} = \dfrac{X}{n}$ is a random variable. In fact X is binomial with parameters n and p. The sample size n may be thought of as the number of trials and p, as the probability of success. Before the sample is actually taken, the probability that a given item will have the characteristic of interest (like being defective) is p. Remember, we do not know the value of p. The value of p is our main interest. We found when we studied the binomial, the mean equals np and the variance equals npq. Or

$$E(X) = np \text{ and } Var(X) = npq.$$

Applying the properties of expectation and variance, we have, dividing both sides by n $E\left(\dfrac{X}{n}\right) = E(\hat{p}) = \dfrac{np}{n} = p$ and we see that $\hat{p} = \dfrac{X}{n}$ is an unbiased estimator of p. Similarly,

$$Var\left(\frac{X}{n}\right) = \frac{pq}{n}.$$

We also found when $np > 5$ and $nq > 5$, the distribution of X can be approximated by a normal distribution. The distribution of $\hat{p} = \dfrac{X}{n}$ can also be approximated by a normal distribution. The variable $Z = \dfrac{\hat{p} - p}{\sqrt{pq/n}}$ has a standard normal distribution. Recalling that $P(-1.96 < Z < 1.96) = 0.95$ and substituting for Z, we have $P\left(-1.96 < \dfrac{\hat{p} - p}{\sqrt{pq/n}} < 1.96\right) = 0.95$. Solving the inside part for p, we have $\left(\hat{p} - 1.96\sqrt{\dfrac{pq}{n}}, \hat{p} + 1.96\sqrt{\dfrac{pq}{n}}\right)$ as a 95% confidence interval for p. Note that we have a problem. The unknown p and q under the square root keeps us from being able to find a numerical value for our interval. We use the estimates for p and q obtained from the sample to finally arrive at our 95% confidence interval.

$$\left(\hat{p} - 1.96\sqrt{\frac{\hat{p}\hat{q}}{n}}, \hat{p} + 1.96\sqrt{\frac{\hat{p}\hat{q}}{n}}\right)$$

197

EXAMPLE 8.1 Cell phones are mass produced. A sample of 200 is taken and 10 have a small scratch on the body of the phone. Estimate the proportion of the day's production p that have a small scratch on the body with a 90% confidence interval. The point estimate of p is $\hat{p} = \dfrac{X}{n} = \dfrac{10}{200} = 0.05$ or 5%. First question: Is the normal approximation valid? np is estimated by $n\hat{p} = 200(0.05) = 10$ and $n\hat{q} = 200(0.95) = 190$ and both exceed 5. So the normal approximation is valid. For 90% confidence we need to put 0.05 in the right and left tails of the standard normal curve. Using EXCEL to find the standard normal values, we find =NORMSINV(0.05) = −1.645 and =NORMSINV(0.95) = 1.645. The lower confidence limit is $0.05 - 1.645\sqrt{\dfrac{0.05(0.95)}{200}} = 0.025$ and the upper confidence limit is $0.05 + 1.645\sqrt{\dfrac{0.05(0.95)}{200}} = 0.075$. We are 90% confident that the percentage with a small scratch is between 2.5% and 7.5%. Generalizing, a $(1 - \alpha)$ confidence interval for p is $\left(\hat{p} - z_{\alpha/2}\sqrt{\dfrac{\hat{p}\hat{q}}{n}},\ \hat{p} + z_{\alpha/2}\sqrt{\dfrac{\hat{p}\hat{q}}{n}} \right)$.

The MINITAB solution to the above problem uses the pulldown menu **Stat ⇒ Basic Statistics ⇒ 1-proportion.** The dialog box given in Figure 8.1 is filled in as shown. Under options, choose 90% confidence level and choose use test and interval based on the normal distribution.

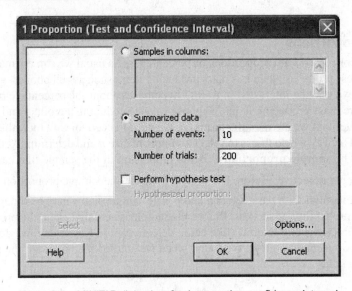

Figure 8.1   MINITAB dialog box for 1-proportion confidence interval.

The following output is produced by the dialog box shown in Figure 8.1. Since we are interested only in the 90% confidence interval at this point, the perform hypothesis test box is left blank. The output is as follows:

```
Test and CI for One Proportion

Sample   X    N    Sample p    90% CI
1        10   200  0.050000    (0.024651, 0.075349)

Using the normal approximation.
```

The 90% confidence interval is the same as was reached using EXCEL. The quantity $|\hat{p} - p|$ is the error made when p is estimated by $\hat{p}$. Another way of expressing the confidence interval is $|\hat{p} - p| < z_{\alpha/2}\sqrt{\dfrac{p(1 - p)}{n}}$. The quantity $E = z_{\alpha/2}\sqrt{\dfrac{p(1 - p)}{n}}$ is called the **maximum error of estimate**. If the error is specified, then the sample size needed to give that error is determinable. Simply solve the equation for n. When the equation is solved for n, we get $n = p(1 - p)[z_{\alpha/2}/E]^2$. If a previous study involving p has been conducted, then p may be replaced by $\hat{p}$ to obtain $n = \hat{p}(1 - \hat{p})[z_{\alpha/2}/E]^2$ as the needed sample size. If nothing is known concerning the possible size of p, then p may be replaced by $p = 0.5$ to obtain a conservative estimate of n. The reason for this is that $p(1 - p)$ is maximum when $p = 0.5$. The function $f(p) = p(1 - p) = p - p^2$ is a quadratic function in p. It reaches its maximum value when $p = 0.5$. If the derivative is set equal to 0 and the resulting equation solved for p, the solution is $p = 0.5$. Therefore, if no estimate of p is available, $n = 0.25[z_{\alpha/2}/E]^2$ is the sample to use to give error of estimate equal to E.

EXAMPLE 8.2 The Department of Health wishes to estimate the percentage of teens p who currently smoke. They wish to estimate p with a 95% maximum error of estimate equal to 3%. (a) If the estimate one year ago was 25%, what is the sample size needed today? (b) If no previous figure exists to use for p, what is the sample size?

(a) Using the last year's figure for p, the sample size is $n = \hat{p}(1 - \hat{p})[z_{\alpha/2}/E]^2 = 25(75)[1.96/3]^2 = 800.33$. Round up to 801.

(b) No previous estimate exists. The sample size is $n = 0.25[z_{\alpha/2}/E]^2 = 0.25[1.96/0.03]^2 = 1067.11$. Round up to 1068.

Note: Use percentage for both E and your estimate of p or use decimal values for both. In part (a), percentages were used and in part (b), decimal values were used. Also, note that if no previous estimate for p is known, the sample size is considerably larger.

Now we consider the case of two proportions, p_1 and p_2. Suppose we are comparing two machines with respect to the percentage of defectives that the two produce. Two samples of sizes n_1 and n_2 are taken. The binomial random variables X_1 and X_2 represent the number of defectives in the two samples. The sample proportions, $\hat{p}_1 = \dfrac{X_1}{n}$ and $\hat{p}_2 = \dfrac{X_2}{n}$, are random variables that represent the proportion of successes in each of the samples. The difference $\hat{p}_1 - \hat{p}_2$ is an unbiased point estimate of $p_1 - p_2$, that is $E(\hat{p}_1 - \hat{p}_2) = p_1 - p_2$. It can be shown that the standard error of the difference in sample proportions is the following: $\sigma_{(\hat{p}_1 - \hat{p}_2)} = \sqrt{\dfrac{p_1 q_1}{n_1} + \dfrac{p_2 q_2}{n_2}}$. This standard error is approximated by $\sqrt{\dfrac{\hat{p}_1 \hat{q}_1}{n_1} + \dfrac{\hat{p}_2 \hat{q}_2}{n_2}}$. When the sample sizes are large enough, $Z = \dfrac{\hat{p}_1 - \hat{p}_2 - (p_1 - p_2)}{\sqrt{\hat{p}_1 \hat{q}_1/n_1 + \hat{p}_2 \hat{q}^2/n_2}}$ has an approximate standard normal distribution. The probability is $1 - \alpha$ that $Z = \dfrac{\hat{p}_1 - \hat{p}_2 - (p_1 - p_2)}{\sqrt{\hat{p}_1 \hat{q}_1/n_1 + \hat{p}_2 \hat{q}^2/n_2}}$ is between $-z_{\alpha/2}$ and $z_{\alpha/2}$. If the resulting inequality is solved for

$p_1 - p_2$ we obtain $(\hat{p}_1 - \hat{p}_2) - z_{\alpha/2}\sqrt{\dfrac{\hat{p}_1 \hat{q}_1}{n_1} + \dfrac{\hat{p}_2 \hat{q}_2}{n_2}} < p_1 - p_2 < (\hat{p}_1 - \hat{p}_2) + z_{\alpha/2}\sqrt{\dfrac{\hat{p}_1 \hat{q}_1}{n_1} + \dfrac{\hat{p}_2 \hat{q}_2}{n_2}}$ as a $(1 - \alpha)$ confidence interval for $p_1 - p_2$.

EXAMPLE 8.3 Two machines are to be compared with respect to the percentage of defectives that they give. A sample of 200 is taken from each machine. The number of defectives from machine 1 is 20 and the number defectives from machine 2 is 15. Use EXCEL to set a 99% confidence interval on $p_1 - p_2$.

Figure 8.2 gives an EXCEL method of setting a 99% confidence interval on $p_1 - p_2$.

Numerical value	EXCEL command	Book notation
A	B	C
0.1	=20/200	$\hat{p}_1$
0.075	=15/200	$\hat{p}_2$
0.9	=1-A1	$1-\hat{p}_1$
0.925	=1-A2	$1-\hat{p}_2$
0.028229	=SQRT((A1*A3/200)+(A2*A4/200))	$\sqrt{\dfrac{\hat{p}_1 \hat{q}_1}{n_1} + \dfrac{\hat{p}_2 \hat{q}_2}{n_2}}$
0.025	=A1-A2	$\hat{p}_1 - \hat{p}_2$
2.575829	=NORMSINV(0.995)	$Z_{0.005}$
-0.04771	=A6-A7*A5	$(\hat{p}_1 - \hat{p}_2) - z_{\alpha/2}\sqrt{\dfrac{\hat{p}_1 \hat{q}_1}{n_1} + \dfrac{\hat{p}_2 \hat{q}_2}{n_2}}$
0.097713	=A6+A7*A5	$(\hat{p}_1 - \hat{p}_2) + z_{\alpha/2}\sqrt{\dfrac{\hat{p}_1 \hat{q}_1}{n_1} + \dfrac{\hat{p}_2 \hat{q}_2}{n_2}}$

Figure 8.2 EXCEL routine for confidence interval for $p_1 - p_2$.

The MINITAB pulldown **Stat ⇒ Basic Statistics ⇒ 2-proportions** is used to set the confidence interval on $p_1 - p_2$. The results are

```
Test and CI for Two Proportions
Sample    X     N     Sample p
1        20    200    0.100000
2        15    200    0.075000

Difference = p (1) - p (2)
Estimate for difference: 0.025
99% CI for difference: (-0.0477130, 0.0977130)
```

Hypotheses Concerning One Proportion

We have considered the estimation of a proportion in the first section of this chapter. Much of the same theory is used to test hypotheses about a proportion. An outline of the statistical test concerning a population proportion is given first, followed by an example. The test is based on the normal distribution. The sample size n will be assumed to be large enough so that $\hat{p} = \dfrac{X}{n}$ may be assumed to have an approximate normal distribution. The null hypothesis is $H_o: p = p_o$. When the null is true, the mean of the random variable $\hat{p} = \dfrac{X}{n}$ will be p_o and the standard deviation of $\hat{p} = \dfrac{X}{n}$ will be $\sqrt{\dfrac{p_o q_o}{n}}$. **Remember, the null is assumed true until we get results that are not very likely to occur if the null is true. When this happens, we reject the null hypothesis.** Assuming the null hypothesis is true, the variable $Z = \dfrac{\hat{p} - p_o}{\sqrt{p_o q_o/n}}$ will have a standard normal distribution.

Large-sample test of hypothesis about a population proportion or percentage:

One-tailed test   Two-tailed test

$H_o: p = p_o$ $H_o: p = p_o$

$H_a: p < p_o$ $H_a: p \neq p_o$

$H_a: p > p_o$

$$\text{Test statistic: } Z = \frac{\hat{p} - p_o}{\sqrt{\dfrac{p_o q_o}{n}}}$$

Rejection region   Rejection region

$Z < -z_\alpha$ $Z < -z_{\alpha/2}$ $Z > z_{\alpha/2}$

$Z > z_\alpha$

Assumption: The sample is large enough so that the normal approximation is valid.

EXAMPLE 8.4 A traffic engineer claims that over 20% of all trucks that cross a bridge exceed the weight limit for the bridge. One hundred trucks are stopped and weighed after crossing the bridge. Twenty-five of the 100 are over the weight limit. Test $H_o: p = 20\%$ versus $H_a: p > 20\%$ at $\alpha = 0.05$. Test using all the three methods of testing hypotheses.
The MINITAB output is as follows:

```
Test and CI for One Proportion
Test of p = 0.2 vs p > 0.2
                                  95%
                                  Lower
Sample   X     N    Sample p    Bound      Z-Value   P-Value
1       25   100    0.250000    0.178776   1.25      0.106
```

METHOD 1   The rejection region is $Z > 1.645$. Since the computed test statistic ($Z = 1.25$) does not fall in the rejection region, we are unable to reject the null.

METHOD 2   The p-value is 0.106. Since the p-value is not less than α, we are unable to reject the null.

METHOD 3   The one sided confidence interval is $(0.179, \infty)$. The null value is in the interval. We are unable to reject the null.

Test the hypothesis using EXCEL.

From Figure 8.3, the rejection region is $Z > 1.645$, the test statistic is 1.25 which does not exceed 1.645. Therefore, do not reject the null.

Value	EXCEL command	Book terminology
1.64485	=NORMSINV(0.95)	Critical value
1.25	=(0.25-0.2)/SQRT(0.2*0.8/100)	Test statistic
0.10565	=1-NORMSDIST(1.25)	p-value
0.17876	=0.25-1.645*SQRT(0.25*0.75/100)	95% lower bound

Figure 8.3   EXCEL solution to test of hypothesis about p.

The p-value is not less than $\alpha = 0.05$. Therefore, do not reject the null.

The 95% lower bound is $(0.179, \infty)$. The $p_o = 0.2$ is in this interval. Therefore, do not reject the null. The EXCEL and the MINITAB solutions are the same.

Hypotheses Concerning Two Proportions

When making decisions about two populations with respect to proportions, we use the random variables $\hat{p}_1 = \frac{X_1}{n_1}$ and $\hat{p}_2 = \frac{X_2}{n_2}$. Two cases present themselves and they require slightly different theory. The two cases are $H_o: (p_1 - p_2) = D_o$ and D_o is either 0 or not 0. In both cases, the same test statistic will be used: $Z = \frac{\hat{p}_1 - \hat{p}_2 - D_o}{\sigma_{(\hat{p}_1 - \hat{p}_2)}}$. However, our estimate of $\sigma_{(\hat{p}_1 - \hat{p}_2)}$, the standard error of the difference in sample proportions, will differ depending on the value of D_o. First, consider the case where the hypothesized difference in proportions is non-zero. An outline of this case is:

One-tailed test   Two-tailed test

$H_o: (p_1 - p_2) = D_o \neq 0$ $H_o: (p_1 - p_2) = D_o \neq 0$

$H_a: (p_1 - p_2) < D_o$ $H_a: (p_1 - p_2) \neq D_o$

$H_a: (p_1 - p_2) > D_o$

Test statistic: $Z = \dfrac{\hat{p}_1 - \hat{p}_2 - D_o}{\sigma_{(\hat{p}_1 - \hat{p}_2)}}$, where $\sigma_{(\hat{p}_1 - \hat{p}_2)} = \sqrt{\dfrac{\hat{p}_1 \hat{q}_1}{n_1} + \dfrac{\hat{p}_2 \hat{q}_2}{n_2}}$

Rejection region   Rejection region

$Z < -z_\alpha$ $Z < -z_{\alpha/2}$ $Z > z_{\alpha/2}$

$Z > z_\alpha$

Assumption: Samples are large enough so that binomial distribution is approximated by normal.

$n_1 \hat{p}_1 > 5$, $n_1 \hat{q}_1 > 5$, $n_2 \hat{p}_2 > 5$, $n_2 \hat{q}_2 > 5$

EXAMPLE 8.5   It is hypothesized that the percentage of women engineers who are considered to be of a healthy weight exceeds men engineers who are considered to be of a healthy weight by 10%. The alternative hypothesis is that the percentage is less than 10%. That is, the null hypothesis is $H_o: (p_1 - p_2) = 0.10$ and the alternative hypothesis is

H_a: $(p_1 - p_2) < 0.10$ at $\alpha = 0.05$. The data in random samples were as follows: Women: $n_1 = 200$, $x_1 = 78$. Men: $n_2 = 200$, $x_2 = 63$. Give an EXCEL solution followed by a MINITAB solution.

Figure 8.4 contains the EXCEL solution to the example. The number 0.39 is in cell A1.

Value	EXCEL command	Book notation
A	B	C
0.39	=78/200	$\hat{p}_1$
0.315	=63/200	$\hat{p}_2$
0.075	=78/200-63/200	$\hat{p}_1 - \hat{p}_2$
0.0476	=SQRT(A1*(1-A1)/200+A2*(1-A2)/200)	$\sigma_{(\hat{p}_1-\hat{p}_2)}$
-0.5249	=(A3-0.1)/A4	Test statistic
-1.6448	=NORMSINV(0.05)	$Z < -z_\alpha$
0.2998	=NORMSDIST(A5)	*p*-value
0.1533	=A3 + NORMSINV(0.95)*A4	Upper confidence bound

Figure 8.4   EXCEL solution to test of hypothesis about $(p_1 - p_2) = D_0$.

The MINITAB solution is as follows:

```
Sample   X     N      Sample p
1        78    200    0.390000
2        63    200    0.315000

Difference = p (1) - p (2)
Estimate for difference: 0.075
95% upper bound for difference: 0.153340
Test for difference = 0.1 (vs < 0.1): Z = -0.52 P-Value = 0.300
```

More of the work is done for the engineer in MINITAB. A better understanding of how the tests work is required for EXCEL.

Now, consider the case where the hypothesized difference in proportions is zero. An outline of this case is:

One-tailed test	Two-tailed test
H_o: $(p_1 - p_2) = 0$	H_o: $(p_1 - p_2) = 0$
H_a: $(p_1 - p_2) < 0$	H_a: $(p_1 - p_2) \neq 0$
H_a: $(p_1 - p_2) > 0$	

Test statistic: $Z = \dfrac{\hat{p}_1 - \hat{p}_2 - 0}{\sigma_{(\hat{p}_1-\hat{p}_2)}}$, where $\sigma_{(\hat{p}_1-\hat{p}_2)} = \sqrt{\hat{p}\hat{q}\left(\dfrac{1}{n_1} + \dfrac{1}{n_2}\right)}$ and $\hat{p} = \dfrac{x_1 + x_2}{n_1 + n_2}$

Rejection region	Rejection region
$Z < -z_\alpha$	$Z < -z_{\alpha/2}$ $Z > z_{\alpha/2}$
$Z > z_\alpha$	

Assumption: Samples are large enough so that binomial distribution is approximated by the normal distribution.

$n_1\hat{p}_1 > 5$, $n_1\hat{q}_1 > 5$, $n_2\hat{p}_2 > 5$, $n_2\hat{q}_2 > 5$

In this case, the null says the two proportions are equal. Assuming the null to be true, we pool our two samples to get an estimate of the common proportion p. This pooled estimate $\hat{p} = \dfrac{x_1 + x_2}{n_1 + n_2}$ is our best estimate of that common proportion. In the case where D_o was not 0, we did not have a common proportion in the two populations.

EXAMPLE 8.6 Engineers were interested in comparing the percentage of drivers who drive over the 75 mph limit on Interstate 10 with the percentage who drive over the 75 mph limit on Interstate 20 in 10 counties in west Texas. The null hypothesis is $H_o: (p_1 - p_2) = 0$ and the alternative hypothesis is $H_a: (p_1 - p_2) \neq 0$ with $\alpha = 0.10$. The data were: $n_1 = 400$, $x_1 = 235$ and $n_2 = 400$, $x_2 = 205$. Solve using EXCEL and MINITAB.

The EXCEL solution is shown in Figure 8.5.

Value	EXCEL command	Book notation
A	B	C
0.5875	=235/400	$\hat{p}_1$
0.5125	=205/400	$\hat{p}_2$
0.55	=(235+205)/(400+400)	$\hat{p} = \dfrac{x_1 + x_2}{n_1 + n_2}$
0.035178	=SQRT(A3*(1-A3)*(1/400+1/400))	$\sigma_{(\hat{p}_1 - \hat{p}_2)}$
2.132014	=(A1-A2)/A4	Test statistic
−1.64485	=NORMSINV(0.05)	$Z < -z_{0.05}$
1.644854	=NORMSINV(0.95)	$Z > z_{0.05}$
0.033006	=2*(1-NORMSDIST(A5))	p-value
0.017137	=(A1-A2)-1.644854*A4	Lower limit
0.132863	=(A1-A2)+1.644854*A4	Upper limit

Figure 8.5   EXCEL solution to test of hypothesis about $(p_1 - p_2) = 0$.

The MINITAB solution is as follows:

```
Sample   X     N      Sample p
1        235   400    0.587500
2        205   400    0.512500

Difference = p (1) - p (2)
Estimate for difference: 0.075
90% CI for difference: (0.0173018, 0.132698)
Test for difference = 0 (vs not = 0): Z = 2.13 P-Value = 0.033
```

Using either EXCEL or MINITAB, the null hypothesis is rejected.

Hypotheses Concerning Several Proportions

In this chapter we have been concerned with testing one and then two proportions. Our next step is testing several proportions. How do we decide whether it is reasonable to treat several proportions as if they have a common value? There are at least two techniques for testing hypotheses about several proportions that are equivalent. The second technique has certain advantages that we shall see in the next section entitled the analysis of $r \times c$ tables. We shall now discuss both techniques.

Technique 1 for testing $H_o: p_1 = p_2 = \dots = p_k = p$ versus the alternative that not all the proportions are equal. Let X_i be the number of successes in samples of sizes n_i. Remember that even though the variables X_i are binomial, if the sample size is large enough, X_i are approximately normal with mean $n_i p_i$ and variance $n_i p_i q_i$, and the variable $Z_i = \dfrac{X_i - n_i p_i}{\sqrt{n_i p_i q_i}}$ is approximately a standard normal. It can be shown that if a standard normal variable is squared, the result is a chi-square. If you add k chi-square variables, you obtain a chi-square variable having k degrees of freedom. However, you loose 1 degree of freedom when you estimate the common p-value. In conclusion, we have $\chi^2 = \sum_{i=1}^{k} \frac{(x_i - n_i p_i)^2}{n_i p_i q_i}$ as chi-square with $(k-1)$ degrees of freedom and the common p_i value is estimated by $\hat{p} = \frac{x_1 + x_2 + \cdots + x_k}{n_1 + n_2 + \cdots + n_k}$. (This is an upper-tailed test only. The null hypothesis is rejected only if the observed values and their mean values differ which causes the test statistic to be large.)

EXAMPLE 8.7 Three assembly lines are assembling cell phones. Samples of sizes $n_1 = 100$, $n_2 = 80$, and $n_3 = 90$ are selected from the three lines. The number with a defect in the samples are $x_1 = 7$, $x_2 = 10$, and $x_3 = 6$. Use EXCEL to perform the test $H_o: p_1 = p_2 = p_3 = p$ for $\alpha = 0.05$. Calculate the test statistic, the critical value, and the p-value. Figure 8.6 shows the computation of the test statistic, the critical value, and the p-value (Example 8.7).

Value	EXCEL command	Book notation
A	B	C
0.085185	=(7+10+6)/(100+80+90)	Pooled est. of p
0.295899	=(7-100*A1)^2/(100*A1*(1-A1))	First term of χ^2
1.627354	=(10-80*A1)^2/(80*A1*(1-A1))	Second term of χ^2
0.396057	=(6-90*A1)^2/(90*A1*(1-A1))	Third term of χ^2
2.31931	=SUM(A2:A4)	Test statistic
5.991465	=CHIINV(0.05,2)	Critical value
0.313594	=CHIDIST(A5,2)	p-value

Figure 8.6   EXCEL test of $H_o: p_1 = p_2 = p_3 = p$ (Technique 1).

Technique 2 for testing $H_o: p_1 = p_2 = \ldots = p_k = p$ versus the alternative that not all the proportions are equal. Two tables are prepared called the observed and the expected tables. The test statistic is then computed using the following: $\chi^2 = \sum_{i=1}^{2} \sum_{j=1}^{k} \frac{(o_{ij} - e_{ij})}{e_{ij}}$. This test statistic has $(k-1)$ degrees of freedom.

EXAMPLE 8.8 Observed table

	Line 1	Line 2	Line 3
Defect	7	10	6
Non-defect	93	70	84

The estimated value of a defect is $p = 23/270 = 0.085185$.
Expected table

	Line 1	Line 2	Line 3
Defect	0.085185(100)	0.085185(80)	0.085185(90)
Non-defect	(1 − .085185)(100)	(1 − .085185)(80)	(1 − .085185)(90)

	Line 1	Line 2	Line 3
Defect	8.5185	6.548	7.66665
Non-defect	91.4815	73.452	82.33335

$$\chi^2 = \sum_{i=1}^{2} \sum_{j=1}^{k} \frac{(o_{ij} - e_{ij})^2}{e_{ij}} = \frac{(7 - 8.5185)^2}{8.5185} + \frac{(10 - 6.548)^2}{6.548} + \frac{(6 - 7.66665)^2}{7.66665} + \frac{(93 - 91.4815)^2}{91.4815}$$

$$+ \frac{(70 - 73.452)^2}{73.452} + \frac{(84 - 82.33335)^2}{82.33335}$$

$$= 2.31931$$

Figure 8.7 contains the EXCEL test of $H_o: p_1 = p_2 = p_3 = p$ (Technique 2). The data is contained in A1:C2. The column totals are in A3:C3 and the row totals are in D1:D2. The grand total is in D3. The cell A5 contains the expression =A3*$D1/$D3. The $ symbol keeps the D1 and D3 from changing when a click-and-drag is performed from left to right. The cell A6 contains the expression =A3*$D2/$D3. Again a click-and-drag is performed left to right. This saves a lot of keyboarding when entering the expected values in the space A5:C6. (Try it. You will like it.) Now enter =(A1-A5)^2/A5 in cell A8 and click-and-drag to the right and down until you cover the space A8:C9. In A11 enter =SUM(A8:C9). This will compute your test statistic, $\chi^2 = \sum\limits_{i=1}^{2}\sum\limits_{j=1}^{k} \dfrac{(o_{ij} - e_{ij})^2}{e_{ij}}$. In A12, the entry =CHIDIST(A11,2) will give the area to the right of 2.31931, which is the p-value. Go through the steps in the past paragraph. It will make a lot more sense if you do it while you have EXCEL in front of you.

	A	B	C	D	E
	7	10	6	23	Observed
	93	70	84	247	Data
	100	80	90	270	
	8.518519	6.814814815	7.666667		Expected
	91.48148	73.18518519	82.33333		Data
	0.270692	1.488727858	0.362319		
	0.025206	0.138626481	0.033738		
	2.31931				Test statistic
	0.313594				p-value

Figure 8.7   EXCEL test of $H_o: p_1 = p_2 = p_3 = p$ (Technique 2).

The MINITAB solution to this example is below. Enter the data in the worksheet as follows:

Line1	Line2	Line3
7	10	6
93	70	84

Execute the pulldown **Stat** $\Rightarrow$ **Tables** $\Rightarrow$ **Chi-square test**. The output is as follows:

```
Chi-Square Test: line1, line2, line3
Expected counts are printed below observed counts
Chi-Square contributions are printed below expected counts

          line1     line2     line3     Total
    1     7         10        6         23
          8.52      6.81      7.67
          0.271     1.489     0.362

    2     93        70        84        247
          91.48     73.19     82.33
          0.025     0.139     0.034

Total     100       80        90        270

Chi-Sq = 2.319, DF = 2, P-Value = 0.314
```

Study and compare the two sets of output. The conclusion is that there is no difference in the assembly lines with respect to their proportion of defective cell phones output.

Technique 1 and Technique 2 are equivalent to the method given in the previous section for testing the null hypothesis that p_1 and p_2 were equal. Recall the example we gave there.

EXAMPLE 8.9 Engineers were interested in comparing the percentage of drivers who drive over the 75 mph limit on Interstate 10 with the percentage who drive over the 75 mph limit on Interstate 20 in 10 counties in west Texas. The null hypothesis is H_o: $(p_1 - p_2) = 0$ and the alternative hypothesis is H_a: $(p_1 - p_2) \neq 0$ with $\alpha = 0.10$. The data was: $n_1 = 400$, $x_1 = 235$ and $n_2 = 400$, $x_2 = 205$.

The test using the standard normal distribution in Figure 8.5 gave $Z = 2.132007$ and p-value = 0.033006. The test using the chi-square with 1 degree of freedom in Figure 8.8 gave $\chi^2 = 4.5454$ and p-value = 0.033006. Note that the square of the Z value, 2.132007^2, gives the chi-square value 4.5454. Both tests give the same p-value. This shows the equivalence of the two tests when only two proportions are involved.

A	B	C	D
235	205	440	Observed
165	195	360	Data
400	400	800	
220	220		Expected
180	180		Data
1.022727273	1.022727		
1.25	1.25		
4.545454545			Test statistic
0.033006262			p-value

Figure 8.8   Showing the equivalence of tests for $p_1 = p_2$.

The MINITAB solution is:

Chi-Square Test: interstate10, interstate20

```
Expected counts are printed below observed counts
Chi-Square contributions are printed below expected counts
            interstate10     interstate20     Total
    1       235              205              440
            220.00           220.00
            1.023            1.023

    2       165              195              360
            180.00           180.00
            1.250            1.250

Total       400              400              800
    Chi-Sq = 4.545, DF = 1, P-Value = 0.033
```

Analysis of $r \times c$ Tables

A company assembles video game consoles. The engineer in charge of the training program is concerned about the connection between the employees' performance in the training program and their success on the job. In particular, she is interested in whether the factors are independent of one another or whether there is a dependency. In statistics, this question is answered by performing a contingency table analysis. An outline of such a test is as follows:

H_o: The two classifications are independent of one another.

H_a: The two classifications are dependent upon one another.

$$\text{Test statistic: } \chi^2 = \sum_{j=1}^{c} \sum_{i=1}^{r} \frac{[o_{ij} - e_{ij}]^2}{e_{ij}}$$

where

$e_{ij} = \dfrac{n_i \cdot n_j}{n}$, n_i = total for row i, n_j = total for column j, and n = observed counts.

Rejection region: $\chi^2 > \chi_\alpha^2$ where χ_α^2 has $(r-1)(c-1)$ degrees of freedom.

Assumptions:

1. The n observed counts are a random sample from the population of interest. This is a multinomial experiment with $r \times c$ possible outcomes.

2. For the χ^2 approximation to be valid, it is required that the estimated expected counts be ≥ 5 in all cells.

EXAMPLE 8.10 The company discussed above classified 200 employees according to performance in training program and success on the job. The results are shown in Figure 8.9. Test for independence at $\alpha = 0.05$.

		Performance in training program		
		Below avg.	Average	Above avg.
Success	Poor	25	15	10
in	Average	25	55	20
job	Very good	10	25	15

Figure 8.9 Does success in the job depend on performance in training program?

Figure 8.10 shows the file structure needed for analysis by STATISTIX.

Category	Row	Column
25	1	1
25	2	1
10	3	1
15	1	2
55	2	2
25	3	2
10	1	3
20	2	3
15	3	3

Figure 8.10 Data file for STATISTIX.

The pulldown **Statistics** $\Rightarrow$ **Association tests** $\Rightarrow$ **Chi-square** is used in STATISTIX. The output is as follows:

```
Chi-Square Test for Heterogeneity or Independence
for category = row column

                                 column
row                   1            2            3
            +----------+----------+----------+
1    Observed |   25    |   15    |   10    |    50
     Expected |  15.00  |  23.75  |  11.25  |
   Cell Chi-Sq |  6.67   |   3.22  |   0.14  |
            +----------+----------+----------+
2    Observed |   25    |   55    |   20    |   100
     Expected |  30.00  |  47.50  |  22.50  |
   Cell Chi-Sq |  0.83   |   1.18  |   0.28  |
            +----------+----------+----------+
3    Observed |   10    |   25    |   15    |    50
     Expected |  15.00  |  23.75  |  11.25  |
   Cell Chi-Sq |  1.67   |   0.07  |   1.25  |
            +----------+----------+----------+
                  60          95          45        200
```

```
Overall Chi-Square     15.31
P-Value                0.0041
Degrees of Freedom     4
```

The two variables are dependent, since the *p*-value is less than α.

Estimation of proportions

8.1. Engineers analyzed the breaking strength of carbon fibers used in fibrous composite materials. They found that out of a sample of 100, 6 had breaking strengths of less than 1.2 GPa (gigapascals). They wished to set a 99% confidence interval on the true proportion having breaking strengths less than 1.2 GPa. Use MINITAB to set a 99% confidence interval on the true proportion of nonconforming fibers *p*.

SOLUTION

When the dialog box is obtained for the 1-proportion test, fill it out as shown below. When the options are selected, choose 99% as the confidence level and select use interval based on the normal distribution.

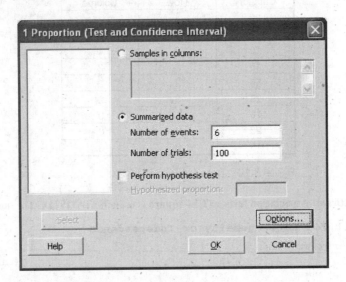

This will produce the following output:

Test and CI for One Proportion

Sample X	N	Sample p	99% CI
6	100	0.060000	(0.000000, 0.121173)

Using the normal approximation.

    We can be 99% confident that the true proportion of nonconforming fibers is somewhere between 0% and 12 %.

8.2. An engineer studies a cylinder boring process for an engine block. Specifications require that these bores be 3.5199 ± 0.0004 in. Management is concerned that the true proportion of cylinder bores outside the specifications is excessive. Current practice is willing to tolerate up to 10% outside the specifications. Out of a random sample of 165, 36 were outside the specifications. Construct a 99% confidence interval for the true proportion of bores outside the specifications.

SOLUTION

```
Test and CI for One Proportion

Sample X      N     Sample p   99% CI
1       36   165    0.218182   (0.135361, 0.301002)
```

Using the normal approximation.

The true proportion of bores outside the specifications is very likely greater than 10%.

8.3. Resistors labeled as 50 Ω are presently purchased from two vendors, vendor A and vendor B. The specification for this type of resistor is that its actual resistance should be within 3% of its labeled resistance. In a sample of 100 resistors from vendor A, 75 met the specification. In a sample of 100 from vendor B, 83 met the specification. Find a 95% confidence interval on $P_B - P_A$.

SOLUTION

The MINITAB pull-down that is used is **Stat $\Rightarrow$ Basic Statistics $\Rightarrow$ 2-proportions**.
   The output is as follows:

```
Test and CI for Two Proportions

Sample  X    N      Sample p
1       83   100    0.830000
2       75   100    0.750000

Difference = p (1) - p (2)
Estimate for difference: 0.08
95% CI for difference: (-0.0323523, 0.192352)
```

A 95% confidence interval for the difference in true proportions is (−0.032, 0.192).

Hypotheses concerning one proportion

8.4. A supplier of semiconductor wafers states that no more than 5% of the company's wafers are defective. In order to test the claim, an engineer selects a sample of 200 and finds that 27 are defective. He uses the results of the sample to test the hypothesis that the true defective is 5% versus it is greater than 5%. The test is performed at level of significance equal to 0.05.

SOLUTION

The result of the test as given by MINITAB is:

```
Test of p = 0.05 vs p > 0.05

                                95% Lower
Sample  X    N      Sample p    Bound      Z-Value  P-Value
1       27   200    0.135000    0.095255   5.52     0.000
```

Using the normal approximation.

   The value of the test statistic is Z-value = 5.52 with p-value = 0.000. Using either method of testing, the null hypothesis is rejected. That is, 5.52 exceeds the critical value which is 1.65 or the p-value is less than 0.05.

8.5. A survey engineer measured the orthometric height above sea level for a sample of 775. The engineer found that 590 of the 775 gave results that were within the class C tolerance limits. Test the hypothesis that the true proportion within class C tolerance limits is 70% at $\alpha = 0.05$.

SOLUTION

```
Test of p = 0.7 vs p not = 0.7

Sample  X     N     Sample p   95% CI                 Z-Value  P-Value
1       590   775   0.761290   (0.731277, 0.791303)   3.72     0.000
Using the normal approximation.
```

The null hypothesis is rejected.

8.6. The parts in a large warehouse will be purchased if it is concluded that fewer than 3% do not meet specifications. A quality engineer selects a random sample of 350 parts from the warehouse. It is found that 6 parts do not meet specifications. Test the null hypothesis that $p = 0.03$ versus the research hypothesis that $p < 0.03$ at $\alpha = 0.05$.

SOLUTION

```
Test of p = 0.03 vs p < 0.03

                               95% Upper
Sample  X   N    Sample p    Bound       Z-Value   P-Value
1       6   350  0.017143    0.028555    -1.41     0.079
Using the normal approximation.
```

From the output we cannot conclude that the warehouse contains less than 3% defective.

Hypotheses concerning two proportions

8.7. An engineer studied newly hatched larvae of the common carp for exposure to copper or lead during the embryonic development. The environmental engineer was interested in testing whether the percentage of defective larvae differed for the two metal solutions. The eggs and sperm were obtained from artificially induced spawning at an inland fisheries institute. The engineer placed 100 eggs in each solution. The number of defective larvae was 38 with copper and 22 with lead. Let P_1 be the proportion of defective larvae in the copper solution and P_2 be the proportion of defective larvae in the lead solution. Test at 0.05 significance level that $P_1 = P_2$ versus the proportions are not equal.

SOLUTION

The MINITAB solution is:

```
Test and CI for Two Proportions

Sample  X     N      Sample p
1       38    100    0.380000
2       22    100    0.220000

Difference = p (1) - p (2)

Estimate for difference: 0.16

95% CI for difference: (0.0349303, 0.285070)

Test for difference = 0 (vs not = 0): Z = 2.47 P-Value = 0.014
```

Because the *p*-value is smaller than the significance level, reject that the two proportions are the same.

8.8. Two extrusion machines that manufacture steel rods are being compared. In a sample of 1000 rods from machine 1935 met specifications regarding length and diameter. In a sample of 1000 rods from machine 2914 met specifications regarding length and diameter. Can it be concluded that the proportions meeting specifications are the same? Use level of significance equal to 0.01.

SOLUTION

The MINITAB output is as follows:

```
Test and CI for Two Proportions
Sample  X     N      Sample p
1       935   1000   0.935000
2       914   1000   0.914000

Difference = p (1) - p (2)
Estimate for difference: 0.021
95% CI for difference: (-0.00213912, 0.0441391)
Test for difference = 0 (vs not = 0): Z = 1.78 P-Value = 0.076
```

Since the *P*-value is not less than 0.01, do not conclude that the proportions are different.

8.9. Resistors labeled as 50 ohms are purchased from two different vendors. The specification for this type of resistor is that its actual resistance be within 5% of its labeled resistance. In a sample of 150 resistors purchased from vendor A, 125 of them met the specification. In a sample of 250 resistors purchased from vendor B, 215 of them met the specification. Vendor A is the current supplier, but if the data demonstrates convincingly that a greater proportion of the resistors from vendor B meet the specification, a change will be made. Test that the proportions are equal versus they are not equal.

SOLUTION

```
Test and CI for Two Proportions
Sample    X          N          Sample p
1         125        150        0.833333
2         215        250        0.860000

Difference = p (1) - p (2)
Estimate for difference: -0.0266667
95% CI for difference: (-0.100199, 0.0468653)
Test for difference = 0 (vs not = 0): Z = -0.72 P-Value = 0.470
```

The data is not convincing that the proportions are unequal.

Hypotheses concerning several proportions

8.10. Three production lines are compared with respect to the proportion of defectives they produce. Line 1 has 150 items selected and 15 of the items are found to be defective. Line 2 has 125 selected and 9 of the items are found to be defective. Line 3 has 175 selected and 20 are found to be defective. Use MINITAB to test the null hypothesis that all the three lines produce the same proportion of defectives versus the three proportions are not the same at $\alpha = 0.05$.

SOLUTION

```
Chi-Square Test: line 1, line 2, line 3
Expected counts are printed below observed counts
Chi-Square contributions are printed below expected counts

          line 1   line 2   line 3   Total
   1      15       9        20       44
          14.67    12.22    17.11
          0.008    0.849    0.488

   2      135      116      155      406
          135.33   112.78   157.89
          0.001    0.092    0.053

Total     150      125      175      450

Chi-Sq = 1.491, DF = 2, P-Value = 0.475
```

Since the *P*-value > 0.05, the null hypothesis of equal proportions cannot be rejected.

8.11. Refer to the Problem 8.7. Show that the test for two equal proportions based on the standard normal distribution given in the problem is equivalent to the test based on the chi-square distribution given in the section on hypotheses concerning several proportions.

SOLUTION

In Problem 8.7 we found the *Z*-test statistic was equal to 2.47 with a *p*-value equal to 0.014. Using MINITAB, the chi-square test gives the following chi-square test statistic (6.095) with the *p*-value equal to 0.014.

```
Chi-Square Test: copper, lead
Expected counts are printed below observed counts
Chi-Square contributions are printed below expected counts
          copper lead  Total
    1      38     22     60
           30.00  30.00
           2.133  2.133
    2      62     78    140
           70.00  70.00
           0.914  0.914
Total     100    100    200
Chi-Sq = 6.095, DF = 1, P-Value = 0.014
```

Note that 2.47 squared equals 6.100. The fact that Z squared equals chi-squared shows the equivalence of the two procedures.

8.12. Airplanes approaching the runway for landing are required to stay within the localizer (a certain distance to the left and right of the runway). When an airplane deviates from the localizer, it is referred to as an exceedence. Consider three airlines at a large airport. During a 3-week period, airline 1 had 14 exceedences out of 156 flights. Airline 2 had 11 exceedences out of 198 flights. Airline 3 had 13 exceedences out of 175 flights. Test the hypothesis that the proportion of exceedences is the same for the three airlines at $\alpha = 0.05$.

SOLUTION

```
Chi-Square Test: airline 1, airline 2, airline 3
Expected counts are printed below observed counts
Chi-Square contributions are printed below expected counts
          airline 1    airline 2    airline 3    Total
    1      14           11           13            38
           11.21        14.22        12.57
           0.697        0.730        0.015
    2      142          187          162           491
           144.79       183.78       162.43
           0.054        0.057        0.001
Total     156          198          175           529
Chi-Sq = 1.553, DF = 2, P-Value = 0.460
```

There is not a significant difference in the number of exceedences for the three airlines.

Analysis of $r \times c$ tables

8.13. Using the data in Figure 8.9 and MINITAB, test that performance in the training program and success in the job are independent of each other. Compare the output with the output given by the package STATISTIX in the section on analysis of $r \times c$ tables.

SOLUTION

The data must be entered in the MINITAB worksheet as follows:

C1	C2	C3
25	15	10
25	55	20
10	25	15

The pulldown **Stat ⇒ Tables ⇒ Chi-square test** gives the following output.

Chi-Square Test: C1, C2, C3

```
Expected counts are printed below observed counts
Chi-Square contributions are printed below expected counts
           C1       C2       C3      Total
    1      25       15       10       50
         15.00    23.75    11.25
         6.667    3.224    0.139

    2      25       55       20       100
         30.00    47.50    22.50
         0.833    1.184    0.278

    3      10       25       15       50
         15.00    23.75    11.25
         1.667    0.066    1.250

Total      60       95       45       200
Chi-Sq = 15.307, DF = 4, P-Value = 0.004
```

Compare this output with that given by STATISTIX.

Chi-Square Test for Heterogeneity or Independence

```
for category = row column

                            column
row                 1           2           3

             +---------+---------+-------+
1    Observed |   25    |   15    |   10  |      50
     Expected |  15.00  |  23.75  | 11.25 |
   Cell Chi-Sq |   6.67  |   3.22  |  0.14 |
             +---------+---------+-------+
2    Observed |   25    |   55    |   20  |     100
     Expected |  30.00  |  47.50  | 22.50 |
   Cell Chi-Sq |   0.83  |   1.18  |  0.28 |
             +---------+---------+-------+
3    Observed |   10    |   25    |   15  |      50
     Expected |  15.00  |  23.75  | 11.25 |
   Cell Chi-Sq |   1.67  |   0.07  |  1.25 |
             +---------+---------+-------+
                  60        95        45       200

Overall Chi-Square 15.31
P-Value 0.0041
Degrees of Freedom 4
```

8.14. One thousand and twenty-one cylindrical steel pins are subject to length specification as well as diameter specification. The following table gives the results of the measurements.

	Diameter		
	Too thin	OK	Too thick
Length			
Too short	13	117	4
OK	62	664	80
Too long	5	68	8

Test that length and diameter are independent of each other at $\alpha = 0.05$ by the use of MINITAB.

SOLUTION

The MINITAB output is as follows:

```
Chi-Square Test: C1, C2, C3
Expected counts are printed below observed counts
Chi-Square contributions are printed below expected counts
           C1        C2        C3      Total
    1      13        117       4       134
           10.50     111.43    12.07
           0.595     0.279     5.400

    2      62        664       80      806
           63.15     670.22    72.63
           0.021     0.058     0.749

    3      5         68        8       81
           6.35      67.35     7.30
           0.286     0.006     0.067

Total      80        849       92      1021
Chi-Sq = 7.461, DF = 4, P-Value = 0.113
```

The null hypothesis is that diameter measurement is independent of length measurement. Since the p-value > 0.05, the hypothesis of independence is not rejected.

8.15. Specifications for the diameter of a roller are 1.90–2.10 centimeters. Rollers that are too thick can be reground, while those that are too thin must be scrapped. Three machinists grind these rollers. Samples of rollers were collected from each machinist, and their diameters were measured. The results were as given in the below table.

Machinist	Good	Regrind	Scrap
A	328	58	14
B	231	48	21
C	409	73	18

Test that the classification of the roller is independent of the machinist at $\alpha = 0.10$.

SOLUTION

The output from MINITAB is as follows:

Chi-Square Test: Good, Regrind, Scrap

Expected counts are printed below observed counts

Chi-Square contributions are printed below expected counts

	Good	Regrind	Scrap	Total
1	328	58	14	400
	322.67	59.67	17.67	
	0.088	0.047	0.761	
2	231	48	21	300
	242.00	44.75	13.25	
	0.500	0.236	4.533	
3	409	73	18	500
	403.33	74.58	22.08	
	0.080	0.034	0.755	
Total	968	179	53	1200

Chi-Sq = 7.033, DF = 4, P-Value = 0.134

The *p*-value indicates that we can not reject independence.

SUPPLEMENTARY PROBLEMS

8.16. The temperature of a process is important to the engineer in charge. The daily temperatures are determined and it is found that on 8 days out of 35, the temperature is below a critical value. The proportion of days p that the temperature is below the critical value is to be estimated. It is hypothesized that the proportion is 0.1 versus it is greater than 0.1. Give the MINITAB output for testing this hypothesis.

8.17. A sample of 100 fuses from company A had 10 defectives and a sample of 125 fuses from company B had 8 defectives. Test that $p_1 = p_2$ versus $p_1 \neq p_2$ using the standard normal distribution. Use $\alpha = 0.05$. Give the MINITAB output for testing this hypothesis.

8.18. Do Problem 8.17 above using the chi-square distribution instead of the standard normal distribution. Show that it is equivalent to the standard normal distribution solution.

8.19. Four age groups of engineers were compared with respect to their attitude toward the Internet. The number of engineers with positive attitudes toward the Internet in the four age groups is as follow: age group 1:68 out of 125, age group 2:75 out of 125, age group 3:80 out of 125, and age group 4:60 out of 125. The null hypothesis is that the four proportions are equal and the research hypothesis is that the proportions differ. Give the MINITAB solution and your conclusion at $\alpha = 0.05$.

8.20. Mechanical engineers tested a new welding technique. They classified welds with respect to appearance and also by use of X-ray inspection. They were interested in finding out if the two classifications were independent of one another. The data is shown in the following table.

X-ray	Appearance		
	Bad	Normal	Good
Bad	15	8	5
Normal	10	35	6
Good	8	13	25

Use MINITAB to test for independence at $\alpha = 0.05$.

CHAPTER 9

Statistical Process Control

Introduction to Statistical Process Control (SPC)

Two names associated with SPC are Dr. Walter Shewhart and Dr. W. Edwards Deming. Dr. Walter Shewhart developed the techniques of quality control while at Bell Telephone Laboratories. He was trying to improve Bell telephones. Dr. W. Edwards Deming introduced quality control in Japan and is credited with making Japan competitive after World War II in automobile production.

This section will introduce the topic of SPC and the software that is used to begin the practical use of some of the more useful control charts associated with SPC.

MINITAB will be used to illustrate many of the SPC techniques. SPC is a set of graphs and techniques that are used to help improve the quality of goods produced.

Pareto Charts

One of the first graphs that we shall study is a **Pareto chart**. This is a graph that lists defects and their counts.

EXAMPLE 9.1 The table that follows gives a list of defects and their counts for a production process that produces new automobiles.

Defects and Counts

Defect	Count
Scratch	5
Discoloration	3
Alignment	6
Missing parts	10
Defective parts	8
Mismatching	15
Dents	7
Malfunction meter	2

The Pareto chart lists the defects that occur and how often they occur. By looking at the bar graph we will be able to see the defects that occur the most often and that we need to work on reducing the most. The data is entered in C1 and C2 of the MINITAB worksheet. The MINITAB pulldown to use is **Graph ⇒ Barchart**. The dialog box is filled as shown in the following dialog box.

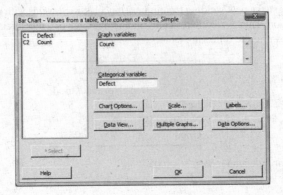

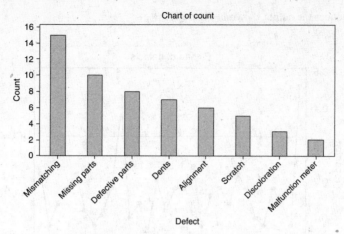

Figure 9.1 MINITAB bar chart showing the defects that occur
the most often.

Figure 9.1 is called a **Pareto chart** of defects. We would try to eliminate the defects which occur the most often such as Mismatching and Missing parts.

P Charts

The **P chart** is a control chart that is used to graph the fraction of a sample that is nonconforming. P charts are used for attributes, quality characteristics that can be classified as conforming or nonconforming.

EXAMPLE 9.2 Suppose cardboard cans are being manufactured for lemon juice concentrate. Thirty samples are selected out of 50 cardboard cans each. The data is shown in the following table:

Sample	Defects	Sample	Defects	Sample	Defects
1	23	11	8	21	20
2	4	12	4	22	9
3	7	13	12	23	6
4	9	14	9	24	4
5	4	15	9	25	6
6	5	16	6	26	11
7	9	17	10	27	9
8	11	18	3	28	7
9	7	19	7	29	4
10	9	20	5	30	11

Use the pulldown **Stat ⇒ Control Charts ⇒ Attribute Charts ⇒ P** to get the following dialog box.

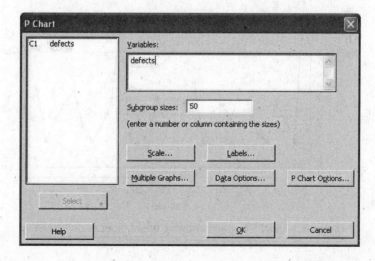

Selecting the OK produces the output shown in Figure 9.2.

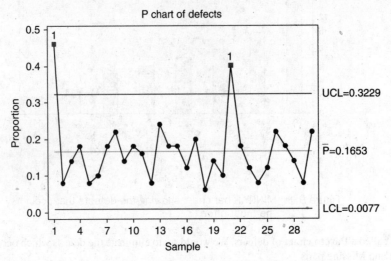

Figure 9.2   MINITAB graph showing the proportion defective in
30 samples of size 50.

The P chart for this process would suggest that we look closer at the process during the time period that the first and twenty-first samples were taken.

NP Charts

The **NP chart** is the control chart for the number of nonconforming units. NP charts are used for attributes. The only difference between the P chart and the NP chart is that the proportion of defectives is plotted on the vertical axis of the P chart, whereas the number of defectives is plotted on the vertical axis of the NP chart.

EXAMPLE 9.3   Suppose the data in the 30 samples are entered in column C1 of MINITAB. The pulldown **Stat ⇒ Control charts ⇒ Attribute charts ⇒ NP** gives the output shown in Figure 9.3. The lower control limit (LCL) is 0.39 and the upper control limit (UCL) is 16.15.

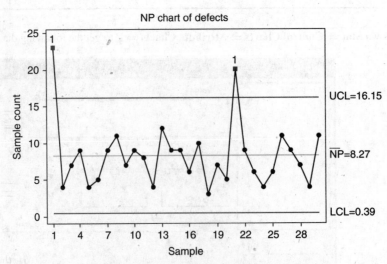

Figure 9.3   MINITAB graph showing the number of defectives in
30 samples of size 50.

C Charts

The **C chart** is the control chart for nonconformities or defects. A product may contain one or more defects and not be defective. If, for example, a computer has a flaw in its cabinet, we would not necessarily wish to reject the computer. In these situations we are more interested in the number of defects per inspection unit. The center line and the control limits are computed from c, the number of nonconformities per inspection unit. You can enter a standard or historical value for c or you can use the average value of c computed from the data.

EXAMPLE 9.4 The following table gives the data. Printed circuit boards are inspected for defects. The inspection unit is defined as 100 circuit boards. The number of defects per 100 circuit boards for 26 samples is listed in the table.

The Number of Defects per 100 Circuit Boards Given for 26 Samples

Sample	Defects	Sample	Defects
1	21	14	19
2	24	15	10
3	16	16	17
4	12	17	13
5	15	18	22
6	5	19	18
7	28	20	39
8	20	21	30
9	31	22	24
10	25	23	16
11	20	24	19
12	24	25	17
13	16	26	15

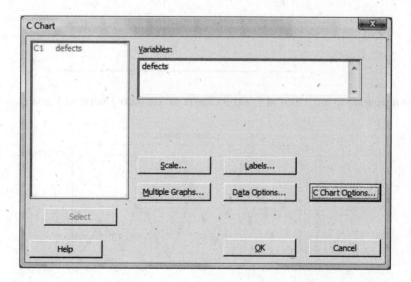

After the 26 defect values are entered into column C1 of the work sheet of MINITAB, the pulldown sequence **Stat ⇒ Control Charts ⇒ Attribute Charts ⇒ C** is given. The C chart dialog box given above is filled as shown and executed to give Figure 9.4.

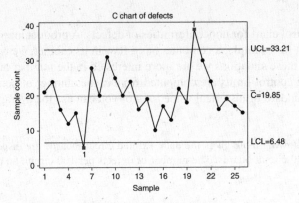

Figure 9.4   MINITAB C chart for the data.

Figure 9.4 gives the UCL at 33.2 and the LCL at 6.5. The center line is given at 19.8. The process is out of control for sample 6 and sample 20.

U Charts

The **U chart** is the attribute control chart for nonconformities per unit. It is used in controlling the noncon-formities per unit when the sample size is not one inspection unit. The center line and control limits are computed from u, the number of nonconformities per unit. You can enter a standard or historical value for u or you can use the average value of u from the data.

EXAMPLE 9.5   Five personal computers were periodically sampled from the final assembly line and inspected for defects. This was repeated 20 times. The inspection unit was one computer. The sample size is 5.

Sample	Defects	Sample	Defects
1	10	11	9
2	12	12	5
3	8	13	7
4	14	14	11
5	10	15	12
6	16	16	6
7	11	17	8
8	7	18	10
9	10	19	7
10	15	20	5

The pulldown sequence to use is **Stat ⇒ Control Charts ⇒ Attribute Charts ⇒ U** and gives the output shown in Figure 9.5.

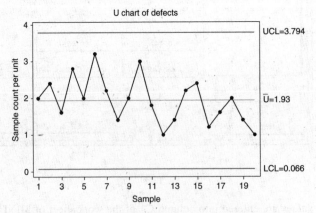

Figure 9.5   U chart for data in the above table.

Since the U chart limits are between the LCL and the UCL, we can say that this process is in control.

R, S, and Xbar Charts

Thus far, our control charts have been concerned with controlling defectives, number of defectives, or counts. These are attributes that we are controlling. When measurements are made, we are interested in controlling the variability or the mean of these measurements. The charts that are used for controlling the variability are called **R charts** or **S charts**. The chart that is used for controlling the mean is called the **Xbar chart**.

EXAMPLE 9.6   A process is being controlled by measuring 25 samples of size 5 of tensile strength in pounds per square inch. The data is shown in Table 9.1. If for each case, or sample, the sample mean, the sample range, and the sample standard deviation is calculated, we obtain Table 9.2. For the ith case or ith sample, the sample mean is the sum of the values in the ith sample divided by 5. For the ith case or sample, the sample standard deviation is S_i which is equal to the standard deviation of the 5 sample values. The range for any sample is the largest value in the sample minus the smallest value in the sample. All the sample means, sample standard deviations, and sample ranges for all 125 samples are shown in Table 9.2. These values are used to construct the Xbar chart, the R chart, and the S chart. The charts for variability are measured first. The R chart is constructed first and then the S chart is measured. Both show that variability is in control. Then the Xbar chart for measuring the control of the mean is constructed.

Table 9.1   Twenty-five Samples of Tensile Strength

Case	Strength	Case	Strength	Case	Strength	Case	Strength	Case	Strength
1	1515	6	1519	11	1499	16	1511	21	1500
	1518		1522		1503		1509		1498
	1512		1523		1507		1503		1503
	1498		1517		1503		1510		1504
	1511		1511		1501		1507		1508
2	1504	7	1498	12	1507	17	1508	22	1511
	1511		1497		1503		1511		1514
	1507		1507		1502		1513		1509
	1499		1511		1500		1509		1508
	1502		1508		1501		1506		1506
3	1517	8	1511	13	1500	18	1508	23	1505
	1513		1518		1506		1509		1508
	1504		1507		1501		1512		1500
	1521		1503		1498		1515		1509
	1520		1509		1507		1519		1503
4	1497	9	1506	14	1501	19	1520	24	1501
	1503		1503		1509		1517		1498
	1510		1498		1503		1519		1505
	1508		1508		1508		1522		1502
	1502		1506		1503		1516		1505
5	1507	10	1503	15	1507	20	1506	25	1509
	1502		1506		1508		1511		1511
	1497		1511		1502		1517		1507
	1509		1501		1509		1516		1500
	1512		1500		1501		1508		1499

Table 9.2   Sample Mean, Sample Standard Deviation, and Sample Range for Each Sample

Case	$\overline{X}_i$	S_i	R_i
1	1510.8	7.661593	20
2	1504.6	4.615192	12
3	1515	6.892024	17
4	1504	5.147815	13
5	1505.4	5.94138	15
6	1518.4	4.774935	12
7	1504.2	6.300794	14
8	1509.6	5.549775	15

(continued)

Table 9.2   Sample Mean, Sample Standard Deviation, and Sample Range for Each Sample (*continued*)

Case	$\overline{X}_i$	S_i	R_i

The **R chart** is formed by entering the 125 observations in column C1and then giving the pulldown **Stat ⇒ Control Charts ⇒ Variable Charts for subgroups ⇒ R**. The result is shown in Figure 9.6.

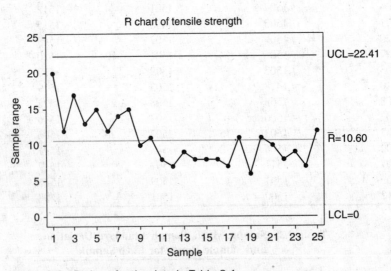

Figure 9.6   R chart for the data in Table 9.1.

The tensile strength readings are in control with respect to variability as measured by the R chart. The **S chart** is formed by entering the 125 observations in column C1 and then giving the pulldown **Stat ⇒ Control Charts ⇒ Variable Charts for subgroups ⇒ S**. The result is shown in Figure 9.7. This chart, like the R chart, shows that variability of tensile strength is in control.

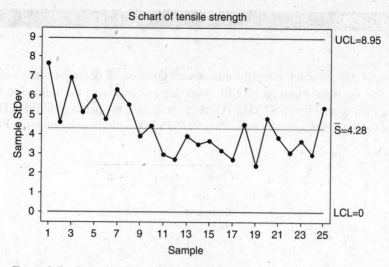

Figure 9.7   S chart for the data in Table 9.1.

Figure 9.8 shows that mean tensile strength is not in control. There are, according to our samples, three places where the chart is not in control.

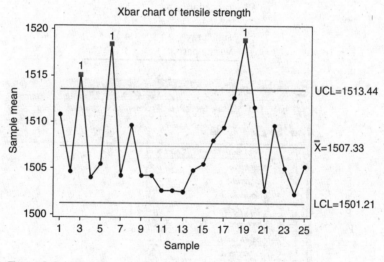

Figure 9.8   Xbar chart for the data in Table 9.1.

The control charts are used to check if a process is in control. The control chart that is used depends upon the level of the characteristic that is being measured. The characteristic being measured may give rise to continuous data, binary data, or count data. Continuous data means that either an R chart, an S chart, or an Xbar chart will be used. A P chart is used for binary data (defective and non-defective). The C chart is used for count data. All the charts consists of a center line, a lower limit that is 3 standard deviations below the center line and a upper limit that is 3 standard deviations above the center line. The process is in control as long as the statistic is between the lower limit and the upper limit.

SOLVED PROBLEMS

Pareto charts

9.1. The reasons due to which aircraft tanks were classified as defective were collected over several months and are placed into an EXCEL worksheet as shown in the following EXCEL worksheet. They are then copied into a STATISTIX data sheet. Use the package STATISTIX to form a Pareto chart for the defective aircraft tanks. This STATISTIX data sheet is shown in Figure 9.9.

SOLUTION

Cause	Count
adhesive	6
alignment	2
alodine	1
cast voids	2
damaged	34
dalam comp	2
dimensions	36
fairing	3
film	5
machining	29
masking	17
out order	4
paint dam	1
paint spec	2
primer dam	1
procedure	1
rusted	13
salt spray	4
wrong part	3

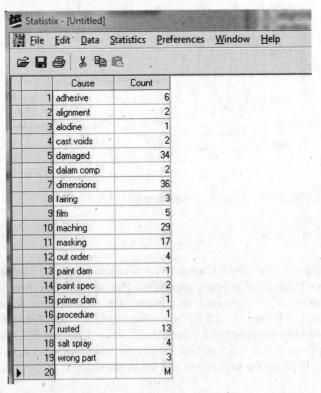

Figure 9.9   Aircraft tanks Causes and Counts copied into STATISTIX data page.

The Pareto chart in Figure 9.10 is given by the pulldown **Statistics** $\Rightarrow$ **Quality Control** $\Rightarrow$ **Pareto Chart**.

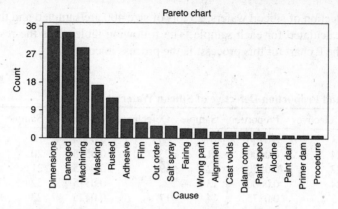

Figure 9.10   Pareto chart showing causes and counts.

9.2. Explain how to construct a Pareto chart for the data in Problem 9.1 using EXCEL.

SOLUTION

First sort the data by the variable COUNT and get the ordered data as follows:

Cause	Count
alodine	1
paint dam	1
primer dam	1
procedure	1
alignment	2
cast voids	2
dalam comp	2
paint spec	2
fairing	3
wrong part	3
out order	4
salt spray	4
film	5
adhesive	6
rusted	13
masking	17
maching	29
damaged	34
dimensions	36

Now, use EXCEL to bar chart the sorted data to get Figure 9.11.

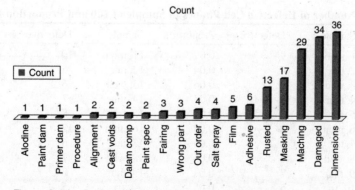

Figure 9.11   Pareto chart showing causes and counts.

P charts

9.3. In the production of silicon wafers, 25 lots of size 400 are sampled and the proportion of defective wafers is calculated for each sample. The following table gives the results. Use STATISTIX to construct the P chart for this process. Is the process in control?

SOLUTION

Number and Proportion Defective of Silicon Wafers

Sample	Defectives	Proportion	Sample	Defectives	Proportion	Sample	Defectives	Proportion
1	18	0.0450	10	20	0.0500	19	25	0.0625
2	24	0.0600	11	23	0.0575	20	17	0.0425
3	22	0.0550	12	19	0.0475	21	18	0.0450
4	24	0.0600	13	23	0.0575	22	24	0.0600
5	15	0.0375	14	17	0.0425	23	17	0.0425
6	22	0.0550	15	26	0.0650	24	17	0.0425
7	23	0.0575	16	16	0.0400	25	15	0.0375
8	29	0.0725	17	22	0.0550			
9	13	0.0325	18	20	0.0500			

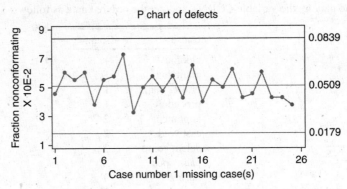

Figure 9.12   P chart for proportion defective of silicon wafers in 400.

The STATISTIX P chart is shown in Figure 9.12. It shows that this process is in control.

9.4. A process produces cell phones. A sample of size 100 is selected daily over a 20-day period. The following table gives the number of defective cell phones produced in the sample of 100 over the 20-day period. Construct a P chart using the data and MINITAB. Comment on the control over the 20-day period.

SOLUTION

Number of Defective Cell Phones in Sample of 100 and Proportion Defective

Sample	Defectives	Proportion	Sample	Defectives	Proportion
1, mon	9	0.09	11, mon	10	0.10
2, tue	4	0.04	12, tue	6	0.06
3, wed	6	0.06	13, wed	6	0.06
4, thur	5	0.05	14, thur	7	0.07
5, fri	5	0.05	15, fri	8	0.08
6, mon	8	0.08	16, mon	12	0.12
7, tue	6	0.06	17, tue	6	0.06
8, wed	6	0.06	18, wed	7	0.07
9, thur	7	0.07	19, thur	8	0.08
10, fri	5	0.05	20, fri	5	0.05

The MINITAB P chart is shown in Figure 9.13.

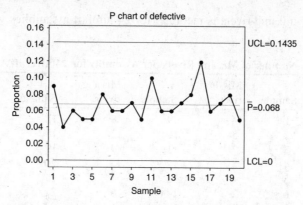

Figure 9.13   P chart for proportion defective cell phones.

The P chart seems to show the "Monday effect." That is, samples 1, 6, 11, and 16 are all samples taken on Monday and are 5-day highs. However the sample proportions are all within the lower and upper control limits.

NP charts

9.5. Refer to Problem 9.3. Use MINITAB to form an NP chart for the number defective.

SOLUTION

The only difference between Figure 9.13 and Figure 9.14 is that the y-axis in Figure 9.13 is the proportion defective in the 400 silicon wafers and in Figure 9.14 the y-axis is the number defective in the 400 silicon wafers.

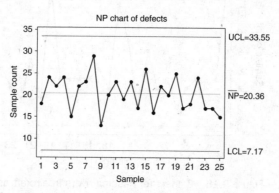

Figure 9.14   NP chart for the number of defects
per 400 silicon wafers.

9.6. Refer to Problem 9.4. Use MINITAB to form an NP chart for the number defective. The MINITAB NP chart is shown in Figure 9.15.

SOLUTION

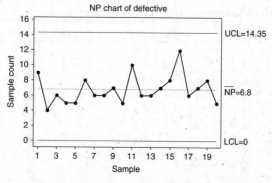

Figure 9.15   NP chart for the number defective
per 100 cell phones.

C charts

9.7. The number of missing rivets is recorded for 24 aircraft assemblies with the results given in the following table.

Number of Missing Rivets per Assembly for 24 Aircraft Assemblies

Sample	Missing rivets	Sample	Missing rivets	Sample	Missing rivets
1	14	9	8	17	5
2	5	10	2	18	5
3	7	11	9	19	15
4	5	12	13	20	3
5	4	13	4	21	7
6	2	14	6	22	3
7	5	15	5	23	4
8	4	16	5	24	6

Use MINITAB to create a C chart for the above process. Comment on the C chart.

SOLUTION

The MINITAB C chart for missing rivets is shown in Figure 9.16.

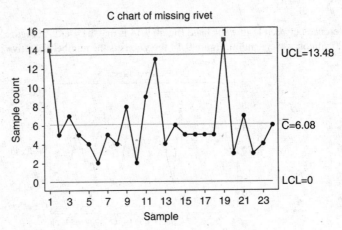

Figure 9.16   C chart for missing rivets in aircraft assemblies.

Sample 1 and sample 19 show a lack of control.

9.8. Rolls of sheet aluminum, used to manufacture cans, are examined for surface flaws. The following table gives the number of flaws in 40 samples of 100 meters squared each. Use the STATISTIX software to create the C chart for this process.

Flaws in 40 Samples of 100 Meters Squared Each

Sample	Flaws	Sample	Flaws	Sample	Flaws	Sample	Flaws
1	11	11	8	21	10	31	10
2	6	12	10	22	8	32	7
3	9	13	13	23	13	33	13
4	9	14	12	24	11	34	13
5	8	15	9	25	10	35	14
6	10	16	11	26	8	36	7
7	11	17	8	27	10	37	17
8	9	18	7	28	12	38	7
9	5	19	7	29	4	39	8
10	9	20	13	30	11	40	8

SOLUTION

The STATISTIX C chart for the number of flaws 40 samples of 100 square meters is shown in Figure 9.17.

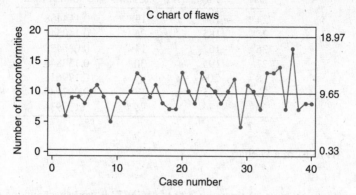

Figure 9.17   STATISTIX C chart for surface flaws in 40 samples of 100 meters squared.

The C chart shows that the process is in control.

U charts

9.9. The director of operations for a large airline was interested in improving the process of baggage handling at the airport. Records were available that reported the number of claims for lost baggage that were processed each day for a period of one month. The airlines' traffic varied over the month. The number of daily flights was also recorded. The exact number of bags carried per flight was not available. The number of flights per day was considered as the area of opportunity. The data is shown in the following table:

Number of Airline Flights per Day and Baggage Claims for a 30-Day Period

Day	Flights	Claims	Claims/Flight
1	172	14	0.081395
2	181	23	0.127072
3	168	17	0.101190
4	188	25	0.132979
5	157	27	0.171975
6	203	42	0.206897
7	191	35	0.183246
8	169	29	0.171598
9	180	30	0.166667
10	165	23	0.139394
11	185	15	0.081081
12	162	27	0.166667
13	200	41	0.205000
14	196	50	0.255102
15	172	23	0.133721
16	179	28	0.156425
17	164	20	0.121951
18	186	13	0.069892
19	160	2	0.012500
20	198	42	0.212121
21	195	38	0.194872
22	170	23	0.135294
23	175	28	0.160000

(continued)

Number of Airline Flights per Day and Baggage Claims for a 30-Day Period (*continued*)

Day	Flights	Claims	Claims/Flight
24	166	19	0.114458
25	185	26	0.140541
26	162	14	0.086420
27	195	30	0.153846
28	206	37	0.179612
29	175	17	0.097143
30	174	24	0.137931

Use MINITAB to form a U chart.

SOLUTION

The data is put into columns C1 through C4 of the MINITAB worksheet and the dialog box is filled out as in Figure 9.18.

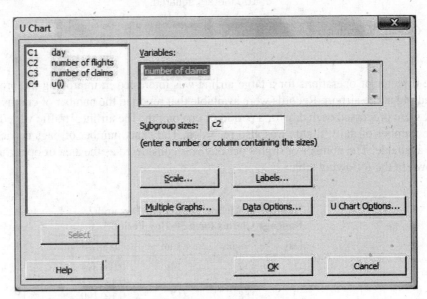

Figure 9.18   Dialog box for forming U chart.

Figure 9.19 shows the MINITAB U chart for claims.

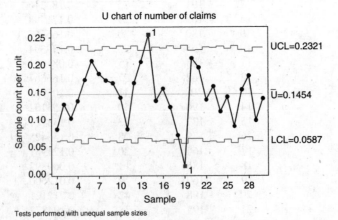

Figure 9.19   U chart for number of claims.

The U chart shows that the process is out of control on days 14 and 19.

9.10. Work Problem 9.9, using the software package STATISTIX.

SOLUTION

The data in the table is entered in a STATISTIX worksheet as shown in Figure 9.20.

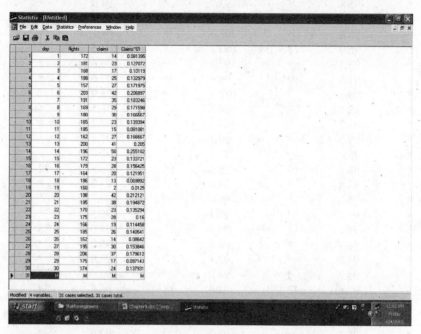

Figure 9.20 Baggage claims data in STATISTIX data worksheet.

The pulldown menu **Statistics ⇒ Quality Control ⇒ U Chart** gives the output.

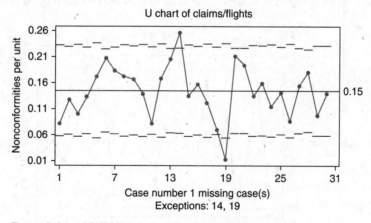

Figure 9.21 STATISTIX U chart for baggage claims data.

Figure 9.21 gives the STATISTIX U chart that corresponds to Figure 9.19.

R, S, and Xbar charts

9.11. R charts check to make sure that variability is in control. An Xbar chart assumes that the variability is in control. Therefore, these two charts are usually executed at the same time. At least the R chart is run to check on variability before the Xbar chart is run. Copper wires are coated with a thin plastic coating. Samples of four wires are taken every half an hour and the thickness of the coating (in mils) is measured. The data from the last 24 samples is given in the below table. Use MINITAB to create an R chart using the data.

Thickness of Coating of Wires for 24 Samples

Sample	mils	Sample	mils	Sample	mils	Sample	mils
1	94.1	7	97.6	13	95.0	19	91.5
*	95.8	*	93.7	*	94.0	*	94.3
*	97.2	*	95.3	*	97.2	*	100.0
*	95.0	*	95.3	*	94.1	*	93.9
2	95.9	8	100.2	14	93.5	20	96.6
*	93.3	*	92.1	*	96.6	*	92.8
*	93.5	*	96.6	*	97.6	*	101.7
*	95.3	*	102.2	*	95.9	*	97.1
3	89.5	9	94.3	15	98.5	21	99.9
*	96.7	*	94.2	*	94.2	*	94.2
*	99.5	*	90.7	*	97.4	*	96.1
*	95.0	*	93.1	*	92.7	*	95.4
4	98.0	10	96.3	16	96.8	22	95.2
*	93.8	*	94.6	*	93.2	*	95.1
*	96.0	*	99.6	*	94.7	*	92.5
*	97.1	*	98.1	*	98.5	*	96.3
5	95.1	11	99.4	17	94.0	23	96.1
*	94.9	*	98.6	*	95.0	*	98.1
*	94.8	*	92.4	*	93.9	*	94.6
*	94.7	*	94.0	*	93.9	*	92.8
6	92.6	12	94.4	18	96.6	24	98.6
*	93.6	*	97.3	*	99.6	*	99.0
*	99.8	*	95.2	*	98.4	*	103.0
*	96.1	*	91.2	*	92.0	*	102.5

SOLUTION

The R chart shown in Figure 9.22 shows that variability of the coating process is in control.

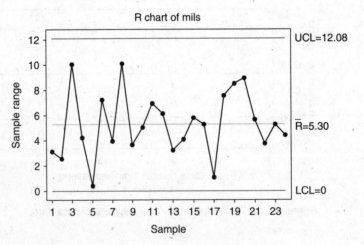

Figure 9.22   R chart for the thickness of wires.

9.12. Use an S chart to check the variability of the data in Problem 9.11.

SOLUTION

The pulldown **Stat** ⇒ **Control Charts** ⇒ **Variable charts for subgroups** ⇒ **S Chart** gives the S chart shown in Figure 9.23. This S chart is in control. If the measurements in sample 1 are put into C3 of MINITAB, sample 2 are placed into C4, and sample 3 are placed into C5 and the command Display Descriptive Statistics C3,C4, C5 is given we get

```
Descriptive Statistics: C3, C4, C5
Variable    N      N*     StDev
C3          4      0      1.315
C4          4      0      1.296
C5          4      0      4.21
```

Note in Figure 1.26 the standard deviation of samples 1, 2, and 3 are 1.315, 1.296, and 4.21 respectively. The MINITAB software calculates and plots all 24 of the standard deviations. The S chart shows that the variability of the coating on the wires is in control.

The S chart to check variability is shown in Figure 9.23.

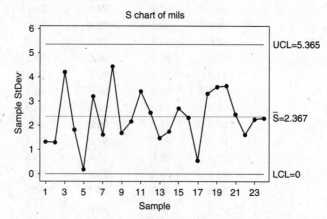

Figure 9.23   The S chart for checking the variability of the 24 subgroups.

9.13. In Problems 9.11 and 9.12 the variation in thickness of coating of the wires was seen to be in control by considering the R chart or the S chart. Now use MINITAB to make sure that the average of the thickness of coating is in control. That is, do an Xbar chart on the data in Problem 9.11.

SOLUTION

Figure 9.24 shows the MINITAB Xbar chart for the thickness of the coating on the wires.

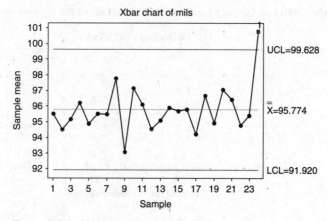

Figure 9.24   Xbar chart for thickness of coating on the wires.

The Xbar chart goes out of control in the 24th subgroup.

9.14. Refer to Problem 9.11. Find an R chart using STATISTIX.

SOLUTION

First the data is entered in EXCEL as shown in the following table. The first sample of thickness coatings are entered as X1, X2, X3, and X4 or 94.1, 95.8, 97.2, and 95.0. The second sample of thickness is 95.9, 93.3, 93.5, and 95.3. Until the last sample is entered, that is, 98.6, 99.0, 103.0, and 102.5.

Next the EXCEL table is copied into the data sheet of STATISTIX as shown in Figure 9.25.

EXCEL Table Where the 24 Samples of Four Coatings of Wires are Entered

x1	x2	x3	x4
94.1	95.8	97.2	95.0
95.9	93.3	93.5	95.3
89.5	96.7	99.5	95.0
98.0	93.8	96.0	97.1
95.1	94.9	94.8	94.7
92.6	93.6	99.8	96.1
97.6	93.7	95.3	95.3
100.2	92.1	96.6	102.2
94.3	94.2	90.7	93.1
96.3	94.6	99.6	98.1
99.4	98.6	92.4	94.0
94.4	97.3	95.2	91.2
95.0	94.0	97.2	94.1
93.5	96.6	97.6	95.9
98.5	94.2	97.4	92.7
96.8	93.2	94.7	98.5
94.0	95.0	93.9	93.9
96.6	99.6	98.4	92.0
91.5	94.3	100.0	93.9
96.6	92.8	101.7	97.1
99.9	94.2	96.1	95.4
95.2	95.1	92.5	96.3
96.1	98.1	94.6	92.8
98.6	99.0	103.0	102.5

Figure 9.25   STATISTIX data for R Chart.

The pulldown **Statistics ⇒ Quality Control ⇒ R Chart** gives the R chart shown in Figure 9.26.

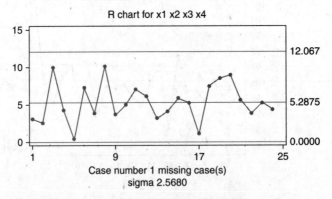

Figure 9.26   R chart using software STATISTIX.

The R chart in Figure 9.22 and Figure 9.26 are the same. The data is set up differently for MINITAB and STATISTIX.

9.15. Set up the S chart for the data in Problem 9.11 using STATISTIX.

SOLUTION

Set up your data as shown in Figure 9.25. The pulldown **Statistics ⇒ Quality Control ⇒ S Chart** gives the S chart shown in Figure 9.27.

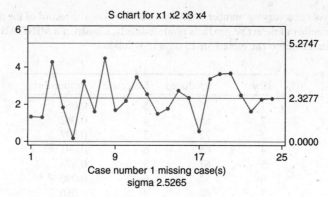

Figure 9.27   S chart for the wire coatings.

9.16. Use the software STATISTIX to form the Xbar chart for the wire coatings.

SOLUTION

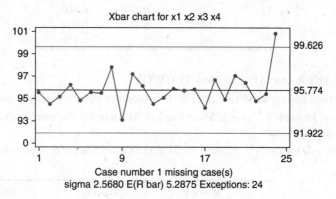

Figure 9.28   Xbar chart for the wire coatings.

Compare Figure 9.28 with Figure 9.24.

SUPPLEMENTARY PROBLEMS

9.17. In Problem 9.1 of solved problems in this chapter, construct a Pareto chart using MINITAB. Include cumulative percentage. Give the percentage accounted for by the top cause. Give the percentage accounted for by the top two causes. Give the percentage accounted for by the top three causes.

9.18. An engineering process has defectives caused by five different causes. The five causes and their counts are as follows over a one-week period.

Cause	Count
Caulking	198
Connecting	103
Gapping	72
Fitting	25
Torque	18
All causes	416

Construct a cumulative Pareto chart using SPSS. Use the pulldown menu **Analyze ⇒ Quality Control ⇒ Pareto Charts**. Give the percentage accounted for by the top cause. Give the percentage accounted for by the top two causes.

9.19. A company produces a varying number of products daily. It keeps a record of the number of products produced as well as the number of defective products produced daily. Construct a MINITAB P chart for the proportion of defectives produced daily. The record for 15 days is as follows:

Day	Produced	Defective	Proportion
1	75	5	0.067
2	65	4	0.062
3	72	7	0.097
4	66	6	0.091
5	67	5	0.075
6	67	6	0.090
7	60	3	0.050
8	59	8	0.140
9	65	7	0.108
10	71	6	0.085
11	73	9	0.123
12	74	10	0.135
13	60	6	0.100
14	68	5	0.074
15	66	9	0.136

9.20. In Problem 9.19, construct a P chart using STATISTIX.

9.21. Give the details in finding an SPSS P chart for the proportion defective in Problem 9.19.

9.22. Refer to solved Problem 9.3. Use SPSS to form an NP chart for the number defective. Give the SPSS steps needed to form the NP chart.

9.23. The number of missing rivets is recorded for 24 aircraft assemblies with the results given below.

Sample	Missing rivets	Sample	Missing rivets	Sample	Missing rivets
1	14	9	8	17	5
2	5	10	2	18	5
3	7	11	9	19	15
4	5	12	13	20	3
5	4	13	4	21	7
6	2	14	6	22	3
7	5	15	5	23	4
8	4	16	5	24	6

Build the SPSS data editor, and supply all the details needed to form a C chart for this process.

9.24. R charts check to make sure that variability is in control. An Xbar chart assumes that the variability is in control. Therefore, these two charts are usually executed at the same time. At least the R chart is run to check on variability before the Xbar chart is run. Copper wires are coated with a thin plastic coating. Samples of four wires are taken every half an hour and the thickness of the coating (in mils) is measured. The data from the last 24 samples is given in Problem 9.11. Set up the data file in SPSS needed to check for control in variability using an R chart. Also use SPSS to set up an Xbar chart and check for control of the mean.

Answers to Supplementary Problems

1.16.

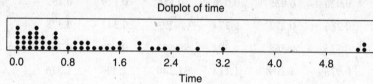

Dotplot of time

The distribution has a shape that is skewed to the right. Three of the fifty times exceed 4.8 microseconds. Twenty-six of the times are less than 0.8 microseconds.

1.17. Stem-and-Leaf Display: Time

```
Stem-and-leaf of Time   N = 50
Leaf Unit = 0.10
  10     0    0000001111
  20     0    2222223333
  25     0    45555
  25     0    67
  23     0    9999
  19     1    11
  17     1    233
  14     1    55
  12     1    6
  11     1    89
   9     2    11
   7     2    2
   6     2    5
   5     2    7
   4     2
   4     3    1
```

3	3	
3	3	
3	3	
3	3	
3	4	
3	4	
3	4	
3	4	
3	4	
3	5	
3	5	2
2	5	44

The three largest times are 5.2, 5.4, and 5.4 microseconds.

1.18. Sorted data (times)

0.000	**0.006**	**0.013**	**0.026**	**0.035**	**0.070**	**0.115**	**0.129**	**0.157**
0.184	**0.208**	**0.222**	**0.266**	**0.272**	**0.274**	**0.293**	**0.347**	**0.378**
0.390	**0.391**	**0.486**	**0.535**	**0.568**	**0.574**	**0.592**	**0.619**	**0.771**
0.902	**0.935**	**0.956**	**0.975**	**1.109**	**1.114**	**1.240**	**1.305**	**1.376**
1.530	**1.588**	**1.614**	**1.868**	**1.922**	**2.103**	**2.152**	**2.250**	**2.525**
2.793	**3.192**	5.296	5.417	5.448				

Mean of the times = 0.190 Standard deviation of the times = 1.346

Mean − 3 Standard deviations = −3.848

Mean + 3 Standard deviations = 4.228

Of the 50, 47 values are within 3 standard deviations of the mean or 94%.

1.19.

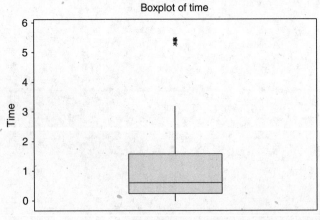

Boxplot of time

The box contains the middle 50% of the times. The symbols are located at 5.296, 5.417, and 5.448. They are more than 3 standard deviations away from the mean. They are called **outliers**.

1.20.

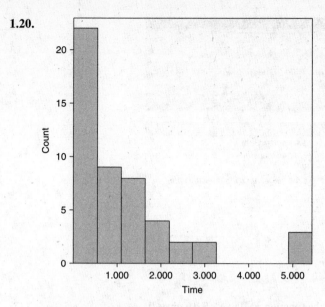

An SPSS plot of a histogram of the data is shown above.

1.21. Suppose the descriptive statistics requested in the following dialog box are requested:

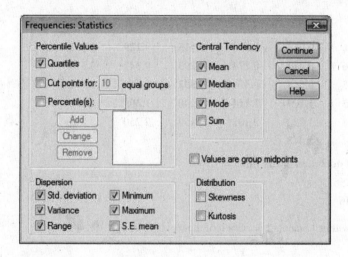

Statistics

		Time
N	Valid	50
	Missing	0
Mean		1.15062
Median		.60550
Mode		.000[a]
Std. Deviation		1.345668
Variance		1.811
Range		5.448
Minimum		.000
Maximum		5.448
Percentiles	25	.25500
	50	.60550
	75	1.59450

[a]Multiple modes exist. The smallest value is shown.

1.22.

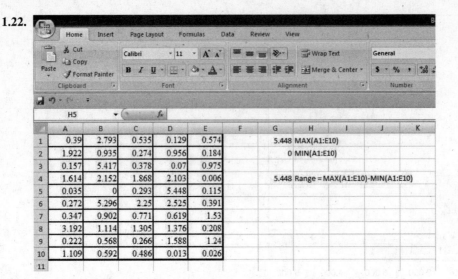

The function =MAX(A1:E10) gives 5.448 and the function =MIN(A1:E10) gives 0.

The function =MAX(A1:E10) – MIN(A1:E10) gives the range, 5.448. The range gives a measure of variation which plays an important role in statistical process control.

1.23. Sorted data (times)

0.000	**0.006**	**0.013**	**0.026**	**0.035**	**0.070**	**0.115**	**0.129**	**0.157**
0.184	**0.208**	**0.222**	**0.266**	**0.272**	**0.274**	**0.293**	**0.347**	**0.378**
0.390	**0.391**	**0.486**	**0.535**	**0.568**	**0.574**	**0.592**	**0.619**	**0.771**
0.902	**0.935**	**0.956**	**0.975**	**1.109**	**1.114**	**1.240**	**1.305**	**1.376**
1.530	1.588	1.614	1.868	1.922	2.103	2.152	2.250	2.525
2.793	3.192	5.296	5.417	5.448				

Mean of the times = 0.190 **Standard deviation of the times** = 1.346

Mean – 1 Standard deviation = –1.156

Mean + 1 Standard deviation = 1.536

The data in bold in the sorted data is within 1 standard deviation of the mean. $(37/50) \times 100 = 74\%$.

1.24. The 50 times given in Table 1.7 are stacked in column A of the EXCEL worksheet. Data analysis is chosen in the data sheet. This provides the following dialog box:

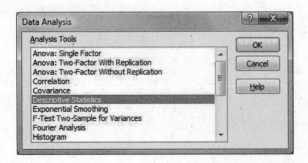

This provides the following descriptive statistics dialog box which is filled in as shown:

Descriptive Statistics		? X
Input		
Input Range:	$A$1:$A$50	OK
Grouped By:	⦿ Columns	Cancel
	○ Rows	Help
☐ Labels in first row		
Output options		
⦿ Output Range:	$C$1	
○ New Worksheet Ply:		
○ New Workbook		
☑ Summary statistics		
☐ Confidence Level for Mean:	95 %	
☐ Kth Largest:	1	
☐ Kth Smallest:	1	

This gives the following EXCEL output:

Column 1	
Mean	1.15062
Standard Error	0.190306
Median	0.6055
Mode	#N/A
Standard Deviation	1.345668
Sample Variance	1.810823
Kurtosis	3.980657
Skewness	1.988476
Range	5.448
Minimum	0
Maximum	5.448
Sum	57.531
Count	50

1.25. The sorted 50 ignition times are:

0.000	0.006	0.013	0.026	0.035	0.070	0.115	0.129	0.157
0.184	0.208	0.222	0.266	0.272	0.274	0.293	0.347	0.378
0.390	0.391	0.486	0.535	0.568	0.574	0.592	0.619	0.771
0.902	0.935	0.956	0.975	1.109	1.114	1.240	1.305	1.376
1.530	1.588	1.614	1.868	1.922	2.103	2.152	2.250	2.525
2.793	3.192	5.296	5.417	5.448				

The 25th value in the sorted array is 0.592 and the 26th value in the sorted array is 0.619. The average of the two middle values in the sorted array is (0.592 + 0.619)/2 = 0.6055.

The built-in median function gives =MEDIAN(A1:E10) = 0.6055.

1.26. Percentiles

Variable	Cases	10.0	20.0	30.0	40.0	50.0
strength	30	53.010	54.940	56.120	57.220	59.400

1.27. SPSS output

Statistics

		Time
N	Valid	30
	Missing	1
Percentiles	10	53.0100
	20	54.9400
	30	56.1200
	40	57.2200
	50	59.4000

1.28. EXCEL output

A	B	C
55.5	61.4	63.8
53.1	49.8	60.8
56.4	61.4	55.7
63.1	60.2	60.3
61.2	53.2	56.7
70.9	52.7	56
56.7	58	66.1
59.3	64.6	53
63.5	60.5	67.9
59.5	54.8	59.2

53.09 PERCENTILE(A1:C10,0.1)
55.36 PERCENTILE(A1:C10,0.2)
56.28 PERCENTILE(A1:C10,0.3)
57.48 PERCENTILE(A1:C10,0.4)
59.4   PERCENTILE(A1:C10,0.5)

Note that EXCEL does not give the same values as STATISTIX or SPSS for P_{10}, P_{20}, P_{30}, and P_{40}.

1.29. Descriptive Statistics: strength

```
Variable    Mean      StDev
strength    59.177    4.857
```

1.30. Descriptive Statistics: strength

```
Variable    Mean      Median
strength    59.177    59.400
```

CHAPTER 2

2.25. S is the set of ordered pairs (resistance of the first resistor, resistance of the second resistor).

$S = \{(9, 18), (9, 19), (9, 20), (9, 21), (10, 18), (10, 19), (10, 20), (10, 21), (11, 18), (11, 19), (11, 20), (11, 21),$
$(12, 18), (12, 19), (12, 20), (12, 21)\}$

2.26. $A = \{(11, 18), (11, 19), (11, 20), (11, 21), (12, 18), (12, 19), (12, 20), (12, 21)\}$
$B = \{(9, 18), (10, 18), (11, 18), (12, 18)\}$
$C = \{(9, 19), (10, 18)\}$

2.27. $A \cap B = \{(11, 18), (12, 18)\}$
$A \cup B = \{(11, 18), (11, 19), (11, 20), (11, 21), (12, 18), (12, 19), (12, 20), (12, 21), (9, 18), (10, 18)\}$
$B \cap C^c = \{(9, 18), (11, 18), (12, 18)\}$

2.28. There are $3 \times 2 \times 3 \times 2 = 36$ ways to order your computer.

2.29. The combination of 12 taken 3 at a time is COMBIN(12,3) = 220.

2.30. The EXCEL solution is =36^8 or 2.82111E+12.

2.31. The relative frequency definition gives the probability as $83/125 = 0.664$.

2.32. The subjective probability is 0.10.

2.33. The number of pairs of nuts and bolts possible is $(10)(15) = 150$. The number of pairs of thick nuts and bolts is $(4)(5) = 20$, the number of medium nuts and bolts is $(3)(5) = 15$, and the number of pairs of thin nuts and bolts is $(3)(5) = 15$. The probability that the selected nut fits the selected bolt is $(20 + 15 + 15)/150 = 50/150 = 1/3$.

2.34. True.

2.35. True.

2.36. True.

2.37. The listed events are mutually exclusive. (a) The probability that the customer testing service will rate the antipollution device as very poor or poor or fair is $0.04 + 0.16 + 0.20 = 0.40$. (b) Since the events are mutually exclusive, the probability that it will be rated as very good and excellent is 0.

2.38. Let A be the event that the family owns a flat screen TV and B be the event that the family owns a computer. $(A$ and $B)$ is the event that the family owns both. $P(A$ or $B) = P(A) + P(B) - P(A$ and $B) = 0.8 + 0.7 - 0.65 = 0.85$.

2.39. A is the event that the student is enrolled in Advanced Statistics, B is the event that the student is enrolled in Operations Research.

$(A \cup B) = (A \cap B^c) \cup (A \cap B) \cup (B \cap A^c)$

The number of students in $A \cup B$ is the number in $A \cap B^c$ plus the number in $A \cap B$ plus the number in $B \cap A^c$.

The number in $A \cup B$ is $50 + 40 + 20 = 110$. The number in $(A \cup B)^c$ is $150 - 110$ or 40. That is, the number in neither course is 40.

2.40. The probability (flaw on the side | flaw on the top) $= P(A|B) = \dfrac{P(A \cap B)}{P(B)} = \dfrac{0.01}{0.03} = 0.33$.

2.41. The probability (flaw on the top | flaw on the side) $= P(A|B) = \dfrac{P(A \cap B)}{P(B)} = \dfrac{0.01}{0.02} = 0.50$.

2.42. $P(\text{all 4 work}) = (1 - 0.05)^4 = (0.95)^4 = 0.81$.

2.43. Let B be the event that the Ford fails an emissions test within two years. Let A_1 be the event the chosen Ford has the smallest engine, let A_2 be the event the chosen Ford has the medium engine, and A_3 be the event that the chosen Ford has the largest engine.

$P(B) = P(B|A_1)P(A_1) + P(B|A_2)P(A_2) + P(B|A_3)P(A_3) = (0.15)(0.50) + (0.10)(0.40) + (0.05)(0.10) = 0.12$.

2.44. Refer to the notation in the solution to the previous problem $P(A_1|B)$ is given by

$$\frac{P(B|A_1)P(A_1)}{P(B|A_1)P(A_1) + P(B|A_2)P(A_2) + P(B|A_3)P(A_3)} = \frac{(0.15)(0.50)}{(0.15)(0.50) + (0.10)(0.40) + (0.05)(0.10)}$$

Or, $0.075/0.12 = 0.625$.

2.45. $P(A_3|B)$ is given by

$$\frac{P(B|A_3)P(A_3)}{P(B|A_1)P(A_1) + P(B|A_2)P(A_2) + P(B|A_3)P(A_3)} = \frac{(0.05)(0.10)}{(0.15)(0.50) + (0.10)(0.40) + (0.05)(0.10)}$$

Or, $0.005/0.12 = 0.0417$.

2.46. $E(X) = (2)(0.3) + (1)(0.6) + (0)(0.1) = 1.2$

2.47. $E(X) = (0)(0.6) + (1)(0.2) + (2)(0.15) + (3)(0.05) = 0.65$

2.48. $E(X) = (1)(0.1) + (2)(0.1) + (3)(0.2) + (4)(0.2) + (5)(0.4) = 3.7$

CHAPTER 3

3.21.

First face	Second face	x = sum	x	$p(x)$
1	1	2	2	0.0625
1	2	3	3	0.125
1	3	4	4	0.1875
1	4	5	5	0.25
2	1	3	6	0.1875
2	2	4	7	0.1250
2	3	5	8	0.0625
2	4	6		
3	1	4		
3	2	5		
3	3	6		
3	4	7		
4	1	5		
4	2	6		
4	3	7		
4	4	8		

3.22. $\mu = \Sigma\, xp(x) = 2(0.0625) + 3(0.125) + 4(0.1875) + 5(0.25) + 6(0.1875) + 7(0.125) + 8(0.0625) = 5$

$\sigma^2 = \Sigma\, x_i^2 p(x_i) - \mu^2 = 4(0.0625) + 9(0.125) + 16(0.1875) + 25(0.25) + 36(0.1875) + 49(0.125) + 64(0.0625) - 25$
$= 27.5 - 25 = 2.5$

$\sigma = 1.58$

3.23. According to Chebyshev's theorem, there is at least 75% of the distribution between 1.84 and 8.16. There is 100% of the distribution between 1.84 and 8.16.

3.24. The EXCEL expression =BINOMDIST(2,20,0.01,1) gives the probability of at most 2 failures in the 20. The probability is 0.999. It is almost certain that at most 2 failures occur in the 20 selected.

3.25. (a) The probability that all cards are from the same suit is equal to the probability that all 5 cards are clubs or all 5 are diamonds or all 5 are spades or all 5 are hearts. The probability that all 5 are from the same suit is P(all 5 are clubs) = $(1/4)^5$ or P(all 5 are diamonds) = $(1/4)^5$ or P(all 5 are spades) = $(1/4)^5$ or P(all 5 are hearts) = $(1/4)^5$. The probability that all 5 are from the same suit is $(1/4)^5$ added 4 times, which is $4(1/4)^5 = 4/1024 = 0.0039$.

(b) $P(x) = (0.75)^{x-1}(0.25), x = 1, 2, \ldots$

(c) $b(x) = \binom{6}{x}\left(\dfrac{12}{52}\right)^x\left(\dfrac{40}{52}\right)^{6-x}, x = 0, 1, 2, 3, 4, 5, 6$

(d) $P(X = x) = \dfrac{\binom{4}{x}\binom{52-4}{4-x}}{\binom{52}{4}}, x = 0, 1, 2, 3, 4$

(e) $NB(x) = \binom{x-1}{3-1}(0.25)^3(0.75)^{(x-3)}\ x = 3, 4, \ldots$

3.26. (a)

x	h(x)
0	0.310563
1	0.431337
2	0.20984
3	0.044177
4	0.003965
5	0.000119

(b) **Probability Density Function**

Hypergeometric with N = 50, M = 10, and n = 5

x	P(X = x)
0	0.310563
1	0.431337
2	0.209840
3	0.044177
4	0.003965
5	0.000119

3.27. (a) $Po(x) = e^{-15}\dfrac{15^x}{x!}$, $x = 0, 1, 2, \ldots$

(b)

x	Po(x)	Cumulative
0	3.05902E-07	3.05902E-07
1	4.58853E-06	4.89444E-06
2	3.4414E-05	3.93084E-05
3	0.00017207	0.000211379
4	0.000645263	0.000856641
5	0.001935788	0.002792429
6	0.00483947	0.0076319
7	0.010370294	0.018002193
8	0.0194443	0.037446493
9	0.032407167	0.069853661
10	0.048610751	0.118464412
11	0.066287387	0.184751799
12	0.082859234	0.267611033
13	0.095606809	0.363217842
14	0.102435867	0.465653709
15	0.102435867	0.568089576
16	0.096033625	0.664123201
17	0.084735551	0.748858752
18	0.07061296	0.819471712
19	0.055747073	0.875218785
20	0.041810305	0.91702909

(c) $P(X > 30) = 1 - \text{POISSON}(30,15,1) = 0.000197$

(d) $P(X \le 15) - P(X \le 4) = \text{POISSON}(15,15,1) - \text{POISSON}(4,15,1) = 0.567$

3.28. (a) =MULTINOMIAL(0,25,0,0)*(0.15)^0*(0.65)^25*(0.15)^0*(0.05)^0 = 2.1E-05

    (b) =MULTINOMIAL(25,0,0,0)*(0.15)^25*(0.65)^0*(0.15)^0*(0.05)^0 = 2.52512E-21
        =MULTINOMIAL(0,25,0,0)*(0.15)^0*(0.65)^25*(0.15)^0*(0.05)^0 = 2.10297E-05
        =MULTINOMIAL(0,0,25,0)*(0.15)^0*(0.65)^0*(0.15)^25*(0.05)^0 = 2.52512E-21
        =MULTINOMIAL(0,0,0,25)*(0.15)^0*(0.65)^0*(0.15)^0*(0.05)^25 = 2.98023E-33

    Add these four probabilities to get the answer. It will be approximately 2.1 E-05.

    (c) =MULTINOMIAL(6,6,6,7)*(0.15)^6*(0.65)^6*(0.15)^6*(0.05)^7 = 6.30352E-08

3.29. The MINITAB pulldown **Calc ⇒ Random Data ⇒ Poisson** gives the following dialog box:

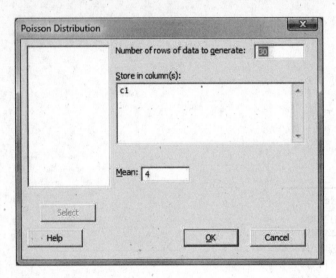

The following simulated sample was produced by MINITAB:

```
Data Display
C1
   5    6    4    6    9    9    4    7    5    9    8    5    8    5    5
  10    6    9    5    6    9    7    5   10    4    5    5    8    8    4
```

3.30. Probability Density Function
    Binomial with n = 500 and p = 0.009
```
x  P( X = x )
0   0.0108850
```

Probability Density Function
    Poisson with mean = 4.5
```
x  P( X = x )
0   0.0111090
```

CHAPTER 4

4.25. (a)

0	0
0.1	0.004
0.2	0.031949
0.3	0.107129
0.4	0.24953
0.5	0.469707
0.6	0.758978
0.7	1.079146
0.8	1.359699
0.9	1.513028
1	1.471518
1.1	1.231367
1.2	0.869062
1.3	0.505244
1.4	0.235537
1.5	0.085451
1.6	0.023347
1.7	0.004636
1.8	0.000644
1.9	6.01E-05
2	3.6E-06

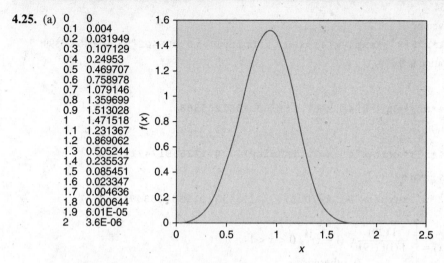

Even though the function is defined from 0 to ∞, it is sufficient to go from 0 to 2. At $x = 2$, the $f(x)$ value is 3.6E-06 and the $f(x)$ values decrease from 2 to ∞.

(b) The antiderivative of $f(x) = 4x^3 e^{-x^4}$ is $-e^{-x^4}$. The antiderivative at the upper limit of integration (∞) is 0. Evaluation at the lower limit of integration gives -1. The value of the integral is $0 - (-1) = 1$.

(c)
```
> evalf(int(4*x^3*exp(-x^4),x=0..infinity)); 1
```

(d) The interval from 0 to 2 is divided into 25 equal intervals. This makes each interval 0.08 wide. Evaluating $f(x) = 4x^3 e^{-x^4}$ at the interval midpoints and multiplying times 0.08 gives rectangle areas that add up to 0.999999 or approximately 1. The details are as follows. This is the approximate evaluation of an integral as a limiting sum of rectangles.

Midpoint	Height	Area of rectangle
0.04	0.000256	2.04799E-05
0.12	0.006911	0.000552845
0.2	0.031949	0.002555907
0.28	0.08727	0.006981595
0.36	0.183516	0.014681249
0.44	0.328201	0.026256101
0.52	0.522777	0.041822122
0.6	0.758978	0.060718239
0.68	1.015614	0.081249104
0.76	1.257797	0.100623789
0.84	1.441037	0.115282942
0.92	1.521591	0.121727289
1	1.471518	0.117721421
1.08	1.292642	0.103411395
1.16	1.021135	0.081690821
1.24	0.717068	0.057365402
1.32	0.441857	0.035348588
1.4	0.235537	0.018842929
1.48	0.106946	0.008555678
1.56	0.040678	0.003254275
1.64	0.012733	0.001018633
1.72	0.003218	0.000257474
1.8	0.000644	5.15131E-05
1.88	9.98E-05	7.98774E-06
1.96	1.17E-05	9.38961E-07
	Sum =	0.999998718

We shall use MAPLE to evaluate (e) through (h). The integrals involved could be evaluated using integration by parts or as the limiting areas of rectangles.

(e)
```
> evalf(int(4*x*x^3*exp(-x^4),x=0..infinity)); 0.9064024772
```
The mean is 0.9064.

The variance is given by
```
> evalf(int(4*x^2*x^3*exp(-x^4),x=0..infinity))- 0.906402^2; 0.0646623399.
```
The standard deviation is 0.254.

(f) $P(X < 1)$ is
```
> evalf(int(4*x^3*exp(-x^4),x=0..1)); 0.6321205588.
```

(g) $P(X > 2)$ is given by
```
> evalf(int(4*x^3*exp(-x^4),x=2..infinity)); 0.112535174710^-6.
```

(h) $P(0.5 < X < 1.5)$ is given by
```
> evalf(int(4*x^3*exp(-x^4),x=(0.5)..(1.5))); 0.9330833471.
```

4.26. (a) The density is $f(x) = \begin{cases} \dfrac{\Gamma(20)}{\Gamma(1)\Gamma(19)}x^0(1-x)^{18}, & 0 < x < 1 \\ 0, & \text{otherwise} \end{cases}$

The density reduces to $19(1-x)^{18}, 0 < x < 1$. The MINITAB plot of the graph is shown below.

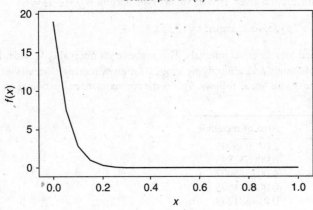
Scatter plot of $f(x)$ vs x

(b)
```
> evalf(int(19*(1-x)^18,x=0..1)); 1
```

(c) The mean is $\mu = \dfrac{\alpha}{\alpha + \beta} = \dfrac{1}{20} = 0.05$. The variance is $\sigma^2 = \dfrac{\alpha\beta}{(\alpha+\beta)^2(\alpha+\beta+1)} = \dfrac{19}{8400} = 0.00226$. The standard deviation is 0.048.

(d)
```
> evalf(int(19*(1-x)^18,x=0..(.1))); 0.8649148282
```

(e) **Data Display**

0.048 0.029 0.043 0.045 0.070 0.035 0.040 0.020 0.005 0.057

4.27. (a) We need to find c such that $\displaystyle\int_0^\infty \int_0^2 \int_0^1 cxy^2 e^{-z}\, dx\, dy\, dz = 1.$

The value of the integral is $\dfrac{4}{3}c$. Set this equal to 1 and solving for c gives $c = 0.75$.

(b)
```
> evalf(int(int((3/4)*x*y^2*exp(-z),y=0..2),z=0..infinity)); 2 x
```

The marginal of X is $f_1(x) = 2x, 0 < x < 1$.
```
> evalf(int(int((3/4)*x*y^2*exp(-z),x=0..1),z=0..infinity)); 0.3750000000y^2
```

The marginal of Y is $f_2(y) = 2y^2/3, 0 < y < 2$.

```
> evalf(int(int((3/4)*x*y^2*exp(-z),x=0..1),y=0..2)); e^(-1,z)
```

The marginal of Z is $f_3(z) = e^{-z}, Z > 0$.

Since $f(x, y, z) = f_1(x) f_2(y) f_3(z)$, X, Y, and Z are independent.

(c)
```
> evalf(int(int(int((3/4)*x*y^2*exp(-z),x=0..1),y=0..1),z=0..1));
0.07901506985
```

$P(X < 1, Y < 1, Z < 1) = 0.07902$

4.28. (a) =NORMDIST(105,100,5,1)–NORMDIST(90,100,5,1), 0.818595

(b) **Cumulative Distribution Function**
```
Normal with mean = 100 and standard deviation = 5

x       P( X <= x )
105     0.841345
```

Cumulative Distribution Function
```
Normal with mean = 100 and standard deviation = 5

x       P( X <= x )
90      0.0227501
```

$0.841345 - 0.0227501 = 0.818595$

(c)
```
> evalf(int(0.0797885*exp(-(x-100)^2/50),x=90..105));  0.8185950647
```

4.29. (a) In MINITAB, α is called the location parameter and β is called the shape parameter in the dialog box for the lognormal distribution. The log-normal is shown in the figure below.

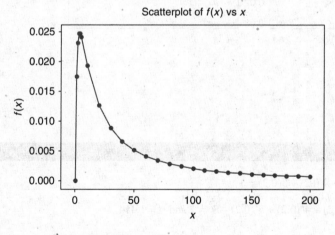

Scatterplot of $f(x)$ vs x

(b) $\mu = e^{\alpha + \beta^2/2} = e^{3.5 + 1.125} = e^{4.625} = 102$

$\sigma^2 = e^{2\alpha + \beta^2}(e^{\beta^2} - 1) = e^{9.25}(e^{2.25} - 1) = 88311$ and $\sigma = 297$

(c)
```
Lognormal with location = 3.5 and scale = 1.5

x       P( X <= x )
150     0.843054
```

4.30. (a) The gamma distribution is $f(x) = \dfrac{1}{\beta^{\alpha}\Gamma(\alpha)} x^{\alpha-1} e^{-x/\beta}$, $x > 0$. Substitute $\alpha = v/2$ and $\beta = 2$. The chi-square

distribution is $f(x) = \dfrac{1}{2^{v/2}\Gamma(v/2)} x^{v/2-1} e^{-x/2}$, $x > 0$.

    (b) The mean of the gamma distribution is $\mu = \alpha\beta$. Substitute $\alpha = v/2$ and $\beta = 2$.

        The chi-square mean is $\mu = (v/2)(2) = v$.

    (c) The variance of the gamma distribution is $\alpha\beta^2$. Substitute $\alpha = v/2$ and $\beta = 2$.

        The chi-square variance is $\sigma^2 = (v/2)(4) = 2v$.

    (d) The figure below shows 4 chi-square distributions. C2 is a chi-square with 2 degrees of freedom. C3 is a chi-square with 5 degrees of freedom. C4 is a chi-square with 10 degrees of freedom. C5 is a chi-square with 15 degrees of freedom.

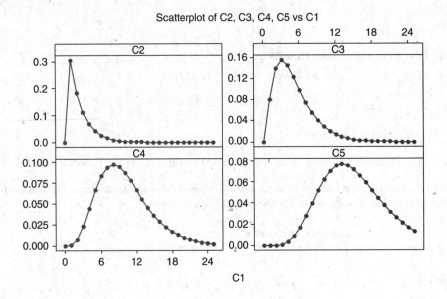

Scatterplot of C2, C3, C4, C5 vs C1

CHAPTER 5

5.11. (a) $\mu = 1(0.2) + 2(0.2) + 3(0.2) + 4(0.2) + 5(0.2) = 3$

        $\sigma^2 = 1^2(0.2) + 2^2(0.2) + 32(0.2) + 4^2(0.2) + 5^2(0.2) - 3^2 = 2$ and $\sigma = 1.414$

    (b) Uniform

    (c)

$\bar{x}$	1	1.5	2	2.5	3	3.5	4	4.5	5
$p(\bar{x})$	0.04	0.08	0.12	0.16	0.20	0.16	0.12	0.08	0.04

        $\mu_{\bar{x}} = 3$ $\sigma_{\bar{x}} = \dfrac{1.414}{\sqrt{2}} = 1$

    (d)

$\bar{x}$	1	1.33	1.67	2	2.33	2.67	3	3.33	3.67	4	4.33	4.67	5
$p(\bar{x})$	.008	.024	.048	.080	.120	.144	.152	.144	.120	.080	.048	.024	.008

        $\mu_{\bar{x}} = 3$ $\sigma_{\bar{x}} = \dfrac{1.414}{\sqrt{3}} = 0.816$

(e)

$\bar{x}$	Count	$p(\bar{x})$
1.00	1	0.0016
1.25	4	0.0064
1.50	10	0.0160
1.75	20	0.0320
2.00	35	0.0560
2.25	52	0.0832
2.50	68	0.1088
2.75	80	0.1280
3.00	85	0.1360
3.25	80	0.1280
3.50	68	0.1088
3.75	52	0.0832
4.00	35	0.0560
4.25	20	0.0320
4.50	10	0.0160
4.75	4	0.0064
5	1	0.0016

$$\mu_{\bar{x}} = 3 \qquad \sigma_{\bar{x}} = \frac{1.414}{\sqrt{4}} = 0.707$$

5.12.

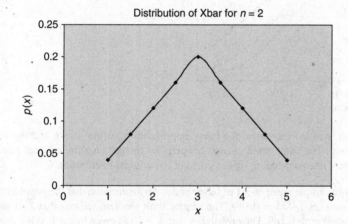

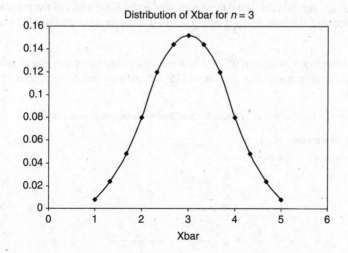

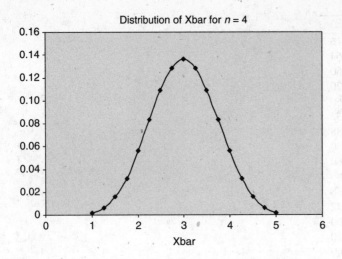

Distribution of Xbar for $n = 4$

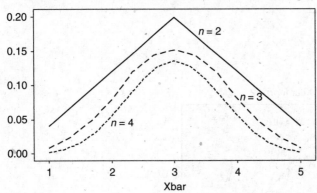

Distribution of Xbar for $n = 2$, 3, and 4 overlaid on the same graph

The plots show that as the sample size (n) increases, the curve approaches a normal curve and the standard deviation of the curve is decreasing. The same phenomenon happens no matter what the original population distribution is. It happens at different rates depending on the shape of the original distribution.

5.13. Calculate the probability of finding a sample mean of 3.1 or larger under the assumption that the population mean is 2.5. The standard error of the mean is $1/\sqrt{35} = 0.169$. The central limit theorem tells us that $\bar{X}$ is normally distributed with mean 2.5 and standard error 0.169. The probability that $\bar{X} \geq 3.1$ is given by the EXCEL command =1-NORMDIST(3.1,2.5,0.169,1) which equals 0.00019. Since this probability is less than 0.05, it is concluded that the population mean fluoride is greater than 2.5. What this result says is that if the mean of the population is really 2.5 and a sample mean is computed, the probability is only 0.00019 that the sample mean would be as large as 3.1 or larger.

5.14. The Anderson–Darling normality test gives a p-value of 0.778, The Ryan–Joiner normality test gives a p-value > 0.1, or the Kolmogorov–Smirnov normality test gives a p-value > 0.15. No matter which of these tests are used, normality is not rejected.

Proceeding, the calculated value for the sample mean, standard error, and standard deviation are:

Descriptive Statistics: fluoride

Variable	Mean	SE Mean	StDev
fluoride	2.740	0.338	1.069

$$t_9 = \frac{\bar{x} - \mu}{\frac{s}{\sqrt{10}}} = \frac{2.74 - 2.5}{0.338} = 0.71$$

The probability of getting a t_9 value this large or larger is found using MINITAB.

Cumulative Distribution Function

```
Student's t distribution with 9 DF
x       P( X <= x )
0.71    0.752156
```

$P(t_9 > 0.71) = 1 - 0.75 = 0.25$

There is a good chance (25%) of getting such a sample if the population mean is 2.5. The population mean, 2.5, is not rejected.

5.15. The two rules usually give the same answer, but not always. I would use the simpler of the two rules and stick with it. That is the second of the two rules.

(a) $n = 150$, $\hat{p} = 0.10$

$n\hat{p} = 15$ and $n\hat{q} = 135$ Yes

$\hat{p} \pm 3\sqrt{\dfrac{\hat{p}\hat{q}}{n}}$ or $0.10 \pm 3\sqrt{\dfrac{0.10(0.90)}{150}}$ or 0.10 ± 0.07 (0.03, 0.17)   Yes

(b) $n = 500$, $\hat{p} = 0.05$

$n\hat{p} = 25$ and $n\hat{q} = 475$ Yes

$\hat{p} \pm 3\sqrt{\dfrac{\hat{p}\hat{q}}{n}}$ or $0.05 \pm 3\sqrt{\dfrac{0.05(0.95)}{500}}$ or 0.05 ± 0.03 (0.02, 0.08)   Yes

(c) $n = 50$, $\hat{p} = 0.15$

$n\hat{p} = 7.5$ and $n\hat{q} = 42.5$ Yes

$\hat{p} \pm 3\sqrt{\dfrac{\hat{p}\hat{q}}{n}}$ or $0.15 \pm 3\sqrt{\dfrac{0.15(0.85)}{50}}$ or 0.15 ± 0.151 (−0.001, 0.301)   No

(d) $n = 1000$, $\hat{p} = 0.001$

$n\hat{p} = 1$ and $n\hat{q} = 999$ No

$\hat{p} \pm 3\sqrt{\dfrac{\hat{p}\hat{q}}{n}}$ or $0.001 \pm 3\sqrt{\dfrac{0.001(0.999)}{1000}}$ or 0.001 ± 0.003 (−0.002, 0.004)   No

5.16. Set the approximate 95% error of estimate equal to 0.05 and solve for n. $2\sigma_{\hat{p}} = 2\sqrt{\dfrac{\hat{p}\hat{q}}{n}} = 2\sqrt{\dfrac{0.15(0.85)}{n}} = 0.05$.

The solution is $n = 204$. The engineer would need to sample an additional 104 units, combine the two results and perform the calculations again.

5.17.

PHAT					PHAT	prob	G*H
A	B	C	D	1	PHAT	prob	G*H
A	B	C	E	0.75	0.5	0.4	0.2
A	B	C	F	0.75	0.75	0.53	0.3975
A	B	D	E	0.75	1	0.07	0.07
A	B	D	F	0.75	Mean of		0.6675
A	B	E	F	0.5	PHAT		
A	C	D	E	0.75			
A	C	D	F	0.75			
A	C	E	F	0.5			
A	D	E	F	0.5			
B	C	D	E	0.75			
B	C	D	F	0.75			
B	C	E	F	0.5			
B	D	E	F	0.5			
C	D	E	F	0.5			

Note that $\mu_{PHAT} = p$. This population is small and the distribution of PHAT is not approximately normal.

5.18. (a) $\sigma^2 = 1^2(0.2) + 2^2(0.2) + 3^2(0.2) + 4^2(0.2) + 5^2(0.2) - 3^2 = 2$ and $\sigma = 1.414$

(b)

S^2	$p(S^2)$
0	0.2
0.5	0.32
2	0.24
4.5	0.16
8	0.08

$E(S^2) = 0(0.2) + 0.5(0.32) + 2(0.24) + 4.5(0.16) + 8(0.08) = 2$

S^2 is an unbiased estimator of σ^2.

S	$p(S)$
0	0.2
0.707	0.32
1.414	0.24
2.121	0.16
2.828	0.08

$E(S) = 0(0.2) + 0.707(0.32) + 1.414(0.24) + 2.121(0.16) + 2.828(0.08) = 1.131$

S is not an unbiased estimator of σ.

(c)

S^2	$p(S^2)$
0.000	0.040
0.333	0.192
1.000	0.144
1.333	0.144
2.333	0.192
3.000	0.096
4.000	0.048
4.333	0.096
5.333	0.048

$E(S^2) = 0(0.040) + 0.333(0.192) + 1(0.144) + 1.333(0.144) + 2.333(0.192) + 3(0.096) + 4(0.048) + 4.333(0.096) + 5.333(0.048) = 2$

S^2 is an unbiased estimator of σ^2.

S	$p(S)$
0.000	0.040
0.577	0.192
1	0.144
1.155	0.144
1.527	0.192
1.732	0.096
2	0.048
2.082	0.096
2.309	0.048

$E(S) = 0(0.040) + 0.577(0.192) + 1(0.144) + 1.155(0.144) + 1.572(0.192) + 1.732(0.096) + 2(0.048) + 2.082(0.096) + 2.309(0.048) = 1.287$

S is not an unbiased estimator of σ.

(d)

S^2	$P(S^2)$
0.00000	0.0080
0.25000	0.0512
0.33333	0.0384
0.66667	0.0576
0.91667	0.1152
1.00000	0.0384
1.33333	0.0288
1.58333	0.0768
1.66667	0.0768
2.00000	0.0768
2.25000	0.1024
2.66667	0.0192
2.91667	0.0768
3.00000	0.0576
3.33333	0.0384
3.58333	0.0384
3.66667	0.0384
4.00000	0.0128
4.25000	0.0384
5.33333	0.0096

$E(S^2) = 2$, S^2 is an unbiased estimator of σ^2.

S	$P(S)$
0	0.0080
0.5	0.0512
0.577347	0.0384
0.816499	0.0576
0.957429	0.1152
1	0.0384
1.154699	0.0288
1.258304	0.0768
1.290996	0.0768
1.414214	0.0768
1.5	0.1024
1.632994	0.0192
1.707826	0.0768
1.732051	0.0576
1.825741	0.0384
1.892969	0.0384
1.914855	0.0384
2	0.0128
2.061553	0.0384
2.3094	0.0096

$E(S) = 1.340$, S is not an unbiased estimator of σ.

5.19.

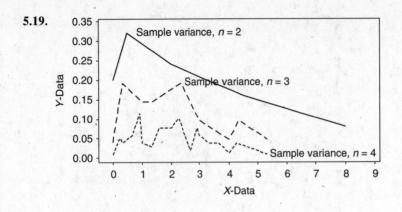

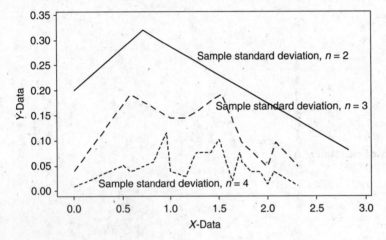

5.20. Suppose we use the package STATISTIX to aid our analysis.

First, test to determine if it is reasonable to assume normality of the data. The results of the Shapiro–Wilk test are:

Shapiro-Wilk Normality Test

Variable	N	W	P
length	10	0.9318	0.4663

The large *p*-value allows us to not reject normality.

Descriptive Statistics

Variable	N	Mean	SD
length	10	99.270	3.3556

The standard deviation of the 10 values is 3.3556. The value of $(n-1)S^2/\sigma^2$ is $9(11.26)/12.25 = 8.27$. This is very close to the mean value of a chi-square distribution with 9 degrees of freedom which is 9. Therefore, assuming that the population has a standard deviation equal to 3.5 does not yield any unusual values for a chi-square random variable. It is reasonable to assume that the process is in control.

CHAPTER 6

6.17. EXCEL

	A	B	C	D
1	temp			
2	65.7		67.895	AVERAGE(A2:A21)
3	64.9		1.951915	STDEV(A2:A21)
4	69.8		0.436462	C3/sqrt(20)
5	70.3		2.093024	TINV(0.05,19)
6	67.3			
7	64.9		**66.98148**	C2−C4*C5
8	67.7		**68.80852**	C2+C4*C5
9	71.5			
10	68.7			
11	67.5			
12	68.4			
13	66.4			
14	69.2			
15	66.6			
16	69.9			
17	70			
18	68.5			
19	65.7			
20	69.2			
21	65.7			

MINITAB

One-Sample T: temp

Variable	N	Mean	StDev	SE Mean	95% CI
temp	20	67.8998	1.9525	0.4366	(66.9860, 68.8136)

SAS

Sample Statistics for temp

N	Mean	Std. Dev.	Std. Error
20	67.90	1.95	0.44

95 % Confidence Interval for the Mean

Lower Limit: 66.98
Upper Limit: 68.81

SPSS

One-Sample Statistics

	N	Mean	Std. Deviation	Std. Error Mean
Temp	20	67.895	1.9519	.4365

One-Sample Test

			Test Value = 0			
					95% Confidence Interval of the Difference	
	t	df	Sig. (2-tailed)	Mean Difference	Lower	Upper
Temp	155.558	19	.000	67.8950	66.981	68.809

STATISTIX

```
Statistix 8.0
Descriptive Statistics

Variable    N    Lo 95% CI    Mean      Up 95% CI    SD
temp        20   66.981       67.895    68.809       1.9519
```

6.18. (a) $\alpha = \text{BINOMDIST}(2,36,0.167,1)+(1-\text{BINOMDIST}(9,36,0.167,1)) = 0.112586$

(b) $\beta = P(\text{not reject null for different values of } p)$

p	β	$1 - \beta$	p	β	$1 - \beta$
0.01	0.005581	0.994419	0.16	0.890654	0.109346
0.02	0.034983	0.965017	0.17	0.884663	0.115337
0.03	0.092775	0.907225	0.18	0.869847	0.130153
0.04	0.173367	0.826633	0.19	0.846847	0.153153
0.05	0.267914	0.732086	0.2	0.816397	0.183603
0.06	0.367779	0.632221	0.21	0.779343	0.220657
0.07	0.46598	0.53402	0.22	0.736647	0.263353
0.08	0.557565	0.442435	0.23	0.689373	0.310627
0.09	0.639433	0.360567	0.24	0.638659	0.361341
0.1	0.709939	0.290061	0.25	0.58568	0.41432
0.11	0.76846	0.23154	0.26	0.531605	0.468395
0.12	0.815018	0.184982	0.27	0.477548	0.522452
0.13	0.849981	0.150019	0.28	0.424536	0.575464
0.14	0.873869	0.126131	0.29	0.373468	0.626532
0.15	0.887241	0.112759	0.3	0.325093	0.674907

The values of p are entered in A1:A30. The function =BINOMDIST(9,36,A1,1)-BINOMDIST(2,36,A1,1) is entered into B1 and a click-and-drag is performed from B1 to B30. The expression $1 - B1$ is entered in C1 and a click-and-drag is performed from C1 to C30.

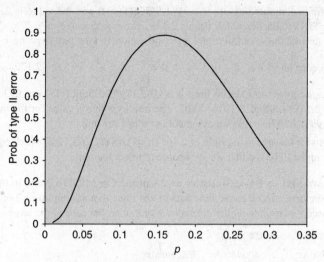

Figure 10.23 Graph of type II errors.

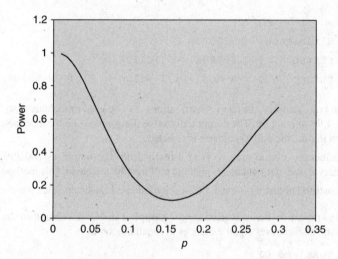

Figure 10.24 Power curve.

Note that in Figures 10.23 and 10.24 the values of p are between 0 and 1. The plot was stopped at $p = 0.3$.

6.19. (a) The test statistic is computed as $=(3.4 - 3)/(1.5/\text{SQRT}(50))$ and the value is 1.886. The critical values are calculated as $=\text{NORMSINV}(0.025) = -1.96$ and $=\text{NORMSINV}(0.975) = 1.96$. The rejection region is $Z < -1.96$ and $Z > 1.96$. The computed test statistic does not fall in the rejection region and the null hypothesis is not rejected.

(b) The 95% confidence interval is $\bar{x} - z_{\alpha/2}\dfrac{s}{\sqrt{n}} < \mu < \bar{x} + z_{\alpha/2}\dfrac{s}{\sqrt{n}}$ or $3.4 - 1.96\dfrac{1.5}{\sqrt{50}} < \mu < 3.4 - 1.96\dfrac{1.5}{\sqrt{50}}$.

Using EXCEL, the lower confidence limit is $=3.4-1.96*1.5/\text{SQRT}(50) = 2.98$ and the upper confidence limit is $=3.4+1.96*1.5/\text{SQRT}(50) = 3.82$. The null hypothesis value for the mean is 3 and 3 falls inside the interval (2.98, 3.82) so that we are unable to reject the null.

(c) The p-value is twice the area to the right of 1.886 or $=2*(1-\text{NORMSDIST}(1.886))$ that equals 0.059, which is larger than the preset alpha (0.05) so that we are unable to reject the null.

Therefore, we are unable to reject the null no matter which of the three methods we use, as will always be the case. The three methods are equivalent.

6.20. (a) The test statistic is computed as $=(3.9-3)/(1.5/\text{SQRT}(13))$ and the value is 2.163. The critical value is calculated as $=\text{TINV}(0.05,12)$ which equals 2.179. The rejection region is $T<-2.179$ and $T>2.179$. The computed test statistic does not fall in the rejection region and the null hypothesis is not rejected.

(b) The 95% confidence interval is $\bar{x}-t_{\alpha/2}\dfrac{s}{\sqrt{n}}<\mu<\bar{x}+t_{\alpha/2}\dfrac{s}{\sqrt{n}}$ or $3.9-2.179\dfrac{1.5}{\sqrt{13}}<\mu<3.9+2.179\dfrac{1.5}{\sqrt{13}}$.

Using EXCEL, the lower confidence limit is $=3.9-2.179*1.5/\text{SQRT}(13)=2.993$ and the upper confidence limit is $=3.9+2.179*1.5/\text{SQRT}(13)=4.807$. The null hypothesis value for the mean is 3 and 3 falls inside the interval (2.993, 4.807) so that we are unable to reject the null.

(c) The p-value is twice the area to the right of 2.163 or $=\text{TDIST}(2.163,12,2)$ which equals 0.051 which is larger than the preset alpha (0.05) so that we are unable to reject the null.

6.21. She uses the pulldown **Stat ⇒ Basic Statistics ⇒ 2-sample t** in MINITAB and in the dialog box she chooses summarized data. She may choose equal variances or not since that assumption is not needed for large samples. She also does not need to worry about the normality assumption because of the large samples.

```
Two-Sample T-Test and CI

Sample  N     Mean     StDev    SE Mean
1       65    16.100   0.250    0.031
2       65    16.350   0.550    0.068


Difference = mu (1) - mu (2)

Estimate for difference: -0.250000

95% CI for difference: (-0.398896, -0.101104)

T-Test of difference = 0 (vs not =): T-Value = -3.34 P-Value = 0.001 DF = 89
```

(a) Because of the large samples, the t and z distributions may be used interchangeably. Therefore, the rejection region is $z<-1.96$ or $z>1.96$. The computed z value is equal to -3.34. Since the computed test statistic is in the rejection region, the null hypothesis is rejected.

(b) The 95% confidence interval may be read directly from the output as $(-0.399, -0.101)$ and since the confidence interval does not contain $D_o=0$, the null may be rejected if this method of testing is used.

(c) If the p-value method of testing is used, the null hypothesis is rejected since the p-value $<\alpha$.

6.22. She uses the pulldown **Stat ⇒ Basic Statistics ⇒ 2-sample t** in MINITAB and in the dialog box she chooses summarized data. She also chooses equal variances in the dialog box.

```
Two-Sample T-Test and CI

Sample  N     Mean     StDev    SE Mean
1       10    16.100   0.250    0.079
2       15    16.350   0.550    0.14


Difference = mu (1) - mu (2)

Estimate for difference: -0.250000

95% CI for difference: (-0.635706, 0.135706)

T-Test of difference = 0 (vs not =): T-Value = -1.34 P-Value = 0.193 DF = 23

Both use Pooled StDev = 0.4567
```

(a) The computed test statistic is $t=-1.34$. The degrees of freedom is $v=10+15-2=23$.

The rejection region is determined as follows:

```
Inverse Cumulative Distribution Function

Student's t distribution with 23 DF

P( X <= x )    x

0.025        -2.06866
```

The rejection region is $t<-2.069$ and $t>2.069$. The computed $t=-1.34$ does not fall in the rejection region.

(b)   The 95% confidence interval is (−0.635706, 0.135706). Since this interval does contain $D_o = 0$, the null is not rejected.

(c)   The p-value is 0.193 which is not less than $\alpha = 0.05$. The null hypothesis is not rejected.

6.23.   The differences pass the normality test. The p-value > 0.15 for the Kolmogorov–Smirnov test. Thus we may assume that the differences are normally distributed.

EXCEL

BE CAREFUL

t-Test: Paired Two Sample for Means

	Variable 1	Variable 2
Mean	26.33333	31.22222
Variance	12.25	9.444444
Observations	9	9
Pearson Correlation	0.79412	
Hypothesized Mean Difference	5	
Df	8	
t Stat	−13.8155	
P(T<=*t*) one-tail	3.64E-07	
t Critical one-tail	1.859548	
P(T<=*t*) two-tail	7.28E-07	
t Critical two-tail	2.306004	

The above analysis is incorrect. The information entered into the dialog box was that the range of variable 1 was A1:A9, the range of variable 2 was B1:B9, and the mean difference was 5 when in fact it would have been −5. However, EXCEL will not all allow a negative value to be put in for D_o. What you have to do is tell the software that the range of variable 1 is B1:B9 and that the range of variable 2 is A1:A9. The mean difference is 5. The output from EXCEL is then shown as below. This is the correct output. The last five rows are now correct. You must peruse your output carefully and ask the question "Is this output reasonable?" If the answer is no, then you have to ask "What have I done that may not be in accord with the programmer who wrote the code?"

t-Test: Paired Two Sample for Means

	Variable 1	Variable 2
Mean	31.22222	26.33333
Variance	9.444444	12.25
Observations	9	9
Pearson Correlation	0.79412	
Hypothesized Mean Difference	5	
Df	8	
t Stat	−0.15523	
P(T<=*t*) one-tail	0.440243	
t Critical one-tail	1.859548	
P(T<=*t*) two-tail	0.880485	
t Critical two-tail	2.306004	

The MINITAB output is

Paired T-Test and CI: highway, stopandgo

Paired T for highway - stopandgo

	N	Mean	StDev	SE Mean
highway	9	31.2222	3.0732	1.0244
stopandgo	9	26.3333	3.5000	1.1667
Difference	9	4.88889	2.14735	0.71578

95% lower bound for mean difference: 3.55786

T-Test of mean difference = 5 (vs > 5): T-Value = -0.16 P-Value = 0.560

Paired T-Test and CI: highway, stopandgo

Paired T for highway - stopandgo

	N	Mean	StDev	SE Mean
highway	9	31.2222	3.0732	1.0244
stopandgo	9	26.3333	3.5000	1.1667
Difference	9	4.88889	2.14735	0.71578

95% upper bound for mean difference: 6.21992

T-Test of mean difference = 5 (vs < 5): T-Value = -0.16 P-Value = 0.440

Note that the *t*-value in the MINITAB output is -0.16. In the EXCEL output it is called *t* Stat and is equal to -0.15523. The *p*-value in MINITAB is called *P*-value and is equal to 0.440. In EXCEL it is written as $P(T <= t)$ one-tail and is 0.440243.

STATISTIX

Statistix 8

Paired T Test for highway - Stop

Null Hypothesis: difference = 5

Alternative Hyp: difference < 5

Mean	4.8889
Std Error	0.7158
Mean - H0	-0.1111
Lower 95% CI	-1.7617
Upper 95% CI	1.5395
T	-0.16
DF	8
P	0.4402

6.24. $v = \dfrac{(s_1^2/n_1 + s_2^2/n_2)^2}{\dfrac{(s_1^2/n_1)^2}{n_1-1} + \dfrac{(s_2^2/n_2)^2}{n_2-1}}$ is used to compute the degrees of freedom.

EXCEL computation of the degrees of freedom:

Process 1	Process 2		
84.3	77	1.300933	VAR(A2:A11)/10
75.3	72.3	75.02334	VAR(B2:B11)/10
76	90.4	5825.395	(D2+D3)^2
75.2	101.9	625.5772	(D2^2/9+D3^2/9)
78.1	60.8	**9.312033**	D4/D5
74.3	89.1		
78.5	70		
70.3	20.8		
75	26		
77.4	91.4		

$v = \dfrac{(s_1^2/n_1 + s_2^2/n_2)^2}{\dfrac{(s_1^2/n_1)^2}{n_1-1} + \dfrac{(s_2^2/n_2)^2}{n_2-1}}$

The MINITAB computation gives

Two-Sample T-Test and CI: process1, process2

Two-sample T for process1 vs process2

	N	Mean	StDev	SE Mean
process1	10	76.44	3.61	1.1
process2	10	70.0	27.4	8.7

Difference = mu (process1) - mu (process2)

Estimate for difference: 6.47377

95% CI for difference: (-13.29480, 26.24233)

T-Test of difference = 0 (vs not =): T-Value = 0.74 P-Value = 0.478 DF = 9

If the value 9.312033 calculated using EXCEL is rounded down to the previous integer as is recommended, the MINITAB, $DF = 9$ is obtained.

6.25. EXCEL computation of pooled standard deviation:

Type 1	Type 2		
2164	1641	6774.444	VAR(A2:A11)
2040	1827	12309.21	VAR(B2:B11)
1926	1847	9541.828	(D2+D3)/2
1911	1800	**97.68228**	SQRT(D4)
1973	1932		
2064	1937		
2018	2023	$S_p^2 = \dfrac{(n_1-1)S_1^2 + (n_2-1)S_2^2}{n_1+n_2-2}$	
2097	1836		
2105	1891		
1972	1715		

When the sample sizes are equal, the pooled variance is just the average of the two sample variances.

Two-Sample T-Test and CI: type1, type2

Two-sample T for type1 vs type2

	N	Mean	StDev	SE Mean
type1	10	2027.0	82.3	26
type2	10	1845	111	35

Difference = mu (type1) - mu (type2)

Estimate for difference: 182.100

95% CI for difference: (90.322, 273.878)

T-Test of difference = 0 (vs not =): T-Value = 4.17 P-Value = 0.001 DF = 18

Both use Pooled StDev = 97.6823

CHAPTER 7

7.10. The symbol χ_α^2 represents the value on the horizontal axis of the chi-square curve with area α to its right. The EXCEL built-in function =CHIINV(0.99,10) gives the chi-square value with area 0.99 to the right of 2.5882 on the horizontal axis for the chi-square distribution having 10 degrees of freedom. $\chi_{0.99}^2 = 2.5882$.

The MINITAB output gives the following output. This tells us that the area to the left of 2.55821 is 0.01.

Inverse Cumulative Distribution Function

Chi-Square with 10 DF

P(X <= x)	x
0.01	2.55821

Therefore, the area to the right of 2.55821 would be $1 - 0.01 = 0.99$. Hence $\chi_{0.99}^2 = 2.55821$.

The EXCEL built-in function =CHIINV(0.01,10) gives the chi-square value with area 0.01 to the right of 23.20925 on the horizontal axis for the chi-square distribution having 10 degrees of freedom. Hence $\chi_{0.01}^2 = 23.20925$.

The MINITAB output gives the following. This tells us that the area to the left of 2.55821 is 0.01.

Inverse Cumulative Distribution Function

Chi-Square with 10 DF

P(X <= x)	x
0.99	23.2093

Therefore, the area to the right of 23.2093 would be $1 - 0.99 = 0.01$. Hence $\chi_{0.01}^2 = 23.2093$.

7.11. The EXCEL output is as follows:

1.6			
1.5	0.024444	VAR(A1:A10)	
1.8	21.9996	9*0.024444/0.01	this is the computed test statistic
1.5			
1.4	0.008879	CHIDIST(22,9)	this is the *p*-value
1.5			
1.6		Since the *p*-value < 0.1 reject the null.	
1.7			
1.9			
1.5			

7.12. The MINITAB pull-down **Stat ⇒ Basic Statistics ⇒ 2 Variances** produced the following output:

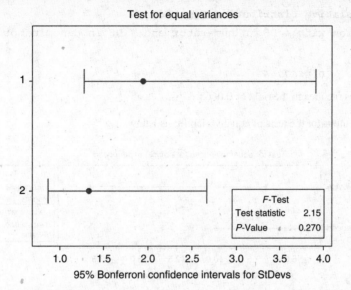

The test statistic F = sample variance 1/sample variance 2 = 2.15 and the p-value corresponding to the test statistic is 0.270. Do not reject the null.

7.13.
```
Group    N       Mean        Std. Dev.     Variance
-------------------------------------------------------
1        10      50.6        7.6333        58.26667
2        12      56.16667    10.744        115.4242
```

Hypothesis Test

```
    Null hypothesis: Variance 1 / Variance 2 = 1
    Alternative: Variance 1 / Variance 2 ^= 1

            - Degrees of Freedom -
    F       Numer.      Denom.       Pr > F
    ------------------------------------------
    0.50    9           11           0.3142
```

The computed test statistic is F = sample variance 1/sample variance 2 = 0.50. The p-value = 0.3142. Because the p-value is greater than 0.05, do not reject the null hypothesis.

7.14. The symbol $F_{0.01}(4,7)$ represents the point on the F distribution curve having 4 and 7 degrees of freedom with area 0.01 to the right of that point.

The EXCEL solution is =FINV(0.01,4,7) = 7.8466. The point 7.8466 has area to the right of 7.8466 equal to 0.01.

The MINITAB inverse cumulative distribution function tells us that area 0.99 is to the left of 7.8466.

Inverse Cumulative Distribution Function
```
F distribution with 4 DF in numerator and 7 DF in denominator

P( X <= x )    x
0.99           7.84665
```

Therefore, 0.01 is the area to the right of 7.8466.

The symbol $F_{0.99}(4,7)$ represents the point on the F distribution curve having 4 and 7 degrees of freedom with area 0.99 to the right of that point.

The EXCEL solution is =FINV(0.99,4,7) = 0.0668. The point 0.0668 has area to the right of that point equal to 0.99.

The MINITAB inverse cumulative distribution function tells us that area 0.01 is to the left of 0.0668.

Inverse Cumulative Distribution Function

```
F distribution with 4 DF in numerator and 7 DF in denominator

P( X <= x )      x
0.01           0.0667746
```

Therefore, 0.99 is the area to the right of 0.0668.

7.15. The MINITAB output for the tests of equal variability is below.

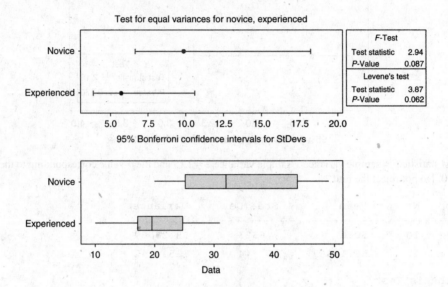

The test that variability is equal for novice and experienced inspectors would not be rejected at $\alpha = 0.05$ by either the F-test or Levene's test.

7.16. The MINITAB test that sigma squared is 0.01 versus it is greater than 0.01 is

```
Null hypothesis Sigma-squared = 0.01
Alternative hypothesis Sigma-squared > 0.01

The chi-square method is only for the normal distribution.
The Bonett method is for any continuous distribution.
Statistics

Variable    N    StDev    Variance
diameter    10   0.156    0.0244

95% One-Sided Confidence Intervals

                       Lower Bound     Lower Bound
Variable    Method     for StDev       for Variance
diameter    Chi-Square  0.114          0.0130
            Bonett      0.100          0.0099

Tests

                                   Test
Variable    Method      Statistic   DF    P-Value
diameter    Chi-Square  22.00        9    0.009
```

7.17. The MINITAB solution is

```
Null hypothesis          Sigma(type1) / Sigma(type2) = 1
Alternative hypothesis   Sigma(type1) / Sigma(type2) not = 1
Significance level       Alpha = 0.05

Statistics

Variable   N      StDev     Variance
type1      10     7.633     58.267
type2      12     10.744    115.424

Ratio of standard deviations = 0.710
Ratio of variances = 0.505

95% Confidence Intervals

                                          CI for
Distribution      CI for StDev            Variance
of Data           Ratio                   Ratio
Normal            (0.375, 1.405)          (0.141, 1.975)
Continuous        (0.347, 1.325)          (0.120, 1.756)

Tests

                                          Test
Method                      DF1    DF2    Statistic   P-Value
F Test (normal)             9      11     0.50        0.314
Levene's Test (any continuous)  1  20     1.62        0.217
```

The F test part of MINITAB corresponds to the following part of the SAS output:

```
        - Degrees of Freedom -
F          Numer.      Denom.        Pr > F
------------------------------------------------
0.50       9           11            0.3142
```

CHAPTER 8

8.16. The MINITAB output for this problem is:

Test and CI for One Proportion

```
Test of p = 0.1 vs p > 0.1

                       95% Lower
Sample  X    N    Sample p    Bound       Z-Value   P-Value
1       8    35   0.228571    0.111823    2.54      0.006

Using the normal approximation.
```

Our conclusion is that we would reject the hypothesis that $P = 0.1$ and accept that $P > 0.1$.

The p-value is 0.006 and we feel fairly confident in our decision.

8.17. The MINITAB output for this problem is:

```
Test and CI for Two Proportions
Sample    X     N     Sample p
1         10    100   0.100000
2          8    125   0.064000
```

```
Difference = p (1) - p (2)
Estimate for difference: 0.036
95% CI for difference: (-0.0367892, 0.108789)
Test for difference = 0 (vs not = 0): Z = 0.99 P-Value = 0.323
Fisher's exact test: P-Value = 0.335
```

We cannot reject the null hypothesis because the *p*-value equals 0.323 which is not less than 0.05.

8.18. The MINITAB output using the chi-square distribution is:

Chi-Square Test: A, B

```
Expected counts are printed below observed counts
Chi-Square contributions are printed below expected counts
```

	A	B	Total
1	10	8	18
	8.00	10.00	
	0.500	0.400	
2	90	117	207
	92.00	115.00	
	0.043	0.035	
Total	100	125	225

Chi-Sq = 0.978, DF = 1, P-Value = 0.323

The *p*-value using the chi-square distribution with 1 degree of freedom is 0.323, the same as with the standard normal distribution in Problem 8.17. Note also, that roughly Z^2 = chi-square.

8.19. The MINITAB output using the pulldown **Stat ⇒ Tables ⇒ Chi-square test**:

Chi-Square Test: group1, group2, group3, group4

```
Expected counts are printed below observed counts
Chi-Square contributions are printed below expected counts
```

	group1	group2	group3	group4	Total
1	68	75	80	60	283
	70.75	70.75	70.75	70.75	
	0.107	0.255	1.209	1.633	
2	57	50	45	65	217
	54.25	54.25	54.25	54.25	
	0.139	0.333	1.577	2.130	
Total	125	125	125	125	500

Chi-Sq = 7.385, DF = 3, P-Value = 0.061

The null hypothesis is not rejected since the *p*-value = 0.061, which is not less than 0.05.

8.20. The MINITAB output is:

Chi-Square Test: bad, normal, good

```
Expected counts are printed below observed counts
Chi-Square contributions are printed below expected counts

          bad       normal    good      Total
1         15        8         5         28
          7.39      12.54     8.06
          7.830     1.646     1.164

2         10        35        6         51
          13.46     22.85     14.69
          0.891     6.463     5.139

3         8         13        25        46
          12.14     20.61     13.25
          1.414     2.809     10.425

Total 33            56        36        125

Chi-Sq = 37.782, DF = 4, P-Value = 0.000
```

The hypothesis of independence is rejected since the p-value is 0.000 and is less than 0.05.

CHAPTER 9

9.17. MINITAB. Pareto chart with cumulative percentage.

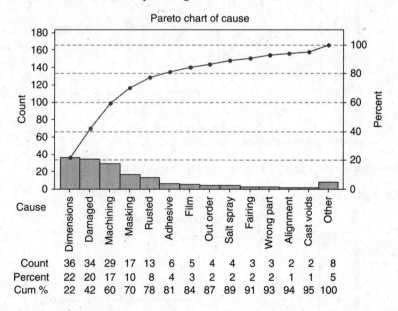

Pareto chart of cause

Cause	Dimensions	Damaged	Machining	Masking	Rusted	Adhesive	Film	Out order	Salt spray	Fairing	Wrong part	Alignment	Cast voids	Other
Count	36	34	29	17	13	6	5	4	4	3	3	2	2	8
Percent	22	20	17	10	8	4	3	2	2	2	2	1	1	5
Cum %	22	42	60	70	78	81	84	87	89	91	93	94	95	100

Total items charted = 166

Percentage accounted for by dimension cause = 36/166 = 21.7%

Percentage accounted for by dimension or damage causes = (36 + 34)/166 = 42.2%

Percentage accounted for dimension or damage or machining causes = (36 + 34 + 29)/166 = 59.6%

9.18. SPSS Cumulative Pareto chart of defects.

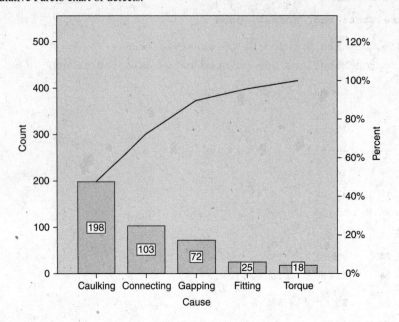

Total items categorized = 416

Percentage accounted for by caulking cause = 198/416 = 47.6%

Percentage accounted for by caulking or connecting causes = (198 + 103)/416 = 72.4%

9.19. MINITAB P chart.

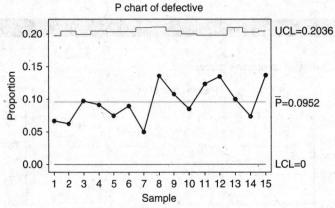

The chart shows that the proportion of defectives produced daily is in control.

9.20. STATISTIX P chart.

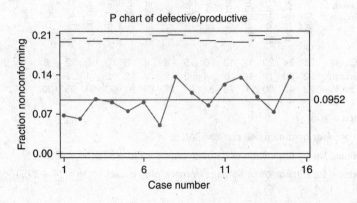

9.21. Enter the data from Problem 9.19 into the SPSS data editor as shown next.

	produced	defective	var
1	75	5	
2	65	4	
3	72	7	
4	66	6	
5	67	5	
6	67	6	
7	60	3	
8	59	8	
9	65	7	
10	71	6	
11	73	9	
12	74	10	
13	60	6	
14	68	5	
15	66	9	

Give the SPSS pulldown **Analyze ⟹ Quality Control ⟹ Control Charts**. Next, select the Attribute Charts p, np.

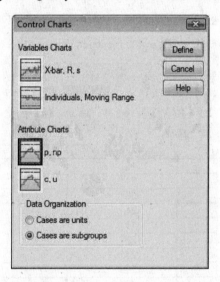

Now, fill in the p, np dialog box as follows:

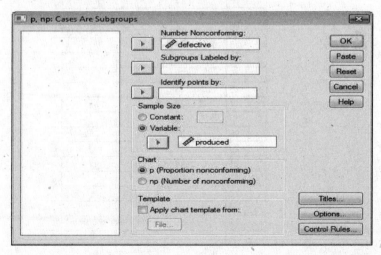

The output is produced as shown next.

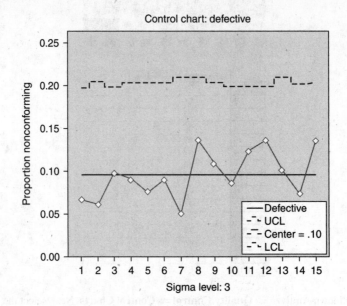

Control chart: defective

This is the SPSS graphic displaying the P chart for the proportion nonconforming.

9.22. The number of defects in each sample of size 400 is entered into the SPSS data editor.

	defects	var	var
1	18		
2	24		
3	22		
4	24		
5	15		
6	22		
7	23		
8	29		
9	13		
10	20		
11	23		
12	19		
13	23		
14	17		
15	26		
16	16		
17	22		
18	20		
19	25		
20	17		
21	18		
22	24		
23	17		
24	17		
25	15		
26			

*Untitled1 [DataSet0] - SPSS Data Editor

File Edit View Data Transform Analyze Gr

27 : defects

Give the SPSS pull down **Analyze ⟹ Quality Control ⟹ Control Charts**. Next, select the Attribute Charts p, np.

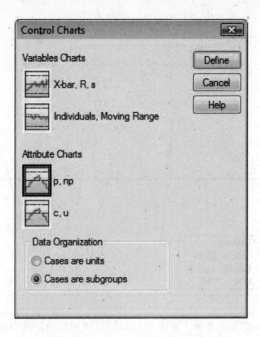

Now, fill in the p, np dialog box as follows:

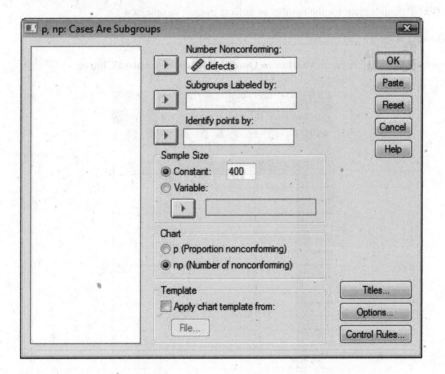

Now form the NP chart.

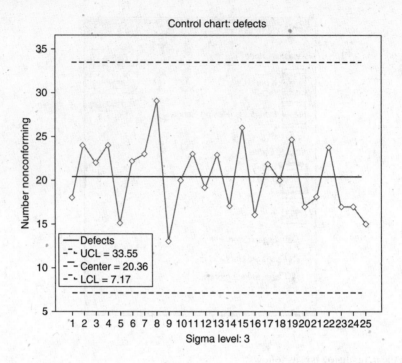

This is the NP control chart for the number of defects in each sample of 400.

9.23. The following is the data as entered into the SPSS data editor.

The following pulldown is given **Analyze ⇒ Quality Control ⇒ Control Charts**.

	Sample	missingrivets	var
1	1	14	
2	2	5	
3	3	7	
4	4	5	
5	5	4	
6	6	2	
7	7	5	
8	8	4	
9	9	8	
10	10	2	
11	11	9	
12	12	13	
13	13	4	
14	14	6	
15	15	5	
16	16	5	
17	17	5	
18	18	5	
19	19	15	
20	20	3	
21	21	7	
22	22	3	
23	23	4	
24	24	6	
25			

The C chart is selected to analyze the process.

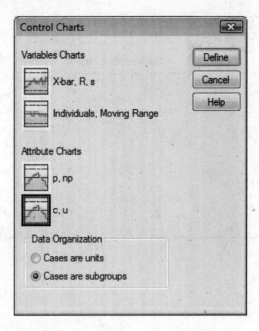

Fill out the following dialog box:

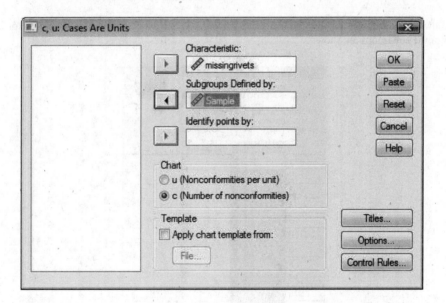

The following C Chart is given by SPSS:

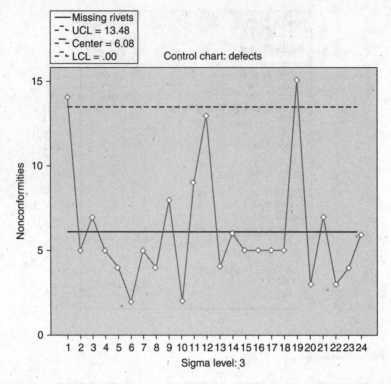

9.24. The following is the data as entered into the SPSS data editor. The data editor contains the data for 24 subgroups. Only 6 of the subgroups are shown.

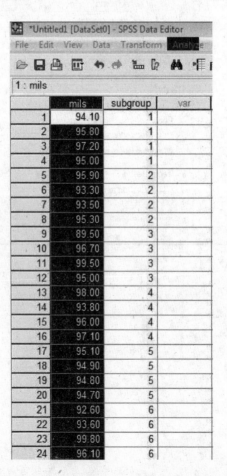

Analyze ⇒ Quality Control ⇒ Control Charts gives the following dialog box which is filled out as shown:

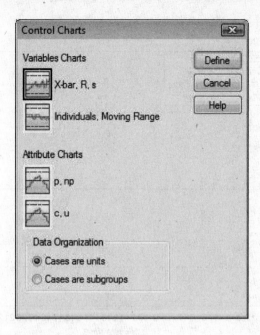

This produces the following dialog box:

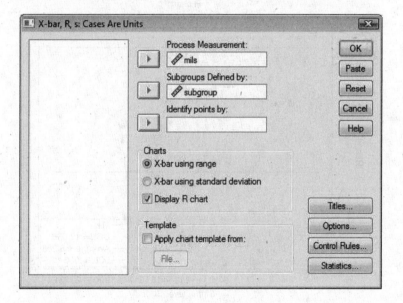

This produces the following Xbar chart and R chart:

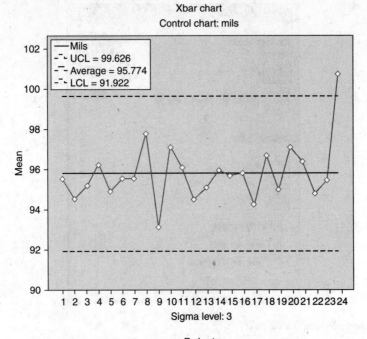

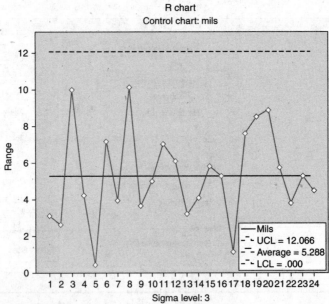

The R chart indicates that variability, as measured by range, is under control. The Xbar chart indicates that the last sample point is out of control.

Index